Dictionary of Mechanical Engineering

D. K. Singh

Dictionary of Mechanical Engineering

Springer

D. K. Singh
Department of Mechanical Engineering
Netaji Subhas University of Technology
New Delhi, India

ISBN 978-981-99-1721-1 ISBN 978-981-99-1722-8 (eBook)
https://doi.org/10.1007/978-981-99-1722-8

This Springer imprint is published by the registered company Springer Nature Singapore Pte Ltd.
The registered company address is: 152 Beach Road, #21-01/04 Gateway East, Singapore 189721, Singapore

Dedicated to my parents, my wife Alka and my daughters Shivangi and Shalvi.

Acknowledgements

I would like to express my sincere thanks to many people who have helped in the publication of this book. First and foremost, my family, who were always supportive of my work and asking about the date of its publication. They made me always in high spirit and confidence. I would also like to thank my publisher Springer for publishing this book and incorporating all my suggestions for the betterment of the book. Also, I am very much thankful to Ms. Priya Vyas, Senior Editor, for taking keen interest in the book. My thanks are also due to all the people connected to the production unit of this work.

I hope this book will prove to be very useful for all the readers in the field of Mechanical Engineering.

D. K. Singh
NSUT, New Delhi, India
dksingh@nsut.ac.in

Contents

About the Author

Dr. D. K. Singh is a Professor and founder Head of Mechanical Engineering Department at Netaji Subhas University of Technology (NSUT), New Delhi, India. He obtained his B.Tech. degree in Mechanical Engineering from National Institute of Technology (NIT) Patna, M.Tech. degree from Birla Institute of Technology (BIT) Mesra, Ranchi, and Ph.D. degree from Uttarakhand Technical University (UTU), Dehradun, India. He has over 30 years of teaching and research experience. He has contributed over 50 papers to various journals of national and international repute and conferences. He has also authored many books on Mechanical Engineering. Two of his books titled *Strength of Materials* and *Fundamentals of Manufacturing Engineering: Concepts and Applications'* are published from Springer. His areas of research include fluid and thermal sciences, mechanics of materials and processes, operations management, and supply chain management. He is a life member of the Indian Society for Technical Education (ISTE), New Delhi.

A

3PL An *Operations Management* term; see **Third-Party Logistics.**

4PL An *Operations Management* term; see **Fourth-Party Logistics.**

5S An *Operations Management* term; see **Five S.**

6σ An *Operations Management* term related to *quality*; see **Six Sigma Approach.**

80/20 Rule An *Operations Management* term; see **Pareto Analysis.**

α-Particle A *Physics* term; see **Alpha Particle.**

β-Particle A *Physics* term; see **Beta Particle.**

γ-Alloy A *Materials Engineering* term; see **Gamma-Alloy.**

ABC Analysis An *Operations Management* term related to *inventory*; a selective inventory control technique which uses *Pareto principle* to classify the inventories into three categories: *A*, *B*, and *C* class items based on the value of consumption of items; *A group inventories* include high-valued items in small volume, represent about 10–20% of the total inventory items but account for about 70–80% of the total inventory cost (e.g. *finished goods*, the most expensive items); *B group inventories* include medium-valued items in moderate volume, represent about 30% of the total inventory items but account for about 15–25% of the total cost (e.g. *semi-finished goods*); *C group inventories* include low-valued items in large volume, represent about 50–60% of the total inventory items but account for only 5% of the total cost (e.g. *spare parts*, the least expensive items); this analysis helps in identifying the few items which are very expensive and require careful watch, scrutiny and follow-up.

ABMA American Bearing Manufacturers Association; an *American* professional body.

Abraham Harold Maslow (1908–1970) A prominent humanistic psychologist who proposed the famous '*Need-hierarchy theory*' of motivation which is based on a hierarchy of five human needs, namely *physiological*, *safety*, *social*, *esteem*, and *self-actualization.*

D. K. Singh, *Dictionary of Mechanical Engineering*,
https://doi.org/10.1007/978-981-99-1722-8_1

Abrasion A *Manufacturing* term; a process in which surfaces of a material are subjected to continuous grinding, rubbing or gouging action due to the presence of a foreign material.

Abrasive Water Jet Machining A *Nonconventional Machining* term, abbreviated as **AWJM**; a nonconventional machining process which uses a high-pressure *abrasive-mixed water jet* to remove materials from a workpiece; offers high material removal rate as compared to water jet machining (WJM); can cut metallic, nonmetallic, and advanced composite materials of various thicknesses; particularly suitable for *heat sensitive materials*, hence is widely used in *aerospace, automotive*, and *electronic industries.*

Abrasive Wear A *Manufacturing* term; a type of wear which occurs when a rough, hard surface slides on a softer surface; appears in the form of grooves on the softer surface; see also **Adhesive Wear** and **Corrosive Wear**.

Absolute Coefficient of Viscosity A *Fluid Mechanics* term; see **Dynamic Viscosity**.

Absolute Entropy A *Thermodynamics* term; the entropy determined relative to the absolute zero temperature (0 K).

Absolute Error A *Metrology* term; the magnitude of the difference between the arithmetic mean and the individual measured value of a physical quantity; given as:

$$|\Delta a_1| = a_m - a_1$$
$$|\Delta a_2| = a_m - a_2$$
$$\ldots\ldots\ldots\ldots\ldots\ldots$$
$$|\Delta a_n| = a_m - a_n$$

where $a_1,\ a_2,\ \ldots a_n$ are the measured values of a physical quantity and a_m is their arithmetic mean, given as:

$$a_m = \frac{a_1 + a_2 + \cdots + a_n}{n}$$
$$= \frac{1}{n} x \sum_{i=1}^{i=n} a_i;$$ see also **Mean Absolute Error**.

Absolute Humidity A *Thermodynamics* term related to *psychrometry*; see **Specific Humidity**.

Absolute Pressure A *Fluid Mechanics* term; the total pressure exerted on a system including atmospheric pressure; equals to the sum of gauge pressure and atmospheric pressure; see also **Gauge Pressure** and **Vacuum**.

Absolute Temperature A *Thermodynamics* term; the temperature measured on the Kelvin scale; based on a hypothetical absolute zero temperature at which matter is devoid of all thermal energy.

Absolute Viscosity A *Fluid Mechanics* term; see **Dynamic Viscosity**.

Absolute Zero Temperature A *Thermodynamics* term; see **Absolute Zero**.

Absolute Zero A *Thermodynamics* term; also called **Absolute Zero Temperature**; a temperature of zero on the Kelvin scale (0 K); the temperature at which all the molecular motion ceases, according to the kinetic theory of gases; the lowest conceivable temperature, and the temperatures lower than this are not defined.

Absorption Dynamometer A *Theory of Machines* term; a type of dynamometer in which work done by a prime mover under test is converted into heat by friction and gets absorbed during measurement; particularly useful for measuring *power* or *torque* developed by *power sources* such as *engines* or *electric motors*; examples include *Prony brake dynamometer* and *rope brake dynamometer*; see also **Transmission Dynamometer**.

Absorption A *Thermodynamics* term; a process to absorb, especially heat; the transfer of one or more constituents from a gaseous mixture to a liquid, where the absorbed material may dissolve physically or react chemically; see also **Adsorption**.

Absorptivity A *Heat Transfer* term; a *radiation property* of the surface of a body which enables it to absorb a fraction of the radiation energy incident on it; depends on the temperature and the wavelength of the radiation; denoted by α (alpha); its value lies between 0 and 1; see also **Emissivity**.

AC An *Electrical* term; see **Alternating Current**.

Acceleration due to Gravity A *Physics* term; the force of gravity which acts on unit mass of a body placed on or near the surface of the earth; the acceleration set up in a body when it falls freely under the effect of gravity alone; a vector quantity which acts towards the centre of the earth; independent of shape, size, and mass of a body, but depends upon mass and radius of the earth; for example, two bodies of different masses, when allowed to fall from the same height in vacuum, will reach the earth at the same time; has maximum value at the surface of the earth (9.81 ms^{-2}), and becomes zero at the centre of the earth; decreases with both *height* and *depth* (the rate of decrease with height is twice as compared to that with depth); denoted by $\boldsymbol{g}$; given as:

$$g = \frac{GM}{R^2}$$

where

G = universal gravitational constant = 6.67×10^{-11} N.m^2.kg^{-2}
M = mass of the earth
R = radius of the earth.

Acceleration A *Physics* term; the time rate of change of velocity; a ratio of the change in velocity to the time taken; a *vector quantity* having the direction of velocity; denoted by *a*; given as:

$$a = \frac{\text{Change in Velocity}}{\text{Time interval}} = \frac{dV}{dt}$$

Acceptance Sampling A *Quality Control* term; a statistical quality control technique used to decide the acceptance or rejection of a shipment of input or output.

Accreditation A *Management* term; a certification issued by a duly recognized body regarding the *facilities, capability, competence,* and *integrity* of any *agency, service* or *operational group* or *individual* to provide the specific service(s) or operation(s) needed; *NBA* in India provides accreditation to educational institutions.

Accuracy A *Metrology* term; the closeness of the measured value of a physical quantity with its true value; used in conjunction with *precision*; see also **Precision.**

Acetone A *Chemistry* term; a liquid chemical having the chemical formula CH_3–CO–CH_3 (dimethyl ketone); has the property of absorbing or dissolving many times of its own volume of acetylene.

Acetylene A *Chemistry* term; a triple-bonded *unsaturated hydrocarbon* discovered by *Edmund Davy* (1785–1857) in the beginning of the nineteenth century; has the chemical formula C_2H_2; the most widely used of all the fuel gases; finds extensive applications in *oxyacetylene welding, heating,* and *cutting*; produced by the chemical reaction between *calcium carbide* (CaC_2) and hot water (H_2O).

Acheson Process A *Chemistry* term; a chemical process used to produce *graphite* artificially in large amounts; involves strong heating of a mixture of *silica* (SiO_2) and *powdered anthracite* (or coke) in an electric furnace continuously for 24–30 hr., which results in the formation of *silicon carbide* (SiC) which decomposes into *graphite* and *silicon,* the latter volatilizes off leaving behind graphite.

Acidic Flux A *Chemistry* and a *Welding* term; a flux used to remove basic impurities like *lime* (CaO) or *metallic oxides* (FeO, MnO, etc.) present in an ore; a few examples include *silica* (SiO_2) and *borax* ($Na_2B_4O_7.10H_2O$); see also **Basic Flux**.

Acidic Refractories A *Materials Engineering* term; refractories which readily combine with bases; *silic*a is their main constituent; a few examples include *quartz, sand,* and *silica brick*; see also **Basic Refractories** and **Neutral Refractories.**

Acoustic Velocity A *Fluid Mechanics* term; see **Sonic Velocity**.

Acoustic-Emission Inspection A *Materials Engineering* term; see **Acoustic-Emission Testing**.

Acoustic-Emission Testing A *Materials Engineering* term; also called **Acoustic-Emission Inspection**; a *nondestructive testing* in which defects are monitored as a result of ultrasonic emissions caused by stressing the flawed material.

Acquisition Cost An *Operations Management* term; the cost of purchasing or producing a unit of a material or product.

Activated Alumina A *Materials Engineering* term; also called **Gamma-Al_2O_3 (γ-Al_2O_3)**; an oxide of *aluminium* (Al_2O_3); one of the two common anhydrous forms of alumina in which it exists in nature; readily dissolves in acids; finds use in column *chromatography*; see also **Corundum**.

Activated Charcoal A *Chemistry* term; an *amorphous* allotropic and active form of carbon; obtained by heating all other forms of charcoal at 1000 °C in a current of superheated steam to eliminate their pores which contain impurities such as residual hydrocarbons, oxygen, etc., thus making the resulting charcoal more active.

Active Pressure A *Civil Engineering* term; a *horizontal earth pressure* exerted by a soil mass against some restraining structures such as a *basement wall* or a *retaining wall*; see also **Passive Pressure**.

Activity Chart A *Motion Study* term; also called **Man–Machine Chart** or **Worker–Machine Chart** or **Operator–Machine Chart**; a chart used to study and improve the utilization of an operator (worker) and a machine or some combinations of operators (a crew) and machines; helpful for minimizing worker and machine delay and for determining the optimal number of machines assigned per worker.

Activity List A *Project Management* term; a list which contains all the activities in a project.

Activity A *Project Management* term; indicates the smallest unit of a work which consumes time, labour and material resources for its completion; see also **Event.**

Activity-on-Arc Network A *Project Management* term; see **AOA Network**.

Activity-on-Arrow Network A *Project Management* term; see **AOA Network**.

Activity-on-Node Network A *Project Management* term; see **AON Network**.

Actuator A *Mechatronics* term; a device used to transfer electrical, pneumatic or hydraulic energy into mechanical energy that moves or displaces something; a device which converts fluid pressure into mechanical motion for the purpose of moving a load.

Acute Triangle A *Mathematics* term; also called **Acute-Angled Triangle**; a triangle whose all the angles are acute, that is, less than 90°; see also **Obtuse Triangle**.

Acute-Angled Triangle A *Mathematics* term; see **Acute Triangle**.

Adam Smith A famous economist who introduced the '*division of labour*' in his book 'The Wealth of Nations' in 1776 to enhance a worker's productivity, which later became part of scientific management.

Adaptive Forecasting An *Operations Management* term related to *forecasting*; see **Exponential Smoothing Method**.

Addendum Circle A *Theory of Machines* term related to *gear*; a circle passing through the *tips* of all the gear teeth; its standard diameter equals to the pitch circle diameter plus 2 modules; see also **Dedendum Circle**.

Addendum A *Theory of Machines* term related to *gear*; the radial distance between the top land and the pitch circle of the gear or the height of a gear tooth *above the pitch circle*; has the standard value of *one module*; see also **Dedendum**.

Addition Polymerization A *Chemistry* term; also called **Chain-Growth Polymerization**; a polymerization process in which the molecules of the same or different monomers add together to form a large polymer molecule; unsaturated organic compounds undergo addition polymerization; a few examples include the formation of *polyethylene from ethylene* and the formation of *polyvinyl chloride (pvc) from vinyl chloride*; see also **Condensation Polymerization**.

Additive A *Materials Engineering* term; a material added to a polymer or other material to provide specific characteristics.

Adhesive Bonding A *Joining* term; a process of joining materials by means of an *adhesive*; the bond is due to molecular attraction between adhesive and the workpiece; useful to join *metals* with *non-metals, plastics,* or *composites.*

Adhesive Force A *Physics* term; also called **Force of Adhesion**; the *intermolecular force* acting between *molecules of different substances*; does not obey the inverse square law of forces; increases rapidly with decrease in distance between the molecules (varies inversely as the seventh power of the distance between the molecules); a few examples include *wetting of the surface of a glass container by water, writing impression left over by a lead pencil (graphite) on a paper*, and *glueing of two surfaces with the use of fevicol, cement,* etc.; see also **Cohesive Force**.

Adhesive Wear A *Manufacturing* term; a type of wear which occurs when *two smooth surfaces* slide over each other and the fragments are pulled off from one surface and adhere to another; see also **Abrasive Wear** and **Corrosive Wear**.

Adhesive A *Chemistry* term; a *polymeric material* used to join *polymers, metals, ceramics, composites* or combinations of these materials; finds applications in areas such as *automotive, aerospace, appliance, electronics, construction,* and *sporting equipment.*

Adiabatic Compression A *Thermodynamics* term; a compression process which does not involve any heat transfer; see also **Isentropic Compression**.

Adiabatic Expansion A *Thermodynamics* term; an expansion process which does not involve any heat transfer; see also **Isentropic Expansion.**

Adiabatic Process A *Thermodynamics* term; a thermodynamic process which does not allow heat transfer through a system (neither heat addition nor heat rejection); see also **Isentropic Process.**

Adiabatic Wall A *Thermodynamics* term; a wall which does not allow the flow of heat across it; see also **Diathermal Wall.**

Adjacent Sides A *Mathematics* term; any two sides starting or originating from the same initial point, (called *vertex*); see also **Opposite Sides.**

Adsorption A *Thermodynamics* term; the removal of impurities from a gas stream by concentration on the surface of a solid or liquid, where the substance that concentrates at the surface is termed adsorbate and the material on the surface of which the adsorption takes place is called adsorbent; see also Absorption.

Aerodynamics A *Fluid Mechanics* term; the study of the effect of air flow on bodies; finds extensive applications in the design of an aircraft.

Aerofoil A *Fluid Mechanics* term; a streamlined body that is designed to produce a streamlined flow pattern, when placed in the free stream of a fluid.

Aerosol A *Chemistry* term; a *colloidal system* in which the dispersion medium is a gas and the dispersed phase is a solid or liquid which is usually smaller than 50 micron in size.

AEV An *Electric Vehicle* term; see **Battery Electric Vehicle.**

AF An *IC Engine* term; see **Air–Fuel Ratio.**

Agate A *Chemistry* term; one of the three *naturally occurring amorphous forms* of silica (SiO_2), other two include *Jasper* and *Onyx.*

Aggregate Plan An *Operations Management* term; a statement which shows the production rates, the work-force levels, and the inventory holdings based on the estimates of customer requirements and capacity limitations of a production system.

Aggregate Planning An *Operations Management* term; see **Aggregate Production Planning.**

Aggregate Production Planning An *Operations Management* term, abbreviated as **APP**; also called simply **Aggregate Planning**; a production plan which helps to arrive at the quantity and timing of resources to be made available to a production system to ensure continuous flow of goods and services to customers; serves the critical role of translating the business plans and strategic intent to operational decisions.

AGMA The short form of *American Gear Manufacturers Association*; an *American* professional body which is the responsible authority for the dissemination of knowledge pertaining to the design and analysis of gearing.

AGV A *Manufacturing* term; see **Automatic Guided Vehicle**.

AI See **Artificial Intelligence**.

Air Preheater A *Power Plant* term related to a *boiler*; an important *boiler accessory* used to extract (absorb) heat from the hot flue gases in the furnace (going as waste) and use it to preheat the air required for combustion; increases the overall boiler efficiency by 5–10%.

Air Standard Cycle A *Thermodynamics* term; a theoretical thermodynamic cycle used in the analysis of internal combustion engines such as petrol, diesel and semi-diesel engines or gas turbine engines; a closed cycle which uses *air* as the working fluid (treated as an ideal gas), and all the processes used in the cycle are reversible; a few examples include *Otto*, *Diesel*, *Dual*, and *Brayton* cycles.

Air-Cooled Condenser A *Power Plant* term; a condenser which uses air as the condensing medium; see also **Water-Cooled Condenser** and **Evaporative Condenser**.

Air-Dried Moulding A *Casting* term; a moulding method in which the surface of a sand mould is dried by leaving the mould in the air.

Air–Fuel Ratio An *IC Engine* term; abbreviated as **AF**; a ratio of the *mass of air* to the *mass of fuel*; see also **Fuel–Air Ratio**.

AISI Short form of American Iron and Steel Institute; an *American* Professional body.

Alclad A *Materials Engineering* term; an alloy of *aluminium* (Al) formed of a thin layer coating (approx. 5% of the thickness) of *pure aluminium* or copper-free aluminium alloys over *duralumin*; produced by *Alcoa* in the form of *sheets* and *plates*; has high corrosion resistant.

All-Electric Vehicle An *Electric Vehicle* term, abbreviated as *AEV*; see **Battery Electric Vehicle**.

Allotropy A *Chemistry* term; the phenomenon of existence of an element in two or more forms which have different *physical properties*, but *identical chemical properties*; for example, carbon (C) exists in both crystalline and amorphous allotropic forms, where carbon's crystalline allotropic forms include *diamond*, *graphite*, and *fullerene,* whereas its amorphous allotropic forms include *coke*, *charcoal*, and *lamp black*.

Allowance Time A *Time Study* term; also called **Fraction Time**: the fraction of time added to the normal time for a job to calculate its standard time; factors such as *worker's fatigue, rest periods* or *unavoidable delays* during working period are included in allowance time, and each is about 20–30% of normal time.

Allowance A *Design* term; the minimum stated clearance or the maximum stated interference for the mating parts; see also **Clearance** and **Interference**.

Alloy Steel A *Materials Engineering* term; a *ferrous* material which contains alloying elements in substantial amounts other than carbon to modify its properties substantially; addition of alloying elements not only overcomes the deficiencies and limitations of plain carbon steels, but also improves their other properties so as to make them extremely useful for industrial applications; classified into *low alloy* and *high alloy* steels, depending upon less than or more than 10% alloy content; a typical example includes *stainless steel.*

Alloy A *Chemistry* term; a material made of two or more elements, one of which is a *metal*; has properties different from its constituents; has metallic characteristics; *steel* is an alloy of *iron* (Fe) and *carbon* (C).

Alloyed Cast Iron A *Materials Engineering* term; the cast iron obtained by adding alloying elements such as *chromium, copper, nickel*, and *molybdenum* in various proportions.

Alpha Error A *Quality Control* term; see **Type I Error**.

Alpha Particle A *Physics* term, abbreviated as **α-Particle**; consists of a large number of *helium ions* (He^{+2}), each having two units of positive charge and 4 times heavier than *hydrogen* (H) atoms; see also **Beta Particle**.

Alpha Ray A *Physics* term; an *electromagnetic radiation* discovered by *Ernest Rutherford* (1871–1937), a New Zealand physicist who is known as the '*father of nuclear physics*'; emitted by radioactive substances such as *uranium* (U), *plutonium* (Pu), and *thorium* (Th); consist of *helium* (He) particles each having two units of positive charge (He^{+2}) and is 4 times as heavy as *hydrogen* (H) atom; moves with a very high velocity of over 16,000 km/s; can ionize the gas through which it passes; see also **Beta Ray** and **Gamma Ray**.

Alpha-Al_2O_3 (α-Al_2O_3) A *Materials Engineering* term; see **Corundum**.

Alternating Current An *Electrical* term, abbreviated as **AC**; a type of electrical current whose polarity changes in a fixed interval of time and has positive and negative values alternatively; flows in *one direction* during any half cycle and in reversed direction during the next half cycle; see also **Direct Current**.

Altitude In *triangle* terms, the perpendicular drawn from any vertex of a triangle to its opposite side; in *mechanics* terms, the height of an object from the earth's surface.

Alumina (Al_2O_3) A *Chemistry* term; an *oxide of aluminium* produced by heating aluminium in oxygen; a white or nearly colourless crystalline substance used as a starting material for smelting of aluminium metal; two common forms of anhydrous alumina include α-Al_2O_3 (also called **Corundum**) and γ-Al_2O_3 (also called **Activated Alumina**).

Aluminium Bronze A *Chemistry* term; a heat-treatable *copper-base* alloy containing up to 12% *aluminium* (Al) along with *copper* (Cu); has *better strength* and *improved corrosion resistance* over brass.

Aluminum A *Chemistry* term; a *metallic element* having symbol *Al* with atomic number 13; a bluish-white *nonferrous metal* having a brilliant lustre which is soon destroyed due to formation of an oxide layer; the most abundant (approx. 8.3% by mass) metal in the earth's crust, and also the third most abundant element in the earth's crust after oxygen (46.6%) and silicon (27.7%); very weak and soft in pure form, but when mixed with small amounts of other alloying elements, its tensile strength, yield strength and hardness can be increased; its principal alloying elements include copper, silicon, manganese, magnesium, and occasionally zinc, nickel and chromium; a very light metal (specific gravity 2.7 as compared to iron's specific gravity of 7.86, thus making it *lighter than iron by one-third for an equivalent volume*); melts at 659 °C and boils at 2450 °C; a malleable and ductile metal which can be rolled into sheets, foils and wires; its important ores include bauxite ($Al_2O_3.H_2O$), cryolite (Na_3AlF_6), and corundum (Al_2O_3), but is generally extracted from *bauxite*; *pure aluminium is not affected by pure water*, but impure form is readily corroded by water containing salts; being lighter in weight, it finds use in making household *utensils*, *aeroplane parts*, and *precision* and *surgical* instruments; also used extensively *in paint as a paint pigment* and for *transporting nitric acid* (HNO_3) as it is not attacked by nitric acid; generally replaces steel and cast iron, where there is a strong need for light weight, corrosion resistance, low maintenance cost or high thermal and electrical conductivity; also used in making body parts, engine blocks, pistons, manifolds and transmission cases in modern motor vehicles in order to ensure fuel economy because of its low weight; its important alloys include *duralumin*, *alclad*, and *γ-alloy*.

Ambient Temperature A *Thermodynamics* term; also called **Sink Temperature**; the temperature (usually of the air) surrounding an operating system; the lowest temperature in the surroundings.

Amethyst A *Chemistry* term; a coloured (purple) *variety of quartz used as a gem*; see also **Jasper** and **Opal**.

Ammonia A *Chemistry* term; also called **Azane**; an inorganic compound of *nitrogen* and *hydrogen* having molecular formula NH_3; industrially produced by *Haber process*; a colourless gas with a characteristic pungent smell and is both caustic and hazardous; extensively used in the manufacture of HNO_3 and important fertilizers such as *ammonium sulphate*, *urea*, *calcium ammonium nitrate*, *ammonium calcium phosphate*; also used as a *cleaning agent*; a widely used *refrigerant* for industrial applications in *ice plants*, *large cold storage*, and *skating rinks*.

Amorphous Materials A *Materials Engineering* term; also called **Noncrystalline Materials**; the materials which do not have a regular periodic arrangement of constituent atoms; for example, *glass* and *clay*; see also **Crystalline Materials**.

Amorphous A *Chemistry* term; an atomic structure in which atoms are not orderly arranged; similar to *noncrystalline structure*.

Amplifier An *Electronics* term; a device used to increase electrical current.

Amplitude A *Physics* term; the distance between the equilibrium position (the mean position) and an extreme position of an oscillation which represents the maximum displacement.

amu A *Chemistry* term; see **Atomic Mass Unit**.

Analog-to-Digital Converter An *Electrical* term, abbreviated as ADC; the conversion of an analog signal to a digital quantity such as binary; see also **Digital-to-Analog Converter**.

Andon An *Operations Management* term; a *warning light system* installed at each assembly line of the workstation; used as a *visible control technique* in *total quality control* (TQC); uses different colours to signify the different types of assistance and support required at the workstation; for example, green colour indicates that the assembly line has no problem, red colour indicates the machine breakdown, white colour indicates the end of a production run, blue colour indicates about a defective unit, and yellow colour indicates about the requirement of a set-up; when an *Andon* light glows, the nearby operators quickly move to assist and solve the problem.

Aneroid Barometer A *Fluid Mechanics* term; a mechanical device used to measure *relatively low pressures* using the elastic deformation of a corrugated diaphragm; see also **Bourdon Gauge**.

Angle of Action A *Cam and Follower* term; the sum of the angle of ascent, angle of descent, and angle of dwell; see also **Angle of Ascent**, **Angle of Dwell,** and **Angle of Descent**.

Angle of Ascent A *Cam and Follower* term; the angle moved by the cam from the instant the follower begins to rise till it reaches its highest position; see also **Angle of Descent**, **Angle of Dwell,** and **Angle of Action**.

Angle of Contact A *Fluid Mechanics* term; the angle made inside a liquid between the tangent to the solid surface and the tangent to the liquid surface at the point of contact between solid and liquid; its value for contact between *pure water* and *clean glass* is zero, for contact between *ordinary water* and *glass* is about 8°, for contact between *water* and *silver* is 90° and for contact between *mercury* and *glass* is 135°.

Angle of Descent A *Cam and Follower* term; the angle moved by the cam during which the follower returns to its initial position; see also **Angle of Ascent**, **Angle of Dwell,** and **Angle of Action**.

Angle of Dwell A *Cam and Follower* term; the angle moved by the cam during which the follower remains in the highest position; see also **Angle of Ascent, Angle of Descent,** and **Angle of Action**.

Angle of Friction A *Physics* term; the angle which the resultant of the force of limiting friction and the normal reaction between any two surfaces in contact makes with the direction of normal reaction; equals to the *angle of repose*; its tangent is equal to the coefficient of static friction; see also **Angle of Repose**.

Angle of Obliquity A *Theory of Machines* term related to *gear*; see **Pressure Angle**.

Angle of Repose A *Physics* term; also called **Angle of Sliding**; the minimum angle of inclination of a plane with the horizontal, such that a body placed over it just begins to slide down; equals to the *angle of friction*; its tangent is equal to the coefficient of static friction; see also **Angle of Friction**.

Angle of Sliding A *Physics* term; see **Angle of Repose**.

Angle Section A *Strength of Materials* term; one of the most commonly used cross sections of structural members; resembles the capital English alphabet *L*.

Angle A *Mathematics* term; a geometrical figure formed by two line segments originating from the same initial point, called *vertex* of the angle; a ratio of the arc length to the radius of a circular sector; denoted by the symbols such as θ (theta), φ (phi), ψ (psi), α (alpha), β (beta), γ (gamma), δ (delta), and others; usually measured in degree; has no unit.

Angular Acceleration A *Physics* term; the time rate of change of angular velocity of a body; a *vector quantity* denoted by α (alpha); the SI unit is rad/s^2; the relationship between linear acceleration (a) and angular acceleration (α) is given as:

$$a = \alpha \cdot r$$

where

r = radius of the circular path.

Angular Deformation A *Strength of Materials* term; the deformation related to change in the angular position of a solid body on being subjected to an external force; see also **Linear Deformation** and **Lateral Deformation**.

Angular Displacement A *Physics* term; the angle traced by the radius vector of a body moving around a circular path at its centre in a given time; a *vector quantity* whose direction depends upon the sense of rotation of the body, and is given by the right-hand rule; usually denoted by *(theta)*, and is expressed in *radian* (*rad*).

Angular Distortion A *Strength of Materials* term; see **Shearing Strain**.

Angular Frequency A *Physics* term; also called **Circular Frequency**; the frequency connected to the rotation of a body; the product of cyclic frequency (n) of oscillation of a body executing periodic motion with a factor 2π ($2\pi n$) or $\frac{2\pi}{T}$ (where T is the time period of oscillation); denoted by ω (Omega); the SI unit is rad/s; see also **Cyclic Frequency**.

Angular Momentum A *Physics* term; also called **Moment of Momentum** or **Rotational Momentum**; a property of a rotating body; the rotational analog of linear momentum; equals to the product of moment of inertia (I) and the angular velocity (ω) of a rigid body about the axis of rotation; a *vector quantity* whose direction is given by the right-handed screw rule or the right hand thumb rule; see also **Linear Momentum**.

Angular Velocity A *Physics* term; the *time rate* (dt) of change of *angular displacement* ($d\theta$) of a body; a *vector quantity* whose direction is same as that of $d\theta$; the SI unit is rad/s; denoted by omega (ω); given as:

$$\omega = \frac{d\theta}{dt}$$

Animal Charcoal A *Chemistry* term; also called **Bone Black**; an *amorphous* allotropic form of carbon obtained by destructive distillation of animal bones which involves strong heating in the absence of air; consists of approximately 10% of *carbon* and remaining *calcium phosphate*.

Anion A *Chemistry* term; a *negatively charged ion* which results due to addition of an electron to a neutral atom or molecule; a few examples include *chlorine ion* (Cl^{-1}), *oxygen ion* (O^{-2}), and *carbonate ion* ($CO_3{}^{-2}$); see also **Cation**.

Anisotropic Material A *Materials Engineering* term; also called **Non-Isotropic Material**; does not exhibit uniform properties in all directions; see also **Isotropic Material**.

Anisotropic A *Materials Engineering* term; the characteristics of having properties (of a material) varying with direction; for example, materials which are single crystals or in which many grains are oriented along a certain direction (occurring naturally or deliberately) have anisotropic *mechanical*, *optical*, *magnetic*, and *dielectric* properties; see also **Isotropic**.

Annealing A *Manufacturing* term related to *heat treatment*; a process of heating a metal above the critical temperature and leaving it in the furnace to cool slowly; offers many changes in the metal like *removes stresses*, *induces softness* for better machining properties, *refines crystalline structure*, and *changes ductility*, *toughness* or *electrical*, *magnetic* or other *physical properties* of the metal.

Anode An *Electrical* term; the negative terminal of a power source (battery) and the positive terminal of an electrolytic cell; see also **Cathode**.

Anodic Reaction A *Chemistry* term; the oxidation reaction which occurs at the *anode* in an electrochemical cell; see also **Cathodic Reaction.**

Anthracite Coal A *Chemistry* term; the *purest type of coal* which contains about 95% of carbon; used as a reducing agent in metallurgical processes and for the manufacture of graphite.

Anticipation Inventory An *Operations Management* term; the inventory which is in demand at some specific periods of time and not required throughout the year, and its demand is accordingly anticipated and met; a few examples include inventory of *fans, air conditioners,* and *air-coolers* before the start of summer; see also **Buffer Inventory, Cycle Inventory,** and **Movement Inventory.**

Anticlockwise Moment A *Physics* term; see **Counterclockwise Moment.**

Anti-Friction Bearing A *Design* term; also called **Rolling Contact Bearing**; a bearing which offers very little friction, much less than the sliding friction; consists of rolling elements such as *balls, cylinders* or *cones* placed between an inner race and an outer race, and has a separator that prevents the rolling elements from rubbing up against one another; involves *rolling friction* between the moving components; the most commonly employed bearing used in *bicycle wheels, robotic joints,* and *automobile transmissions*; see also **Journal Bearing.**

Antoine Chezy (1718–1798) A *French* engineer who is known for the *Chezy equation,* one of the oldest equations used in the area of fluid mechanics; designed a canal for the *Paris water supply system*; formulated *similarity parameter* for predicting flow characteristics of one channel from measurements on another.

Anvil A *Manufacturing* term; the most commonly used *forging tool* used to support a workpiece during its shaping; usually supported on a wooden block or a cast iron stand; made of *wrought iron* or *cast steel.*

AOA Network A *Project Management* term which stands for **Activity-on-Arc Network** or **Activity-on-Arrow Network**; an *event-oriented* network diagram approach in which activities are shown on the *arcs* (or arrows) between two *nodes* (or circles) which represent activity's start or completion time; used in *program evaluation and review technique* (*PERT*); see also **AON Network.**

AON Network A *Project Management* term which stands for **Activity-on-Node Network**; an *activity-oriented* network diagram approach in which activities and activity times are represented by *nodes* (circles) and the *arrows* (also called branches) between the nodes indicate the precedence relationships between activities; used in *critical path method* (*CPM*); see also **AOA Network.**

AOQ A *Quality Control* term; see **Average Outgoing Quality.**

AOQL A *Quality Control* term; see **Average Outgoing Quality Limit.**

APF A *Materials Engineering* term; see **Atomic Packing Factor**.

APICS The short form of American Production and Inventory Control Society; an *American* professional body.

APP An *Operations Management* term; see **Aggregate Production Planning**.

Apparent Viscosity A *Fluid Mechanics* term; the viscosity of a fluid which depends on the shear rate; for example, most non-Newtonian fluids have apparent viscosities; relatively high as compared to the viscosity of water.

Appraisal Cost A *Quality Management* term; a part of *cost of quality* associated with evaluation of quality; includes costs of *inspection*, *testing*, and other related tasks to ensure that the products or processes are acceptable; see also **Prevention Cost**.

Apron A *Lathe* term; a principal part of a lathe which houses the gears and controls for the carriage and feed mechanism.

Arc Voltage A *Welding* term; the voltage generated between the electrode and the workpiece during a welding operation.

Arc Welding A *Welding* term; a *fusion welding process* which uses heat of an arc produced between two electrodes or between the workpiece and an electrode for joining the parts; the temperature produced may be 6000–7000 °C; produces stronger joints because of higher temperatures involved, and hence finds extensive applications; classified into *shielded metal arc*, *submerged arc*, *electroslag*, *electrogas*, and *atomic hydrogen welding processes*; see also **Gas Welding**.

Arc A *Welding* term; see **Electric Arc**.

Archimedes (287–212 B.C.) A *Greek* mathematician and inventor who established elementary principles of *buoyancy* and *floatation* used in *fluid mechanics*.

Archimedes Principle A *Fluid Mechanics* term; named after the *Greek* mathematician and philosopher *Archimedes* (287–212 BC) who discovered and formulated the *principle of buoyancy* in the third century BC; states that a solid body when immersed wholly or partially in a fluid, experiences an upward buoyancy force equal to the weight of the fluid displaced by the body; explains why a floating or sink body appears lighter in weight, while it is under a fluid.

Area Moment of Inertia A *Strength of Materials* term; see **Second Moment of Area**.

Area A *Mathematics* term; a measure of the two-dimensional space that is enclosed by a shape.

Area-Moment Method A *Strength of Materials* term related to *deflection*; see **Moment-Area Method**.

Arithmetic Mean A *Statistics* term; also called simply **Mean**; the most common form of average which takes into account all the numbers, but is affected by extreme values; may not be a whole number even though each item is a whole number; the mean of n numbers from x_1 to x_n can be mathematically expressed as:

$$a_a = \frac{\text{Total of all items}}{\text{Number of items}} = \frac{x_1 + x_2 + \cdots + x_n}{n} = \frac{\sum_{r=1}^{n} x_r}{n}$$

Arthur M. Wahl A person who introduced the *curvature correction factor for helical springs*; famous for his book '*Mechanical Springs*' considered as the *bible* of spring design.

Artificial Diamond A *Chemistry* term; also called **Synthetic Diamond**; a diamond first made artificially in 1955; one of its methods of manufacturing is to subject graphite to a hydrostatic pressure of 14 GPa and a temperature of 3000 °C; has superior properties over natural diamond as it does not contain impurities; used extensively as *abrasives* for *grinding* and *polishing* for industrial applications; *gem-quality synthetic diamond has electrical conductivity 50 times higher than that of natural diamond, and is 10 times more resistant to laser damages*; its possible applications are as heat sinks for computers, in telecommunications, in the integrated circuit (IC) industries and as windows for high-power lasers; see also **Diamond**.

Artificial Intelligence Abbreviated as **AI**; a process which uses machines and computers to replace human intelligence; useful in making decisions which optimize operations and minimize costs.

Artificial Satellite A *Physics* term; a *man-made satellite* which revolves in an orbit around the earth or any other planet; see also **Natural Satellite**.

AS 9100 A *Quality Management* term; a quality standard system which stipulates the requirements for suppliers to an *aerospace industry*; earlier known as AS 9000.

ASHRAE American Society of Heating, Refrigerating and Air-Conditioning Engineers; an *American* professional body.

ASME American Society of Mechanical Engineers; an *American* professional body.

ASRS An *Operations Management* term; see **Automated Storage and Retrieval System**.

Assemble-to-Order An *Operations Management* term; also called **Build-to-Order**; a manufacturing approach or strategy in which products and services are produced in standard modules to which options are added according to customer specifications; the components are made-to-stock and then assembled to order after the customer order comes; has the flexibility of varying the components of the final assembly; no final product inventory exists as in case of make-to-stock and the customer has limited input into the design of the product; a few examples include *computer systems*, *corporate training*, and *industrial equipment*; see also **Make-to-Stock** and **Make-to-Order**.

Assembly Chart An *Operations Management* term; the chart or table which shows the various parts or components of a final product requiring assembly while shaping the product.

Assembly Drawing A *Machine Drawing* term; the *complete drawing* of a given machinery which indicates the relative positions of various component parts assembled together; should not be overcrowded with dimensions and dotted lines.

Assembly Line Technology An *Operations Management* term; a process technology suitable for a narrow range of standardized products in high volume (mass production); uses the product layout.

Assembly Line An *Operations Management* term related to *facilities planning*; see **Product Layout**.

Assembly A *Design* term; a group of machined or handmade parts that fit together to form one unit.

Assignable Variation A *Statistics* term; also called **Special Cause Variation**; arises from external sources which are not inherent in the process, appears sporadically and disrupts the random pattern of common causes; can be traced to a specific reason and hence can be explained or understood and can be eliminated; the potential sources of assignable variations include *machine wear*, *misadjusted equipment*, *fatigued* or *untrained workers*, or *new batches of raw materials*; see also **Natural Variation**.

Assignment Method An *Operations Research* term; a *linear programming solution method* used to assign jobs or personnel to machines or departments.

Assignment Problem An *Operations Research* term; a special case and a completely degenerate form of transportation problem in which the objective is to assign a number of sources to the equal number of destinations at a minimum cost.

Assumptions Applied for Simple Bending A *Strength of Materials* term; include the following:

- Beam's material is *isotropic* and *homogeneous*.
- Transverse section of the beam remains flat before and after bending.
- Beam material's modulus of elasticity remains the same in tension and compression.
- Radius of curvature of the bent beam is very large as compared to its cross-sectional dimensions.

- No resultant push or pull acts on the beam's cross section.
- Load on the beam acts in the plane of bending.

ASTM American Society of Testing Materials; an *American* professional body.

Asymmetric Bending A *Strength of Materials* term; see **Unsymmetrical Bending**.

Atkinson Cycle A *Thermodynamics* term; named after a *British* engineer *James Atkinson* (1846–1914) who invented in 1882; an air-standard cycle which consists of four *reversible processes*, namely *isentropic compression*, *constant volume heat addition*, *isentropic expansion*, and *constant pressure heat rejection*; uses *variable engine strokes* (*shorter compression stroke and longer expansion stroke*) from a complex crankshaft; has greater thermal efficiency than conventional engines.

Atmospheric Air A *Thermodynamics* term; a mixture of dry air and water vapour.

Atmospheric Pressure A *Physics* term; the normal pressure exerted by atmospheric air upon all surfaces in contact; equals to a force of 10.13 N exerted by the gases on each square cm, which is considered as 1 atmosphere pressure, equivalent to 760 mm of Hg; varies with altitude; pressure at a higher altitude is less, as air becomes rarer; the atmospheric pressure is measured by a *barometer*; the standard atmospheric pressure can be expressed in various units as given below:

$$\begin{aligned} \text{1 standard atmospheric pressure} &= 760 \text{ mm of Hg} \\ &= 76 \text{ cm of Hg} \\ &= 10.3 \text{ m of water} \\ &= 1.01325 \text{ bar} \\ &= 1.01325 \times 10^5 \text{ Pa} \\ &= 1.0332 \text{ kgf/cm}^2. \end{aligned}$$

Atom A *Chemistry* term; the smallest as well as a basic unit of a chemical element which takes part in a chemical reaction; has a spherical shape with a radius of 10–8 cm; composed of three fundamental particles, namely *electrons* (negatively charged), *protons* (positively charged) and *neutrons* (no charge); mass of an atom is entirely due to the presence of protons and neutrons which are present in its nucleus, whereas electrons move around the nucleus in different orbits and are held in position by an electrostatic attraction; a few examples include *hydrogen* (H), *nitrogen* (N), *iron* (Fe), and *copper* (Cu); an atom is *electrically neutral* as it contains equal number of electrons and protons.

Atomic Disintegration A *Chemistry* term related to *radioactivity*; a process of disintegration of an atom of a radioactive element; the rate of decay is independent of temperature, pressure, or any external conditions; the whole of the substance never disintegrates completely, although its actual amount may become too small to be measured; the time of complete disintegration of all radioactive elements is thus infinity and does not carry any meaning; a half-life period is commonly used to refer to a time span during which the radioactivity of the element reduces to one-half of its original value.

Atomic Hydrogen Welding A *Welding* term; a fundamentally different *arc welding process* which uses two nonconsumable tungsten electrodes to produce the arc; the heat of the arc is not directly used for welding as it happens in conventional arc welding processes, but is used to break the molecules of hydrogen (H_2) into atoms of hydrogen (H), which recombine after coming out of the arc, producing large amount of heat (temperature around 6000 °C) to be used for welding purpose; does not require shielding atmosphere, as hydrogen itself provides a shielding medium.

Atomic Mass Unit A *Chemistry* term, abbreviated as **amu**; the fundamental unit of *mass of an atom*; approximately equals to the mass of a proton or neutron (1.66 × 10–24 g).

Atomic Mass A *Chemistry* term; the mass of an individual atom expressed in *atomic mass unit* (amu).

Atomic Number A *Chemistry* term: the number of protons in the nucleus of an atom.

Atomic Packing Factor A *Materials Engineering* term, abbreviated as **APF**; a ratio of the *volume of atoms* to the *volume of unit cell* of a crystal; a measure of how dense a metal is; its higher value indicates a denser metal; the *bcc structure* has the APF value of 0.68, the *fcc structure* has the APF value of 0.74, and the *hcp structure* has the APF value of 0.74.

Atomic Radius A *Chemistry* term; the distance of the outermost electron orbital from the atomic nucleus; a parameter used to measure the size of an atom.

Attribute Characteristics A *Statistics* term; also called simply **Attributes**; a product characteristic which can be measured with a discrete response like *yes* or *no*, *good* or *bad*, *success* or *failure*, etc.; see also **Variable Characteristics**.

Attributes A *Statistics* term; see **Attribute Characteristics**.

Audible Wave A *Physics* term; also called **Sound Wave**; the longitudinal wave having frequency varying between 20 and 20,000 Hz, and is audible; produced by vibrating bodies such as vocal cords (human and animal voices), stretched strings (violin, guitar, etc.), stretched membranes (drum, loudspeaker, etc.), air columns (flute, clarinet, etc.), also produced by irregular motion of bodies like doors, tree leaves, etc.; classified into *musical sound waves* and *noise waves* depending on pleasing or irritating effect on the ear.

Audit Findings A *Financial Management* term; the findings or results of the evaluation of the collected audit evidence against audit criteria; can be positive, negative, or neutral.

Audit A *Management* term; a systematic, independent, and documented process for obtaining audit evidence and evaluating it objectively to determine the extent to which audit criteria are fulfilled.

Auditee A *Financial Management* term; a firm (company) which is being audited.

Auditor A *Financial Management* term; an organization or a person with the competence and authorization to conduct an audit.

Augustin Louis de Cauchy (1789–1857) A *French* engineer, mathematician, and hydrodynamicist who contributed to the general field of *theoretical hydrodynamics* and to the *study of wave motion*; well known for the *Cauchy number* (a non-dimensional group) used in the analysis of a *compressible flow*.

Austenite A *Materials Engineering* term; a *solid solution* of *carbon* or *iron-carbide* in face-centered cubic *gamma phase* (γ) of *iron* or *steel*.

Austenitic Stainless Steel A *Materials Engineering* term; also called **Chromium-Nickel Steel**; a corrosion-resistant ferrous alloy which predominantly contains face-centered cubic (γ) phase of iron in which chromium content varies between 16 and 26% and nickel between 3.5 and 22%.

Authority A *Management* term; the right in a position or the right of a person occupying a position to exercise discretion in making decisions affecting others.

Autocratic Leader A *Management* term; a person who commands and expects compliance; he is *dogmatic* and *positive*, and leads by the ability to withhold or give rewards and punishment; see also **Democratic Leader**.

Autocratic Style A *Management* term; a method of functioning which encourages a dictatorial way of doing the work; tends to centralize authority, and limits employee participation.

Autogenous Joining A *Joining* term; see **Autogenous Welding**.

Autogenous Welding A *Joining* term; also called **Autogenous Joining**; a welding process which does not use any filler metal; for example, *solid-state* and *resistance welding processes*; see also **Homogeneous Welding** and **Heterogeneous Welding**.

Autoignition An *IC Engine* term; the premature ignition of a fuel; produces an audible noise, called *engine knock*.

Automated Assembly System An *Operations Management* term related to *automation*; a production system of *automated assembly machines* and *automated materials handling equipment* used to produce major assemblies or completed products.

Automated Storage and Retrieval System An *Operations Management* term related to *automation*; abbreviated as **ASRS**; an automated system of receiving orders for materials, collecting the materials, and delivering them to different workstations for various operations.

Automatic Guided Vehicle A *Manufacturing* term, abbreviated as **AGV**; a material handling system used in a flexible manufacturing system (FMS) to move and transfer materials on different locations of the manufacturing floor.

Automatic Transmission An *Automobile* term; a mechanism used in automobiles to shift gears using a hydraulic torque converter; replaces the clutch pedal found in manual transmission; allows the engine to continue running when the vehicle comes to a stop; currently, in the USA over 84% of the cars have automatic transmission; see also **Manual Transmission**.

Automatic Welding A *Welding* term; a welding process which uses *automatic equipment* to control the entire welding operation including feed, speed, etc. without involving any human intervention.

Automation A *Manufacturing* term; a technological change which involves the replacement of human efforts by mechanical devices; used as a tool to gain competitive advantage through increased productivity and improved quality; see also **Soft Automation** and **Hard Automation**.

Autonomation An *Operations Management* term; the automatic control of defects; the in-built capability of a machine to detect abnormalities or defects in the parts and stop the production line when defects or abnormalities occur; a tool to produce defect-free products and eliminate the need for re-work.

Available Energy A *Thermodynamics* term; also called **Exergy**; the energy which can be converted into work by some means (usually by reversible engines); a few examples include *kinetic energy*, *potential energy*, and *work* itself; *internal energy* and *heat* are partially available energy as they are not convertible into work; see also **Unavailable Energy**.

Average Outgoing Quality Limit A *Quality Control* term, abbreviated as **AOQL**; the maximum average outgoing defectives leaving an inspection process, as the actual percent defectives in lots vary; see also **Average Outgoing Quality**.

Average Outgoing Quality A *Quality Control* term, abbreviated as **AOQ**; the average percentage of outgoing defectives in lots leaving an inspection process; see also **Average Outgoing Quality Limit**.

Average Speed A *Physics* term; a ratio of the total distance travelled by an object to the total time taken by the object.

Avogadro's Hypothesis A *Chemistry* term; see **Avogadro's Law**.

Avogadro's Law A *Chemistry* term; also called **Avogadro's Hypothesis**; named after an Italian scientist *Amedeo Avogadro* (1776–1856); states that equal volumes of all gases under similar conditions of temperature and pressure contain equal number of molecules; see also **Avogadro's Number**.

Avogadro's Number A *Chemistry* term, named after an Italian scientist *Amedeo Avogadro* (1776–1856); indicates the number of molecules in 1 mol of any substance; equals to 6.022×10^{23}; see also **Avogadro's Law.**

AWJM A *Nonconventional Machining* term; see **Abrasive Water Jet Machining.**

AWS American Welding Society; an *American* professional body.

Axial Flow Turbine A *Fluid Mechanics* term; a turbine in which a fluid flows along the axis of the shaft of the turbine; for example, *Pelton turbine* is an axial flow turbine; see also **Radial Flow Turbine** and **Mixed Flow Turbine.**

Axial Flow A *Fluid Mechanics* term; a fluid flow which occurs along the axis of the shaft of a turbine or pump; occurs in *Pelton turbine*; see also **Radial Flow** and **Mixed Flow.**

Axial Force A *Mechanics* term; also called **Axial Load**; a force directed along the axis of a member; produces either *tension* or *compression* in the member; the truss members are subjected to axial forces; the tension test is conducted using an axial force.

Axial Load A *Mechanics* term; see **Axial Force.**

Axial Strain A *Strength of Materials* term; see **Longitudinal Strain.**

Axial-Flow Pump A *Fluid Mechanics* term; a *rotodynamic pump* in which a fluid enters and leaves *axially*, typically along the outer portion of the pump because of blockage by the shaft, motor, hub, etc.; see also **Radial-Flow Pump** and **Centrifugal Pump**;

Axially Loaded Members A *Strength of Materials* term; the structural members (components) subjected to tension or compression force only; solid bars with straight longitudinal axes are the most common type; a few examples include *truss members, connecting rods in engines, spokes in bicycle wheels,* and *columns* in buildings.

Axle A *Design* term; a nonrotating member which carries no torque; used to support rotating wheels and pulleys.

Azane A *Chemistry* term; see **Ammonia.**

B

B2B A *Management* term; see **Business-to-Business.**

B2C A *Management* term; **see Business-to-Consumer**.

Babbitt Metal A *Materials Engineering* term; an alloy of *tin* (Sn), *antimony* (Sb), and *copper* (Cu) consisting of 90% Sn, 7% Sb, and 3% Cu; used in making *bearings for machinery.*

Babcock and Wilcox Boiler A *Power Plant* term related to a *boiler*; the most commonly used *water tube boiler*; categorized as an *externally fired high-pressure* boiler which is horizontally placed, stationary, and involves natural circulation; has high evaporative capacity and produces steam at high pressures; occupies less space as compared to fire tube boilers; see also **Lancashire Boiler**.

Back Order An *Inventory Management* term; an order which is received at a time when there is no inventory in the stock but can be fulfilled after the arrival of new stock; an indication of poor management of the activities in the organization.

Backlash A *Theory of Machines* term related to *gear*; the amount by which width of a tooth space exceeds thickness of the engaging tooth measured on the pitch circles.

Backlog An *Operations Management* term; an accumulation of customer orders which have been promised for delivery at some future date.

Backorder Cost An *Operations Management* term; see **Stockout Cost**.

Backward Extrusion A *Manufacturing* term; see **Indirect Extrusion**.

Backward Scheduling An *Operations Management* term; another method of *job scheduling* in which activities are scheduled once the due date of a job is known; the final operation is scheduled first and then steps within the job are scheduled in reverse order; by subtracting the time needed for each item, the start time is obtained; used in manufacturing environments, as well as service environments such as catering a *banquet* or *scheduling surgery*; see also **Forward Scheduling**.

D. K. Singh, *Dictionary of Mechanical Engineering*,
https://doi.org/10.1007/978-981-99-1722-8_2

Badged Product A *Financial Management* term; an item bought in by a company and then resold to others as its own product without making any change; the product is often badged with the buyer's name so as to be considered its own product.

Bag House An *Air Pollution* term; an air pollution control device which traps the gas-borne particulates by forcing the gas through filter bags.

Bainite A *Materials Engineering* term; an extremely fine *needle-like microstructure* consisting of *ferrite* (α-Fe) and *cementite* (Fe_3C); a structure of steel lying between pearlite and martensite.

Bakelite A *Chemistry* term; a common *phenolic thermoset* used as *adhesives, coatings, laminates,* and *moulded components* for electrical or motor applications.

Balance Sheet A *Financial Management* term; a record sheet (usually an annual record) which presents a summary statement of a firm's financial position at a given point of time; balances the firm's assets (what it owns) against its financing which can be either debt (what it owes) or equity (what was provided by owners).

Baldrige Award A *Quality Management* term; also known as **Malcolm Baldrige National Quality Award**; a quality award started in 1988 and presented every year by the president of the USA to recognize those companies that operate on best business practice.

Band and Block Brake A *Theory of Machines* term; a type of brake which consists of a number of *wooden blocks* secured inside *a flexible steel band*, and both embrace a certain part of the circumference of the drum; more effective than block or band brakes as it produces more friction; see also **Band Brake, Block Brake,** and **Internal Expanding Shoe Brake**.

Band Brake A *Theory of Machines* term; a type of brake which consists of a *rope, belt* or *flexible steel band* which embraces a certain part of the circumference of the drum (or the wheel) and whose one end is connected to one end of a bent lever, while its other end is fixed to a pin on an intermediate point on the same lever; see also **Block Brake, Band and Block Brake,** and **Internal Expanding Shoe Brake**.

Banking of a Road A *Physics* term; a phenomenon of raising outer edge of a curved road above its inner edge; provides the *centripetal force* to a moving vehicle required to balance the *centrifugal force* which may develop at higher speeds of the vehicle, thus preventing its overturning.

Bar Chart A *Statistics* term; see **Bar Graph**.

Bar Code An *information technology* tool; a computer-readable code attached to items like *products, containers, packages*, and *even vehicles* which flow through a supply chain; a *product identifying tool* which includes information such as *product description, item number*, its *source and destination, special handling procedures, costs*, and *order number*.

Bar Graph A *Statistics* term; also called **Bar Chart**; a pictorial representation of the numerical data by a number of bars (*rectangles*) of uniform width (erected horizontally or vertically with equal spacing between them), where each bar represents only one value of the numerical data.

Bar A *Strength of Materials* and a *Thermodynamics* term; in *strength of materials* terms, a circular or rectangular rod used as a test specimen in destructive tests to know the mechanical properties (tensile strength, compressive strength, fatigue, etc.) of materials; in *thermodynamics* terms, the SI unit of pressure commonly used in *meteorology*; 1 bar $= 10^5$ N/m^2.

Bare Electrode A *Welding* term; also called **Uncoated Electrode**; an electrode not coated with any shielding material; normally used with *straight polarity* set-up; suitable for welding of *wrought iron*, and *low* or *medium carbon steel*; used as *consumable* and *nonconsumable* electrodes; as consumable electrodes, they are used as *filler metals* (*MIG welding, submerged arc welding*); as nonconsumable electrodes, they are used to form the electric circuit for the flow of current (*TIG welding, atomic hydrogen arc welding, plasma arc welding*); see also **Coated Electrode**.

Barometer A *Fluid Mechanics* term; an instrument used to measure the *atmospheric pressure* (given as $\rho g h$, where $\rho =$ density of mercury, $h =$ height of mercury column in barometer tube and $g =$ acceleration due to gravity); the barometer reading changes with atmospheric weather and the altitude of the barometer location.

Barotropic Fluid A *Fluid Mechanics* term; a fluid whose bulk modulus (K) is independent of temperature and its density is only a function of pressure.

Barrier-Free Organization A *Management* term; see **Boundaryless Organization.**

Base Metal A *Welding* term; see **Parent Metal.**

Basic EOQ Model An *Operations Management* term; a mathematical formula used to determine the *optimal order size* of inventory items which minimizes the sum of inventory's carrying costs and ordering costs; the following assumptions are considered for basic EOQ model:

- Demand is constant over time.
- No shortages are allowed.
- Lead time for the receipt of orders is constant.
- The order quantity is received all at once.
- Quantity discounts do not exist.
- No safety stock is utilized; see also **Production Quantity Model.**

Basic Flux A *Chemistry* and a *Welding* term; a flux used to remove acidic impurities like *silica* (SiO_2) or *phosphorus pentoxide* (P_2O_5) present in an ore; a few examples include *limestone* ($CaCO_3$) and *magnesite* ($MgCO_3$); see also **Acidic Flux.**

Basic Refractories A *Materials Engineering* term; refractories which readily combine with acids; *magnesium oxide* (MgO) is their main constituent; a few examples include *magnesite* sand *dolomite*; see also **Acidic Refractories** and **Neutral Refractories.**

Batch Production An *Operations Management* term; a type of a production system which involves manufacturing of similar products in batches of a definite number of items; uses general-purpose machine tools arranged on a functional basis; involves more work-in-progress (WIP) inventory; see also **Mass Production** and **Continuous Production.**

Batch Technology An *Operations Management* term; a process technology suitable for producing a variety of products in varying volumes; see also **Batch Production.**

Battery Cell An *Electric Vehicle* term; the basic unit of lithium-ion battery which stores electricity.

Battery Electric Vehicle An *Electric Vehicle* term, abbreviated as *BEV*; also called **All-Electric Vehicle (AEV)**; a 100% battery-powered electric vehicle; a zero emission vehicle which is only powered by on-board rechargeable batteries and no gasoline engine; the batteries of the vehicle are subjected to fast charging using Level 2 charging.

Battery Management System An *Electric Vehicle* term; an electronic system used to manage a rechargeable battery by monitoring and reporting on its state for its optimal performance.

Battery Module An *Electric Vehicle* term; the grouping of battery cells in 6, 8, 10, 12, or up to 24 in numbers.

Battery Pack An *Electric Vehicle* term; a protective battery enclosure which contains batteries in 6, 8, 10, 12, or higher modules.

Battery An *Electrical* term; a device which produces electricity through a chemical reaction between two different metal electrodes separated by a chemical solution called an electrolyte.

BCC Structure A *Chemistry* term; see **Body-Centred Cubic Structure.**

BDC An *IC Engine* term; see **Bottom Dead Centre.**

Beam A *Strength of Materials* term; a structural member in the form of a bar of uniform rectangular or circular cross section which is designed to support vertical loads; essentially a one-dimensional structure as its cross-sectional dimensions are very much smaller compared to its length; classified into *statically determinate* and *statically indeterminate* beams; statically determinate beams include *simple beam, cantilever beam,* and *overhanging beam*; statically indeterminate beams include *fixed beam, continuous beam,* and *propped cantilever beam.*

Bearing Pressure A *Strength of Materials* term; see **Bearing Stress.**

Bearing Stress A *Strength of Materials* term; also called **Bearing Pressure**; a type of pressure exerted between contact surfaces at the area of contact, when two bodies are pressed against each other; a few examples include *soil pressure* beneath a *pier* and contact pressure between a *rivet* and the *side of its hole*; can be reduced by using bearing plates, which distribute the contact forces over a large area.

Bearing A *Design* term; a mechanical device used to reduce friction where two surfaces meet and slide against each other as a result of linear or rotary motion; used to hold a shaft which rotates relative to a fixed support (for instance, *housing of a motor*, *gearbox,* or *transmission*); enables the shaft to rotate with little resistance, while being well-supported against forces acting in the other directions.

Bed of a Lathe A *Manufacturing* term; the base of a lathe on which parts such as *headstock, tailstock, carriage,* etc. are mounted.

Behavioural Science A *Management* term; a science which explores how human behaviour is affected by variables such as *leadership, motivation, communication, interpersonal relationships,* and *attitude change.*

Bell Metal A *Materials Engineering* term; an alloy of *tin* (Sn), and *copper* (Cu) consisting of 25% Sn and 75% Cu; used in making *bells* and *gongs.*

Bellows Meter A *Fluid Mechanics* term; a quantity-measuring device used for gas flow measurements.

Belt Drive A *Theory of Machines* term; a mechanism used to transmit *motion* or *power* between two shafts which are *considerable distance apart*, using a *flat belt* that runs over the pulleys attached to the shafts; a *negative drive* of power transmission as some power is lost due to friction and slippage; classified into *open belt drive* and *closed belt drive*; see also **Rope Drive** and **Chain Drive**.

Belt Friction A *Physics* term; the friction which occurs in a belt drive mechanism; related to the belt tensions as:

$$\frac{T_1}{T_2} = e^{\mu\beta}$$

where

T_1 = tension in the belt on tight side
T_2 = tension in the belt on slack side
μ = coefficient of friction between contact surfaces
β = angle of lap (radian).

Bench Moulding A *Casting* term; the process of making a *small-sized mould* on a bench at a suitable height to the moulder.

Bench Vice A *Manufacturing* term; a *fitting tool* mounted on a bench; used to hold a job between its two jaws while giving the desired shape to the job.

Benchmarking A *Quality Management* term; an approach for setting goals and productivity measures, for example, high standards for the products, practices, and services based on best industry practices; a process of achieving the best level of quality by a company which other companies seek to achieve through continuous improvement; used as a reference point of standard to compare the standards of actual performance; a benchmarked product or service has the highest level of quality standards; see also **Competitive Benchmarking** and **Functional Benchmarking**.

Bend Test A *Strength of Materials* term; a *destructive* test used to know the bending strength of solid materials; the bend test is conducted on a brittle test specimen of rectangular cross section supported at its ends and vertically loaded with one or two point loads.

Bendability A *Manufacturing* term; the ability (property) of a material which enables it to be easily *bent*; *aluminium* and *copper* have good bendability.

Bending Equation A *Strength of Materials* term; see **Flexure Formula**.

Bending Formula A *Strength of Materials* term; see **Flexure Formula**.

Bending Modulus A *Strength of Materials* term; see **Flexural Rigidity**.

Bending Moment Diagram A *Strength of Materials* term related to *bending*; abbreviated as **BMD**; a pictorial representation which shows the distribution of bending moments at various sections of a loaded beam; the positive values of bending moments are shown above the axis of the beam (horizontal axis) and the negative values below the axis (both values as vertical lines), which are connected by a inclined straight line in case of point loads, by a parabolic curve in case of uniformly distributed loads (*udl*) and by a cubic parabolic curve in case of uniformly varying loads acting on the beam; see also **Shear Force Diagram**.

Bending Moment A *Strength of Materials* term; a moment which produces bending in a beam; a restoring couple developed at each cross section of the beam due to its elasticity, when the beam is subjected to an external couple at the elastic limit; the SI unit is N-m.

Bending Spring A *Strength of Materials* term related to a *spring*; see **Laminated Spring**.

Bending Stress A *Strength of Materials* term; also called **Flexure Stress**; a direct stress either tensile or compressive in nature which is produced by a bending moment due to a lateral load applied on a beam; determined by using the *flexure formula.*

Bending A *Manufacturing* term; a deformation (cold working) process used to shape a solid material (*bars*, *wires*, *tubes*, *structural shapes*, and *sheet metal*) into useful products using a bending force; finds extensive applications in various manufacturing operations to make useful products; see also **Roll Bending**.

Benefaction A *Materials Engineering* term related to metallurgy; a process of *increasing the concentration* of an ore by removing the unwanted earthy and siliceous impurities (called *gangue* or *matrix*) from the ore; see also **Pulverization**.

Bernoulli's Principle A *Fluid Mechanics* term; also called **Bernoulli's Theorem** named after *Daniel Bernoulli* (1700–1782), the *Swiss* scientist and mathematician who proposed it in 1738; states that for all points along a streamline in an *incompressible* and *non-viscous* steady flow, the sum of the *pressure head*, the *velocity head*, and the *elevation head* is constant; expressed as:

$$P + \rho g h + \frac{1}{2}\rho v^2 = \text{constant} \tag{1}$$

for a flow in a horizontal plane (say pipe), the equation transforms to

$$P + \frac{1}{2}\rho v^2 = \text{constant} \tag{2}$$

Equation (2) shows that where the velocity is less, the pressure is larger and vice versa; a powerful tool for finding the values of pressure and velocity at two points along a streamline; in the above two equations, the symbols used stand for.

p = pressure of the fluid
ρ = density of the fluid
g = acceleration due to gravity = 9.81 $\frac{m}{s^2}$
v = velocity of fluid flow.

Bernoulli's Theorem A *Fluid Mechanics* term; see **Bernoulli's Principle**.

Beryllium Bronze A *Chemistry* term; a heat-treatable *copper-base* alloy containing about 2% *beryllium* (Be) along with *copper* (Cu); has high *corrosion resistance*, *strength*, *hardness*, and *wear resistance*.

Beta Distribution A *Statistics* term; a *probability* distribution of a random variable; traditionally used in CPM/PERT for estimating the mean and variance of project activity times; see also **Binomial Distribution**.

Beta Error A *Quality Control* term; see **Type II Error**.

Beta Particle A *Physics* term, abbreviated as **β-Particle**; consists of fast moving *electrons*; lighter than an alpha particle; see also **Alpha Particle**.

Beta Ray A *Physics* term; an *electromagnetic radiation* discovered by *Ernest Rutherford* (1871–1937), a New Zealand physicist who is known as the *'father of nuclear physics'*; emitted by radioactive substances such as *uranium* (U), *plutonium* (Pu), and *thorium* (Th); similar to cathode ray and consists of *electrons*; moves much faster than the alpha ray; can cause ionization of air but much lesser than the alpha ray; being more lighter and faster, it is more penetrating than the alpha ray; see also **Alpha Ray** and **Gamma Ray**.

BEV An *Electric Vehicle* term; see **Battery Electric Vehicle**.

Bevel Gear A *Theory of Machines* term related to a *gear*; a type of toothed gear which has teeth (of straight or spiral type) formed on a *conical surface*; used to transmit motion or power between two *intersecting shafts*, or *shafts that lie at right angle to each other*.

Bevel A *Design* term; an angular type of edge preparation of a workpiece.

Bevelling A *Design* term; a type of *chamfering*.

BHP An *IC Engine* term; the short form of **Brake Horse Power**; see **Shaft Power**.

Bias Error A *Metrology* term; see **Systematic Error**.

Biaxial Bending A *Strength of Materials* term; the bending which takes place about both centroidal axes of a beam's cross section; occurs when a load is inclined at an angle with respect to the vertical plane of symmetry of the beam; the applied load is resolved into its components along directions of the two axes of symmetry at the section, of which vertical component causes bending about the horizontal axis and horizontal component causes bending about the vertical axis, and both components produce normal stresses along the longitudinal direction; the method of superposition is used to find the combined stresses.

Biaxial Loading A *Strength of Materials* term; a condition of loading in which a member is subjected to loads in two mutually perpendicular directions; see also **Uniaxial Loading**.

Biaxial Stress A *Strength of Materials* term; a state of stress in which an element of a beam's cross section is subjected to normal stresses acting along two mutually perpendicular directions, and there is no shear stress; for example, *pure biaxial tension* or *pure biaxial compression*; see also **Uniaxial Stress**.

Big M Method An *Operations Research* term; one of the methods for solving a *linear programming problem*; used when the constraints are of '$\geq$' or '=' type.

Bill of Materials An *Operations Management* term, abbreviated as **BOM**; an input of *material requirements planning* (MRP); also called **Product Structure Tree** as it shows that how a product is put together; a list of all *parts, ingredients,* or *materials* required to produce, assemble, or put together one unit of a product.

Billet A *Manufacturing* term; a type of *steel ingot* (square or oblong); considerably smaller than a *bloom*; see also **Bloom** and **Slab**.

Binary Diagram A *Materials Engineering* term related to *phase diagram*; a two-component phase diagram; for example, Fe–C diagram.

Binary Vapour Cycle A *Thermodynamics* term; a vapour power cycle which uses two working fluids, one with good high-temperature characteristics and another with good characteristics at the lower-temperature end of the operating range.

Binding Energy A *Chemistry* term; the energy released in the formation of a nucleus from its constituent nucleons (protons and neutrons); also the energy required to break up a nucleus into its component nucleons; when divided by the number of nucleons, it gives the mean binding energy per nucleon (a measure of the stability of a nucleus); the greater the binding energy per nucleon, the more stable the nucleus is.

Bingham Fluid A *Fluid Mechanics* term; also called **Bingham Plastic Fluid**; a plastic fluid in which a finite stress (yield stress) is required before the fluid begins to flow at all; a few examples include *cream* and *toothpaste.*

Bingham Plastic Fluid A *Fluid Mechanics* term; see **Bingham Fluid**.

Bingham Plastic A *Fluid Mechanics* term; a *viscoplastic material*, named after *Eugene Cook Bingham* (1878–1945) who proposed its mathematical form; behaves as *a rigid body at low stresses*, but flows as *a viscous fluid at high stress.*

Binomial Distribution A *Statistics* term; a *probability* distribution which uses discrete variables such as number of employees or products; see also **Beta Distribution**.

Biological Science The science which deals with *living things*; include subjects like *botany*, *zoology*, *ornithology*, *anthropology*, *entomology*, and *forensic science*; see also **Physical Science**.

Bituminous Coal A *Chemistry* term; a type of coal which contains 88% of *carbon.*

Black Body Radiation A *Physics* term; the radiation emitted by a heated black body; it is higher than the radiation emitted by all real surfaces at the same temperatures.

Black Body A *Physics* term; an idealized surface which completely absorbs the incident thermal radiations of all wavelengths; a *perfect absorber* (absorptivity, $\alpha = 1$) as well as a *perfect emitter* (emissivity, $\varepsilon = 1$).

Black Iron A *Materials Engineering* term; the untreated flat steel which is hot rolled.

Blade Row A *Fluid Mechanics* term related to a *pump*; see **Cascade**.

Blaise Pascal (1623–1662) A seventeenth-century *French* scientist and philosopher who is widely known for *Pascal's law*; has made important contributions in the field of hydrostatics such as *principles of barometer*, *hydraulic press*, and *pressure transmissibility.*

Blank Diameter A *Theory of Machines* term related to *gear*; the diameter of the *addendum circle*; see also **Root Diameter**.

Blanking A *Manufacturing* term; a sheet metal operation used to cut out a useful product from a sheet metal blank using a die-punch assembly; see also **Punching**.

BLDC Motor An *Electrical* term; see **Brushless DC Motor**.

Blind Riser A *Manufacturing* term related to *casting*; also called **Side Riser**; a riser contained entirely *within the mould* and *horizontally located* adjacent to the mould cavity usually along the *parting line.*

Blister A *Casting* term; a type of casting defect and a variation of scar; a certain portion of the casting is covered by a thin layer of metal; see also **Scar**.

Block Brake A *Theory of Machines* term; also called **Shoe Brake**; a type of brake in which motion is retarded or stopped by pressing a *wooden*, *rubber,* or *cast steel block* or *shoe* (replaceable when worn out) against a rotating drum; see also **Band Brake, Band and Block Brake,** and **Internal Expanding Shoe Brake**.

Bloom A *Manufacturing* term; a type of *steel ingot* (square or oblong) with a minimum cross-sectional area of 36 $in.^2$; see also **Billet** and **Slab**.

Blow Hole A *Welding* term; a *welding defect*; occurs in the form of a gas pocket in the weld metal; caused by rapid cooling of the hot metal without escaping of all the gases (making them entrapped) present at the time of solidification.

Blow A *Casting* term; a type of *casting defect* which occurs in the form of *balloon-shaped gas cavities* on the surface of a casting due to pressure of the mould gases; eliminated by improving the permeability of the moulding sand; in *Physics* terms, an impulsive force.

Blower A *Fluid Mechanics* term; a *turbomachine* which is basically a *gas pump*; used to move gases (mostly air) with relatively moderate to high pressure rise (about 25 mm of Hg) and moderate to high flow rate; a few examples include *centrifugal blowers* and *squirrel cage blowers* in *automobile ventilation systems, furnaces*, and *leaf blowers*; see also **Fan** and **Compressor**.

Blow-off Cock A *Power Plant* term related to a *boiler*; a *boiler accessory* used to empty the boiler for the purpose of its cleaning, repair, or inspection and to remove the accumulated mud and sediments deposited at its bottom; fitted at the boiler's lowest part.

BMD A *Strength of Materials* term; see **Bending Moment Diagram**.

Board of Directors A *Management* term; a group elected by a firm's stockholders which has ultimate authority to guide corporate affairs and make general policy; include key corporate personnel as well as outside individuals who typically are the successful business people and executives of other major organizations; forms a part of high level management.

Body Force A *Physics* term; an external force caused by one body on another body without direct physical contact between the bodies; usually expressed as force per unit volume or per unit mass; for example, *gravitational attraction* due to earth on a body or the *electromagnetic force*; see also **Surface Force**.

Body Language A *Management* term; the nonverbal communication cues such as *facial expressions, gestures,* and *other body movements.*

Body A *Physics* term; a certain quantity of matter bounded by a closed surface.

Body-Centred Cubic Structure A *Chemistry* term, abbreviated as **BCC Structure**; one of the three basic types of atomic arrangements found in metals in which the unit cell consists of *eight atoms* one each at every corner of the unit cell and one atom at its centre; the atomic packing factor (APF) equals to 0.68; a few examples include *chromium* (Cr), *manganese* (Mn), *sodium* (Na), *potassium* (K), *tungsten* (W), *barium* (Ba), *iron in α-phase* (below 912 °C), and *iron in δ-phase* (above 1394 °C); see also **Face-Centred Cubic Structure** and **Hexagonal Close-Packed Structure**.

Boiler Accessory A *Boiler* term; an auxiliary or supporting part which is required for the smooth operation of a boiler and increases its overall efficiency; a few examples include *economizer, air preheater, superheater, fuel pump, chimney, air injector, steam separator*, etc.; see also **Boiler Mounting**.

Boiler Mounting A *Boiler* term; an integral fitting or device needed for the operation and safety of a boiler; a few examples include *safety valves, fusible plug, water level indicator, pressure gauge, feed check valve, blow-off cock,* and *steam stop valve*; see also **Boiler Accessory**.

Boiler Operation Mounting A *Boiler* term; the fittings and devices which form integral parts of a boiler and are essentially needed for its operation; a few examples include *pressure gauge, feed check valve, blow-off cock,* and *steam stop valve*; see also **Boiler Safety Mounting**.

Boiler Safety Mounting A *Boiler* term; an integral fitting or device attached to a boiler from a safety point of view; a few examples include *safety valves, fusible plug,* and *water level indicator*; see also **Boiler Operation Mounting**.

Boiler Steel A *Materials Engineering* term; a low-carbon steel with carbon content varying between 0.15 and 0.3%; used in making *boiler sheets* for its body as it has good tensile strength.

Boiler A *Power Plant* term; also called **Steam Generator**; a closed cylindrical vessel in which water or any other liquid is heated to produce steam or vapour at desired pressure; finds use in power plants to generate high pressure and high temperature steam for steam turbines; classified into *low pressure, high pressure, fire tube*, or *water tube* boilers.

Boiling Point A *Physics* term; a temperature at which a liquid substance boils and gets converted into its vapour; for example, the boiling point of water is 100 °C.

Boiling A *Physics* term; a process of formation of *vapour bubbles*, when a liquid is heated at its boiling point.

BOM An *Operations Management* term; see **Bill of Materials.**

Bomb Calorimeter A *Chemistry* term; an instrument used to measure the calorific value (the standard enthalpy of combustion) of *solid fuels*; see also **Continuous Flow Calorimeter.**

Bond Energy A *Chemistry* term; also called **Bond Strength**; the energy required to break *one mole of a particular bond* (Avogadro's number of bonds) in a substance in gaseous state; a measure of bond strength; the higher is the bond energy, the lesser is the reactivity of a molecule; for example, the $N \equiv N$ bond in nitrogen molecule (N_2) is more stable than the $O = O$ bond in oxygen molecule (O_2), indicating that the nitrogen molecule (N_2) is more stable than the oxygen molecule (O_2); the bond energies of some common bonds are given below:

Bonds	**Bond Energy (kJ/mol)**
H–H	458.0
F–F	154.8
Cl–Cl	242.7
H–Cl	430.0
I–Cl	209.2
O=O	494.6
N≡N	945.6
C–C	347.0
C=C	619.0
C≡C	836.0

Bond Length A *Chemistry* term; the distance between the nuclei of two-bonded atoms in a molecule; measured in *angstrom*; for example, the bond length between two hydrogen atoms (H) in hydrogen molecule (H_2) is 0.74 angstrom; the bond lengths of some common bonds are given below:

Bonds	**Bond Length (*Angstrom*)**
H–H	0.74
C–H	1.00
O–H	0.96
F–F	1.42
Cl–Cl	1.99
Br–Br	2.28
I–I	2.67
H–F	0.92
H–Cl	1.27

Bonds	Bond Length (*Angstrom*)
H–Br	1.41
H–I	1.61
C–C	1.54
C=C	1.34
C≡C	1.20

Bond Strength A *Chemistry* term; see **Bond Energy**.

Bone Black A *Chemistry* term; see **Animal Charcoal**.

Borax A *Chemistry* term; also called **Sodium Tetraborate Decahydrate**; a chemical compound of **boron** (B) having the chemical formula of $Na_2B_4O_7 \cdot 10H_2O$; a *white crystalline solid* which is less soluble in cold water but more soluble in hot water; a naturally occurring borax, called *tincal*, is obtained from dried up lakes which contain about 50% borax along with some other insoluble impurities such as clay, sand, etc.; an important *cleansing* and *bleaching agent present in washing powders*; *used in the manufacture of enamels and glazes for earthen wares such as tiles, pottery*, etc. which are resistant to heat, strain, and scratches; also used in the manufacture of *heat-resistant borosilicate glass* (*pyrex glass*), and in the manufacture of medicinal soaps due to its antiseptic properties; finds use as a water softener and *as a flux in soldering*; also used in the laboratory for *borax bead test*.

Borazon A *Materials Engineering* term; see **Cubic Boron Nitride**.

Bore An *IC Engine* term; the diameter of the cylinder of an engine.

Boring A *Manufacturing* term; a process of enlargement of a drilled hole; see also **Drilling** and **Reaming**.

Bottle Glass A *Chemistry* term; an ordinary soda-lime glass which contains MgO to the extent of 3.5% and $Al_2O_3 \cdot Fe_2O_3$ to the extent of 0.5%; used in making *cheap bottles*.

Bottleneck Resource An *Operations Management* term; any resource whose capacity is equal to or less than the demand placed on it.

Bottleneck An *Operations Management* term; a constraint within a production system which limits throughput; may include machine, scarce of highly skilled labour or a specialized tool.

Bottom Dead Centre An *IC Engine* term, abbreviated as **BDC**; a point where the piston is at the bottom of the cylinder of an engine; the position of the piston when it forms the largest volume in the cylinder; occurs when the crankshaft is rotated by 180°; once the piston nears BDC, the intake valve closes so that the cylinder is completely sealed; see also **Top Dead Centre**.

Boundary Layer Thickness A *Fluid Mechanics* term; the distance from a solid boundary within which the flow velocity increases from zero at the wall to 99%, that is, 0.99 of the *free stream velocity* at the edge of the boundary layer.

Boundary Layer A *Fluid Mechanics* term; a very thin layer of fluid adjacent to a surface in which viscosity is important, while the fluid outside this layer is considered frictionless or ideal; the concept of the boundary layer was first introduced by a German physicist *Ludwig Prandtl* (1875–1953) in 1904; one of the important advances in modern fluid mechanics, and is considered important with turbulent flow; the velocity of a fluid particle in a boundary layer is always zero at a fixed wall because of viscosity; see also **Boundary Layer Thickness**.

Boundary Lubrication A *Fluid Mechanics* term; a type of lubrication in which the load-carrying surfaces sliding relative to each other are separated by a *thin film of lubricant* (*water*, *oil*, or *air*); see also **Solid Film Lubrication**.

Boundaryless Organization A *Management* term; also called **Network Organization,** as it functions as a network of a series of project groups or collaborations or **Learning Organization,** as it facilitates the learning of all its members, and continually transforms itself in order to gain a competitive advantage over others or **Virtual Organization** as it virtually eliminates in-house business functions and outsources many activities or **Barrier-Free Organization** or **Modular Organization**; an organization not defined or limited by historical boundaries or categories imposed by traditional structures; a *flatter* organization which eliminates *vertical*, *horizontal*, and *interorganizational* barriers by increasing its interdependence with its environment; provides increased *flexibility* and *adaptability* to cope with rapid technological changes.

Bourdon Gauge A *Fluid Mechanics* term; named after the *French* engineer and inventor *Eugene Bourdon* (1808–1884); a mechanical device used to measure *high pressures* using deflection of an elastic element (a metallic tube, called Bourdon tube); see also **Aneroid Barometer**.

Boyle's Law A *Chemistry* term; named after an *English* scientist *Robert Boyle* (1627–1691) who proposed it; relates *pressure* and *volume* of a gas at *constant temperature*; states that the volume (V) of a given mass of a gas varies inversely proportional to its pressure (P), when temperature is held constant; expressed as:

$$PV = \text{constant}$$

BPR An *Operations Management* term; see **Business Process Reengineering**.

Bragg's Equation A *Physics* term; see **Bragg's Law**.

Bragg's Law A *Physics* term; also called **Bragg's Equation**; an *x-ray diffraction law*, named after *English* physicists *William Henry Bragg* (1862–1942) and his son *William Lawrence Bragg* (1890–1971) who were awarded the Nobel prize for their contribution to *x-ray diffraction* in 1915; gives the condition for x-ray diffraction by a given crystal plane; given as:

$$n\lambda = 2d \sin\theta$$

where

n = an integer number
λ = wavelength of the X-ray
d = interplanar spacing
θ = Bragg angle (one-half of the diffraction angle).

Brake Horse Power An *IC Engine* term, abbreviated as **BHP**; see **Shaft Power**.

Brake Power An *IC Engine* term, abbreviated as **BP**; see **Shaft Power**.

Brake Specific Fuel Consumption An *IC Engine* term; see **Specific Fuel Consumption**.

Brake A *Theory of Machines* term; a device used to offer frictional resistance to a moving body in order to stop or retard its motion or prevent increase in its speed; classified into *block, band, band and block*, and *internal expanding shoe brakes.*

Branch and Bound Technique An *Operations Research* term; a method of optimization to solve problems having finite number of feasible solutions; used to solve *integer linear programming problems, travelling salesman problem*, and *cargo loading problem.*

Brayton Cycle A *Thermodynamics* term related to *air standard cycle*; named after *George Brayton* (1830–1892) who originally developed it; also called **Joule Cycle**; an air standard thermodynamic cycle used in *gas turbine engines*; consists of *four* reversible processes, namely reversible adiabatic (isentropic) compression, constant pressure heat addition, reversible adiabatic (isentropic) expansion and constant pressure heat rejection; the thermal efficiency depends upon its *compression ratio* or *pressure ratio*; for the same compression ratio, both the Otto cycle and the Brayton cycle have same thermal efficiency; see also **Otto Cycle** and **Diesel Cycle**.

Braze Welding A *Joining* term; a process fundamentally different from brazing, where the work material is heated at the joint and molten filler metal fills up the gap between the work parts; see also **Brazing**.

Brazing A *Joining* term; a process of joining metals and alloys without melting them; uses a filler metal (melting point more than 450 °C) which fills the gap between the work parts by *capillary action*; produces a stronger joint as compared to soldering due to higher temperature application: see also **Soldering** and **Braze Welding**.

Breakdown Maintenance An *Operations Management* term; a *reactive maintenance* performed after the machine or equipment has broken down; involves carrying out some repair to restore the machine or equipment to working condition or to replace them with a new machine or equipment; see also **Preventive Maintenance** and **Predictive Maintenance.**

Break-Even Analysis An *Operations Management* term; a graphical technique used to determine a point (called *break-even point*) where the total revenue received from products sold equals to the total cost of manufacturing; commonly used to compare the cost functions of two or more facility alternatives; uses the following formula to calculate the break-even point (BEP):

$$\text{BEP} = \frac{F}{S - V}$$

where

F = fixed cost
S = sale price per unit
V = variable cost per unit.

Brine A *Chemistry* term; the solution of NaCl or $CaCl_2$ having freezing point temperature lower than 0 °C; used as a *coolant* in *cold storage, ice plants,* and *skating rinks* for producing low temperatures.

Brinell Hardness Test A *Strength of Materials* term; see **Brinell Test.**

Brinell Test A *Strength of Materials* term; also called **Brinell Hardness Test**; the oldest destructive method of hardness testing of a material developed by *Johan August Brinell* (1849–1925) in 1900; a commonly employed method which uses a 10 mm diameter hardened steel or tungsten carbide ball that needs to be pressed against the surface of the material being tested with a load of 500 kg, 1500 kg and 3000 kg, and the depth of penetration on the surface is an indication of Brinell hardness number (BHN), given by

$$\begin{aligned}\text{BHN} &= \frac{\text{Load applied in Kgf}}{\text{area of spherical indentation in mm}^2}\\ &= \frac{P}{\left(\frac{\pi D}{2}\right)\left[D - \sqrt{D^2 - d^2}\right]}\end{aligned}$$

where

P = load applied on the test specimen
D = diameter of the ball
d = diameter of indentation produced; see also **Rockwell Test.**

Britannia Metal A *Materials Engineering* term; an alloy of *tin* (Sn), *antimony* (Sb), and *copper* (Cu) consisting of 90% Sn, 8% Sb, and 2% Cu; used in making *table wares.*

British Thermal Unit A *Thermodynamics* term, abbreviated as **Btu**; a unit of heat energy in the English system; the amount of heat required to raise the temperature of 1 lbm (pound mass) of water at 68°F by 1°F; 1 Btu = 1.0551 kJ = 0.252 kcal or 1 kcal equals to 3.968 Btu; a typical match yields about one Btu of energy, if completely burned.

Brittle Material A *Materials Engineering* term; a material which shows very little plastic range beyond elastic limit; the breaking point of brittle materials lies close to the elastic limit; used in making *columns*, but not in making *springs* and *sheets*; a few examples include *cast iron* and *glass*; see also **Ductile Material**.

Brittle A *Manufacturing* term; lacking in deformability; for example, *cast iron is brittle.*

Brittleness A *Strength of Materials* term; the tendency of a material to fail suddenly by breaking without showing any permanent deformation before failure.

Broaching A *Manufacturing* term; a machining process used for mass production of complex geometry parts (complicated hole shapes) using a tool called broach which has a series of cutting teeth along its length.

Broad-Crested Weir A *Fluid Mechanics* term; a type of weir in the form of a *rectangular block* of suitable dimensions; used to measure the liquid flow rate in *open channels*; see also **Sharp-Crested Weir**.

Brushed DC Motor An *Electrical* term; a type of *direct current* (DC) motor which uses *internal commutation, stationary permanent magnets*, and *rotating electrical magnets*; some of its advantages include *low initial cost, high reliability*, and *simple control of motor speed*, whereas its disadvantages include *high maintenance cost* and *low life span*; see also **Brushless DC Motor**.

Brushless DC Motor An *Electrical* term; also called **BLDC Motor**; a type of *direct current* (DC) motor which uses a *rotating permanent magnet* in the rotor and *stationary electrical magnets* on the motor housing; some of its advantages include *simple design, long life span, little* or *no maintenance cost*, and *high efficiency*, whereas its disadvantages include *high initial cost* and *more complicated motor speed*; see also **Brushed DC Motor**.

BSI British Standards Institution; a *British* professional body.

Btu A *Thermodynamics* term: see **British Thermal Unit**.

Bucket In *operations management* terms, the principal *unit of time measurement* in a material requirements planning (MRP) system, which is usually *one week*; in *turbine* terms, the rotating blade of a turbine, which is also called **Runner Blade**; see also **Impeller Blade**.

Buckingham Pi Theorem A *Fluid Mechanics* term; a formalized procedure developed in 1915, named after *Edgar Buckingham* (1867–1940); a theory which organizes steps to ensure dimensional homogeneity; used for deducing the dimensionless groups appropriate for a given fluid mechanics or other engineering problem.

Buckle A *Casting* term; a type of casting defect which occurs in the form of *vee-shaped depressions* on the surface of a casting.

Buckling Load A *Strength of Materials* term related to a *column*; also called **Crippling Load** or **Critical Load**; the load which buckles or tends to buckle a column; used in finding the critical stress for columns.

Buffer Inventory An *Operations Management* term; see **Buffer Stock**.

Buffer Stock An *Operations Management* term; also called **Buffer Inventory, Safety Stock,** or **Safety Stock Inventory**; the inventories kept in stock (in large numbers) to meet the uncertainties in demand or to ensure its continuous flow; needed to take care of unusual product demand and uncertain lead time.

Build-to-Order An *Operations Management* term; see **Assemble-to-Order**.

Bulk Deformation Processes A *Manufacturing* term; the processes in which the surface area of a workpiece changes on the application of deformation forces; a few examples include *forging, rolling*, and *extrusion.*

Bulk Modulus of Elasticity A *Physics* and a *Strength of Materials* term; also called **Volume Modulus of Elasticity** or simply **Bulk Modulus**; an elastic constant denoted by K; defined as the compressive stress per unit volumetric strain; increases with increase in pressure of a fluid, as the resistance of fluid increases with increase in pressure; the bulk modulus for solids are much larger than that for liquids, which in turn, are much larger than that for gases; for a perfectly rigid body (incompressible), the value of K is infinity; the SI unit is pascal (Pa) which is equivalent to N/m^2; has the dimension of [ML^{-1} T^{-2}]; the values of K for some materials are tabulated below:

Materials	***K* (GPa)**
Steel	158
Copper	120
Iron	80
Aluminium	70
Glass	36
Mercury	25
Water	2.2; see also **Young's Modulus of Elasticity** and **Modulus of Rigidity**

Bulk Modulus A *Strength of Materials* term; see **Bulk Modulus of Elasticity**.

Bullwhip Effect A *Supply Chain Management* term; the phenomenon in which fluctuation and distortion of demand information increases (variability magnification) as it moves up the supply chain, from retailers, manufacturers, to suppliers; named so as inaccurate and distorted information travels up the supply chain like a bullwhip uncoiling; occurs due to lack of synchronization among the supply chain members which leads to flow of distorted information among them; tends to reduce the efficiency of a supply chain.

Buoyancy A *Fluid Mechanics* term; a phenomenon in which a solid body immersed partially or wholly in a fluid experiences an upward force exerted by the fluid.

Buoyant Force A *Fluid Mechanics* term; also called **Force of Buoyancy** or **Upward Thrust**; an *upward force* exerted by a fluid on a solid body immersed partially or wholly in the fluid; tends to lift the body and acts at the centre of buoyancy.

Business Ethics A *Management* term; the discipline connected with a systematic study of *morals*, *truth*, and *justice*; a tool for fair business.

Business Plan An *Operations Management* term; a statement which details the income, costs, and profits of a company; also shows a company's *budget*, *sources*, and *allocation of funds*.

Business Process Reengineering An *Operations Management* term, abbreviated as **BPR**; a process that is based on restructuring of the entire business processes needed to cut short the prices, and meet quality and service requirements.

Business Strategy A *Management* term; also called **Business-Level Strategy**; a strategy usually occurring at the business unit or product level; emphasizes on the improvement of the competitive position of a corporation's products or services in the specific industry or market segment served by that business unit; for example, *Mercedes-Benz* takes its skills in *R&D* and positions itself in the *luxury segment of the car market*, whereas *Coca-cola* uses its *marketing skills* to defend its niche against *Pepsi-cola*, and *Apple Computer* uses a *differentiation* competitive strategy which emphasizes *innovative products* with *creative design*.

Business-Level Strategy A *Management* term; see **Business Strategy**.

Business-to-Business A *Management* term, abbreviated as **B2B**; a type of *e-commerce* in which trade transactions take place through electronic communication (over the *Internet*) between companies and their suppliers; see also **Business-to-Consumer**.

Business-to-Consumer A *Management* term, abbreviated as **B2C**; a type of *e-commerce* in which trade transactions take place through electronic communication (over the *Internet*) between businesses and consumers; for example, *online retailing like Amazon* or *online stock brokerage*; see also **Business-to-Business**.

Butt Joint A *Joining* term; a joint produced by placing the two parts end-to-end; see also **Lap Joint**.

C

C Process A *Manufacturing* term related to *casting*; see **Shell Mould Casting**.

Cable A *Design* term; also called **Wire Rope**; an axially loaded flexible member subjected only to tensile forces; consists of a large number of wires wound in some particular manner; a common type of cable is formed by six strands wound helically around a central strand, and each strand, in turn, is constructed of many wires that are also wound helically; used to transmit large forces such as lifting and pulling heavy objects, raising elevators, guying towers, and supporting suspension bridges; unlike springs and prismatic bars, cables cannot resist compression, but can be bent to give any shape.

CAD A *Manufacturing* term; see **Computer-Aided Design**.

Cadmium Sulphide A *Materials Engineering* term; a semiconducting compound consisting of *cadmium* (Cd) and *sulphur* (S) having the chemical formula CdS; used as a relatively low-cost *solar cell* used for the conversion of *solar energy* to *electrical energy*; see also **Gallium Arsenide**.

Calcination A *Materials Engineering* term related to metallurgy; a chemical process of converting an *ore* into its *metallic oxide* by heating it strongly below its melting point either *in absence or limited supply of air*; commonly used to convert metal carbonates and hydroxides to their respective metal oxides; usually carried out in a *reverberatory furnace*; see also **Roasting**.

Calibration A *Metrology* term; a process of checking and adjusting a measuring device against a standard.

Calliper Gauge A *Metrology* term; see **Vernier Calliper**.

Calorie A *Physics* term; the MKS unit of heat; 1 cal = 4.2 J.

Calorific Value A *Chemistry* term; the amount of heat released when a unit mass of a substance (fuel) is completely burnt in presence of oxygen; commonly measured by a *bomb calorimeter*; expressed in kJ/kg or kJ/mol; for example, the calorific value of coal (anthracite) is 36 MJ/kg.

D. K. Singh, *Dictionary of Mechanical Engineering*,
https://doi.org/10.1007/978-981-99-1722-8_3

Calorimetry A *Chemistry* term; the science of measurement of the calorific value of fuels.

Cam and Follower A *Theory of Machines* term; a mechanism used to change continuous rotary motion into intermittent linear motion.

CAM A *Manufacturing* term; see **Computer-Aided Manufacturing**.

Cam A *Theory of Machines* term; a *rotating* or *reciprocating* element of a mechanism which imparts a rotating, reciprocating or oscillating motion to another element (follower) through its direct contact; constitute a *three-link mechanism* of higher pair with follower and frame which supports the cam; used to obtain complex motion which are otherwise difficult to obtain; used in *internal combustion engines for operating valves, clocks, shoe-making equipment*, etc.

Cantilever Beam A *Strength of Materials* term related to a *beam*; a statically determinate beam that is fixed at one end and free at another end; finds use in supporting balconies of houses.

Cantilever Laminated Spring A *Strength of Materials* term related to a *spring*; see **Quarter-Elliptic Leaf Spring**.

Capacitor A *Physics* term; an electrical device involving two electrodes separated by a dielectric; *used to store electrical charge.*

Capacity Flexibility An *Operations Management* term; the ability of a manufacturing unit to rapidly change (increase or decrease) the production levels or to shift production capacity quickly from one product or service to another; can be achieved through *flexible plants, flexible processes*, and *flexible workers.*

Capacity Requirement Planning An *Operations Management* term, abbreviated as **CRP**; a technique used to estimate the labour and equipment resources according to material requirements; see also **Material Requirements Planning**.

Capillarity A *Fluid Mechanics* term; a phenomenon of rise or fall of the level of a liquid in a capillary tube held vertically or inclined in the liquid; the narrower the tube, the greater is the rise or fall; a few common examples include the *absorption of ink by a blotting paper* (has fine pores which act as capillary tubes), *the rising of oil in a wick* (has fine threads which have narrow space), the *swelling of wood in rainy season due to moisture absorption* and the *soaking of water by a towel.*

Capillary Tube A *Fluid Mechanics* term; also called simply **Capillary**; a *glass tube* with a fine and uniform bore throughout its length and open at its both ends; a means to show the capillarity of a liquid.

Capillary A *Fluid Mechanics* term; see **Capillary Tube**.

Capital Assets A *Financial Management* term; see **Fixed Assets**.

Capital Budgeting A *Financial Management* term; a process of evaluating and selecting long-term investments which are consistent with the firm's goal of owner wealth maximization.

Capital Energy A *Power Plant* term; the energy derived from the fuels existing in the earth such as *fossil fuels* and *nuclear fuels.*

Capital Expenditure A *Financial Management* term; an outlay of funds by a firm which is expected to produce benefits over a period of time greater than one year; see also **Operating Expenditure.**

Capital Gain A *Financial Management* term; the amount by which the sale price of an asset exceeds the asset's initial purchase price.

Capital Goods A *Financial Management* term; the high cost items which are bought to improve the efficiency of a company; involve few buyers and are sold after lengthy *negotiations* and *evaluation*; a few examples include *buildings*, *machinery*, *equipment*, *vehicles*, and *tools*; these are not finished goods, instead used in making finished goods.

Capital Productivity A *Management* term; a ratio of the output to the capital input; a measure of *growth* of an organization; see also **Labour Productivity** and **Material Productivity**.

Capital A *Financial Management* term; a manufactured or created product (a physical entity) used solely for the production of goods and services to be consumed by individuals; includes long-lived items such as *land*, *buildings* (*factories* and *warehouses*), and *equipment* (*machines*, *trucks*) which are used in production and the long-term funds of a firm; all the items on the right-hand side of the balance sheet excluding current liabilities; a critical resource in growing economies where labour is combined with capital to produce goods and services; *stocks* and *bonds* are called financial capital.

Capital-Intensive Technology An *Economics* term; a technology which relies heavily on *capital* instead of human labour; see also **Labour-Intensive Technology**.

CAPP An *Operations Management* term; see **Computer-Aided Process Planning**.

Carbon Arc Welding A *Welding* term; an arc welding in which arc is created between a pure graphite (carbon) electrode and the workpiece; suffers from the problem of mixing of carbon disintegrated from the carbon electrode with the workpiece, thus making the joint brittle and weak, hence not used now a days; see also **Twin Electrode Carbon Arc Welding**.

Carbon Black A *Chemistry* term; see **Lamp Black**.

Carbon Dioxide Mould A *Casting* term, abbreviated as **CO_2 Mould**; a sand mould (composed of *clean sand* which is *free of oil*, *resin* or *clay* and *sodium silicate solution*) which uses *carbon dioxide gas* to harden the mould; used to obtain *very smooth and intricate castings.*

Carbon Dioxide A *Chemistry* term having the chemical formula CO_2; a dioxide of carbon obtained by burning carbon in excess of air (oxygen); prepared in laboratory by the action of cold dilute hydrochloric acid (HCl) on marble ($CaCO_3$); a *colourless* gas with a faint pungent odour and slight acidic taste; about 1.5 times heavier than air; dissolves in water to a small extent forming carbonic acid (H_2CO_3), a weak acid; also dissolves in alkalis to form salts; unlike carbon monoxide (CO), CO_2 is not poisonous; however, it does not support life in animals and human beings, but causes suffocation resulting in death due to lack of oxygen.

Carbon Monoxide A *Chemistry* term, abbreviated as **CO**; a neutral oxide of *carbon* (C) obtained by burning carbon in a limited supply of oxygen; a *colourless* and *odourless toxic gas* which is slightly soluble in water; its highly poisonous nature is due to its ability to form a complex with haemoglobin which is about 300 times more stable than the oxygen-haemoglobin complex which can prevent haemoglobin in the red blood corpuscles (RBC) from carrying oxygen round the body, thus resulting in ultimately death; an essential constituent of gaseous fuels like *water gas* ($CO + H_2$) and *producer gas* ($CO + N_2$); *a powerful reducing agent* which can reduce almost all metal oxides other than those of the alkali and alkaline earth metals, aluminium and a few transition metals; its reducing property is used in the extraction of many metals from their oxide ores; readily combines with chlorine (Cl_2) in the presence of sunlight to give *carbonyl chloride*, also called *phosgene* ($COCl_2$), which is an extremely poisonous gas.

Carbon Steel A *Materials Engineering* term; see **Plain Carbon Steel.**

Carbon Tetrachloride A *Chemistry* term, abbreviated as $\mathbf{CCl_4}$; a colourless, nonpolar liquid with high density (1.58 g/cm^3) prepared commercially by passing *chlorine* (Cl) into *carbon disulphide* (CS_2) in the presence of a small amount of *iodine* (I); a common *laboratory* and *industrial solvent* for nonpolar materials; used as a *fire-extinguisher* as it is noncombustible; also used in preparing *Freons* such as $CFCl_3$, CF_2Cl_2, and CF_3Cl, which are used as *refrigerants.*

Carbon A *Chemistry* term; a *nonmetallic chemical element* having symbol *C* with atomic number 6 and atomic weight 12; melts at 3727 °C and boils at 4833 °C; forms a great variety of compounds with other elements and is the major element in *organic compounds*; occurs hardly up to 0.1% in the earth's crust; it occurs in the *combined form* (with oxygen and hydrogen) in all *living tissues* belonging to plant or animal kingdom and in *petroleum* and *coal deposits*; also occurs as carbon dioxide in the atmosphere and as carbonates in rocks and minerals such as *limestone* and *dolomite*, and in pure form (as well as in free state) as diamond and graphite and more abundantly in a contaminated form as coal; exists in various allotropic forms, of which *diamond and graphite are crystalline*, while coal, wood charcoal, animal charcoal, lamp black, coke, and gas carbon are amorphous; exhibits the remarkable property of *catenation.*

Carbonyl Chloride A *Chemistry* term; see **Phosgene.**

Carburizing Flame A *Chemistry* term; also called **Reducing Flame**; a type of oxyacetylene gas flame which contains excess of fuel gas (*acetylene*) as compared to oxygen in the mixture; finds use in gas welding to join *high carbon steel*, *cast iron*, and *cemented carbide*; see also **Oxidizing Flame** and **Neutral Flame**.

Carburization A *Manufacturing* term related to *heat treatment*; see **Carburizing**.

Carburizing A *Manufacturing* term related to *heat treatment*; also called **Carburization**; a *case hardening process* used to produce a *hard* and *wear-resistant surface* in *low carbon steel* parts without affecting their inner core by heating the parts in the presence of a *carbon-rich substance* above *critical temperatures* for a prolonged time.

Cardice A *Chemistry* term; see **Solid Carbon Dioxide**.

Carlo Alberto Castigliano (1847–1884) An *Italian* mathematician and physicist who is well known for his *Castigliano's theorem*, which is used for determining the displacements in a linear-elastic system using the partial derivatives of strain energy.

Carnot Cycle A *Thermodynamics* term; an ideal, hypothetical thermodynamic cycle developed by a French mechanical engineer *Sadi Carnot* (1796–1832); a reversible cycle which consists of two reversible isotherms and two reversible adiabatics; offers the maximum possible thermal efficiency; the efficiency of the Carnot cycle is independent of the working medium, but depends upon the maximum and minimum temperatures used in the cycle; a benchmarked cycle (standard) cycle which other cycles try to approach its thermal efficiency; all other thermodynamic cycles have lower thermal efficiencies than that of a Carnot cycle.

Carnot Efficiency A *Thermodynamics* term; the maximum thermal efficiency of a reversible power cycle operating between two heat-reservoir temperatures; given as:

$$\eta_{\max} = 1 - \frac{T_L}{T_H}$$

where

T_L = absolute temperature of the low-temperature reservoir
T_H = absolute temperature of the high-temperature reservoir.

Carnot Heat Engine A *Thermodynamics* term; a heat engine which operates on the reversible *Carnot cycle*; the thermal efficiency of Carnot heat engine (as well as other reversible engines) is given as:

$$\eta_{\text{th(rev)}} = 1 - \frac{T_L}{T_H}$$

where

T_L = absolute temperature of the low-temperature reservoir

T_H = absolute temperature of the high-temperature reservoir.

Carnot Heat Pump A *Thermodynamics* term; a heat pump which works on the reversed Carnot cycle.

Carnot Principle A *Thermodynamics* term; the two deductions or corollaries derived from the second law of thermodynamics, given below as:

- No engine can be more efficient than a reversible engine operating between the same temperature limits.
- All reversible engines operating between the same temperature limits have the same efficiency.

Carnot Refrigerator A *Thermodynamics* term; a refrigerator which works on the reversed Carnot cycle.

Carriage Spring A *Strength of Materials* term related to a *spring*; see **Laminated Spring.**

Carriage A *Lathe* term; a principal part of a lathe which consists of *tool post, saddle, cross slide, compound slide,* and *apron*; used to carry the cutting tool and control its movements either along the axis of the workpiece (longitudinal direction) or perpendicular to it (transverse direction).

Cascade A *Fluid Mechanics* term related to a *pump*; also called **Blade Row**; an infinite series of two-dimensional blades of a pump.

Case Hardening A *Manufacturing* term related to *heat treatment*; also called **Surface Hardening** or **Skin Hardening**; a heat treatment process used to produce a *hard, wear resistant outer surface* (called *case* or *skin*) on a *mild steel* (*low-carbon steel*) specimen while leaving its interior *soft* and *tough*, using various atmospheres such as *carbon, nitrogen* or their combination up to a depth of about 2 mm; a few examples include *carburizing, nitriding, cyaniding,* and *flame hardening.*

Cast Cobalt Alloy A *Materials Engineering* term; see **Stellite.**

Cast Iron A *Materials Engineering* term; a *ferrous metal* which contains iron between 91 and 94% together with 2–4.5% of *carbon* (C), 0.7–3% of *silicon* (Si) and small quantities of *phosphorus* (P), *sulphur* (S), and *manganese* (Mn); carbon in cast iron is present partly as free carbon in the form of *graphite* and partly in combination with iron as *cementite* (Fe_3C); the least pure form of iron as well as *the least expensive of all metals* because of its abundance occurrence in nature; extracted chiefly from its oxide ore by heating it in the presence of coke and limestone in a blast furnace; melts at relatively low temperature (1200 °C); molten cast iron is easily cast into complex shapes, hence finds extensive applications in casting operations; also can be easily machined to required tolerances; important types of cast iron include *grey, white, malleable,* and *nodular.*

Castability A *Manufacturing* term related to *casting*; the ability of a material which enables it to be easily cast in the shapes of desired products; the *grey cast iron* has the *best castability* among the cast iron group as it has *high fluidity* and *slow solidification rate.*

Castigliano's Theorem A *Strength of Materials* term; a theorem named after an *Italian* mathematician and physicist *Carlo alberto Castigliano* (1847–1884); uses the work-energy principle to find deflection of a straight or curved beam under various loading conditions; states that the partial derivative of strain energy produced (within the elastic limit) with respect to a force producing it, gives the displacement (deflection) in the direction of force, and can be mathematically expressed as:

$$\delta k = \frac{\partial U}{\partial Pk}$$

alternatively, the partial derivative of strain energy produced (within the elastic limit) due to a force acting on a member with respect to any linear displacement, gives force in the direction of displacement, and can be mathematically expressed as:

$$Pk = \frac{\partial U}{\partial \delta k}$$

Casting Defects A *Casting* term; the defects produced in a casting; produced due to *faulty design of pattern, mould, core,* and *gating system* or due to defects in *pouring techniques*; a few examples include *shrinkage, cold shut, misrun, rat tail, hot tear, blow, pin hole,* and *core shift*.

Catalyst A *Chemistry* term; a material used to initiate a chemical reaction.

Catenation A *Chemistry* term; a remarkable property of carbon (C) to make chains of identical atoms through covalent bonds which is due to *small size* and *higher electronegativity* of carbon atom and unique strength of carbon–carbon bonds; helps to form millions of carbon compounds in which carbon atoms are bonded to one another.

Cathode An *Electrical* term; the positive terminal of a power source (battery) and the negative terminal of an electrolytic cell; see also **Anode**.

Cathodic Reaction A *Chemistry* term; the reduction reaction which occurs at the *cathode* in an electrochemical cell; see also **Anodic Reaction**.

Cation A *Chemistry* term; a *positively charged ion* which results due to removal of an electron from a neutral atom or molecule; a few examples include *sodium ion* (Na^{+1}), *calcium ion* (Ca^{+2}), and *ammonium ion* ($NH_4{}^{+1}$); see also **Anion**.

Cauchy Number A *Fluid Mechanics* term; a *non-dimensional* group, named after a *French* engineer, mathematician, and hydrodynamicist *Augustin Louis de Cauchy*

(1789–1857); used in the analysis of a *compressible flow*; denoted by C_a; expressed as:

$$C_a = \frac{\rho v^2}{K}$$

where

ρ = density of the fluid
v = velocity of the fluid
K = bulk modulus of the fluid.

Cauchy Stress A *Strength of Materials* term; see **Engineering Stress.**

Causal Model Method An *Operations Management* term related to *forecasting*; see **Regression Method.**

Cause-and-Effect Diagram A *Quality Management* term; also called **Ishikawa Diagram** named after *Kaoru Ishikawa*, a famous *Japanese* quality expert who evolved this diagram in 1943; a quality improving tool used to trace the customer complaints about quality problems; a graphical description of the elements of a specific quality problem and the relationship between those elements; also called **Fishbone Diagram**, because of its similarity to a fish's body, where the main quality problem is labelled as the fish's 'head', the major categories of potential causes as structural 'bones' and the likely specific causes as 'ribs'.

Cavitation A *Fluid Mechanics* term; a form of water damage associated with the collapse of small vapour bubbles in an adjacent liquid; occurs when the absolute pressure of the liquid becomes equal to or lower than its vapour pressure at a given temperature; for example, cavitation occurs at the tip of a boat's propeller blades, and causes damage to *impeller*, *casing*, and other components of a *centrifugal pump* due to pitting of their surfaces because of erosion; also makes the operation of a pump *noisy*, *produces vibration problems*, and *impairs the hydraulic performance.*

CBN A *Materials Engineering* term; see **Cubic Boron Nitride.**

C-Chart A *Quality Control* term; a *control chart for attributes* used for controlling the number of defects when more than one defect can be present in a product or service; a few examples include scratches on tables, fire alarms in a city, accidents at a particular intersection, complaints at a hotel, and typesetting errors in a newspaper; the control limits for a *c*-chart are defined as: $UCL = c + z\sqrt{c}$ and $LCL = c - z\sqrt{c}$, where '*c*' is the mean of the distribution (the target value), $\sqrt{c}$ is the standard deviation, and '*z*' is the number of standard deviation (usually $z = 3$); distribution of defects for a *c*-chart follows the *Poisson distribution*, which is based on the notion that defects occur over a continuous region and that the probability of two or more defects at any one location is negligible; see also **P-Chart.**

CCl_4 A *Chemistry* term; see **Carbon Tetrachloride.**

Celestial Energy A *Physics* term; the energy derived from the outer atmosphere; for example, *solar energy* and *lunar energy.*

Cell A *Manufacturing* term; see **Manufacturing Cell.**

Cellular Layout An *Operations Management* term related to *facilities planning*; a modern, expensive plant layout in which different jobs in small number are produced by grouping the machines in a cell (a complete manufacturing unit consisting of computer-controlled machines and automatic material handling system designed produce a single parts family, that is, a few parts all with common characteristics); an improved version of product layout as it offers increased manufacturing flexibility to produce a variety of quality products which meet customization; used in *flexible manufacturing system* (FMS).

Cellular Manufacturing A *Manufacturing* term; a manufacturing methodology in which manufacturing activities to process a workpiece are performed in cells consisting of several machines kept together and controlled by a central robot.

Celsius Scale A *Thermodynamics* term; also previously called **Centigrade Scale**; a *temperature scale* used in the SI system and now considered in the English system; named after the Swedish astronomer and inventor *Anders Celsius* (1701–1744); uses the unit *degree Celsius* (°C) for temperature measurement; the ice and steam points on this scale are assigned the values of 0 °C and 100 °C respectively; see also **Fahrenheit Scale, Kelvin Scale,** and **Rankine Scale.**

Cemented Carbide A *Materials Engineering* term; also called **Sintered Carbide** as it is manufactured by powder metallurgy technique using *sintering*; used in making cutting tools to be operated at high cutting speeds without affecting its hardness at increased temperatures; lacks toughness and is brittle; mostly used in the disposable *insert* form in order to replace it easily, when worn out.

Cementite A *Materials Engineering* term; a constituent of *steel* and *cast iron* having the chemical formula Fe_3C; it is *hard* and *brittle.*

Centigrade Scale A *Thermodynamics* term; see **Celsius Scale.**

Centimetre The CGS unit of length; the one-hundredth part of a metre (m); denoted by cm; 1 cm = 10 mm = 10^{-2} m; see also **Metre** and **Millimetre.**

Centipoise A *Fluid Mechanics* term; a smaller unit of viscosity equal to one-hundredth of a poise, that is, 1 centipoise (cP) is equal to 0.01 poise (P); the viscosity of fresh water at room temperature is about 1 cP; see also **Poise.**

Central Force A *Physics* term; a force which always acts towards or away from a fixed point (called the *centre of force*) and whose magnitude depends only on the distance from that point; a few examples include *gravitational force, elastic force,* and *electrostatic force.*

Central Workshop A *Manufacturing* term; a place where manufacturing activities are carried out on a small scale for the purpose of making a job from a study viewpoint; classified into *machine shop, foundry shop, fitting shop, welding shop, sheet metal shop,* and *carpentry shop.*

Centralization A *Management* term; an organizational setup in which the authority to make important decisions is retained by the managers at the top of the hierarchy; see also **Decentralization**.

Centre of Buoyancy A *Fluid Mechanics* term; the centre of gravity of the liquid displaced by a solid body immersed in the liquid; *for a homogeneous and isotropic body, the centre of gravity coincides with the centre of buoyancy.*

Centre of Flexure A *Strength of Materials* term; see **Shear Centre**.

Centre of Gravity A *Physics* term; a point where the entire weight of a body acts, and the total gravitational torque on the body is zero; may or may not lie within the body; the centre of gravity of a body may or may not be the same as its *centre of mass*; for a small size body having its mass uniformly distributed, the centre of gravity coincides with the centre of mass (in case of a solid sphere, where both points lie at its centre); for a large size body having non-uniform mass distribution, the centre of gravity does not coincide with the centre of mass; denoted by G; see also **Centroid**.

Centre of Mass A *Physics* term; a point where the whole mass of a body may be assumed to be concentrated; the centre of mass of regular shaped homogeneous bodies lies at their *geometric centres*; see also **Centre of Gravity** and **Centroid**; the position of the centre of mass of a few *regular shaped bodies* are tabulated below:

Body	Position of the centre of mass
Uniform solid or hollow sphere	Centre of the sphere
Uniform circular ring or disc	Centre of the ring or disc
Uniform rod	Centre of the rod
Rectangular or cubical block	Point of intersection of the diagonals
Hollow and solid cylinder	Middle point of the axis of the cylinder
A triangular plane lamina	Point of intersection of medians of the triangle
Plane lamina (square, rectangle or parallelogram)	Point of intersection of diagonals

Centre of Twist A *Strength of Materials* term; see **Shear Centre**.

Centric Axial Force A *Strength of Materials* term; an axial force acting through the centre of a section; see also **Axial Load**.

Centric Loading A *Strength of Materials* term; a loading condition in which a load acts through the *centroid* of the cross section of a body.

Centrifugal Casting A *Manufacturing* term related to *casting*; a casting process which uses centrifugal force to hold the molten metal against the inside of a rotating mould; produces high-grade cylindrical castings having superior mechanical properties; the typical products made by this method include *cast iron pipes*, *cylinder liners, gun barrels, pressure vessels, brake drums, gears*, and *flywheels*; classified into *true centrifugal casting, semi-centrifugal casting*, and *centrifuging*.

Centrifugal Force A *Physics* term; a force which occurs, when a body is moving along a circular path, by virtue of tendency of the body to regain its natural straight line path; equals in magnitude but opposite in direction to the centripetal force and is directed radially outwards; see also **Centripetal Force**.

Centrifugal Governor A *Theory of Machines* term; a type of governor which works on the principle of balancing a *centrifugal force* (produced due to rotation of governor balls about the axis of a shaft) by a *controlling force* that acts in the *radially inward direction*, and is provided by a *dead weight*, a *spring* or a *combination of the two*; much more frequently used than an inertia governor; a few examples of centrifugal governors include *Watt*, *Porter*, and *Proell* governors; see also **Inertia Governor**.

Centrifugal Pump A *Fluid Mechanics* term; also called **Radial-Flow Pump**; a *rotodynamic pump* in which a fluid enters axially (along the axis of a rotating shaft) in the centre of the pump, but leaves along the outer radius of the pump casing; increases the pressure of fluid by the centrifugal action produced by the rotation of its impeller blades; operates at *high speeds*; involves low initial cost; offers *continuous and smooth delivery*; suitable for *large discharge* and *low heads*; see also **Reciprocating Pump**.

Centrifuge Casting A *Manufacturing* term related to *casting*; see **Centrifuging**.

Centrifuging A *Manufacturing* term related to *casting*; also called **Centrifuge Casting**; *a type of centrifugal casting* used to produce large number of *small size castings* which need not be symmetrical using a rotating mould consisting of several mould cavities on its periphery; uses relatively low rotational speeds; often used to assist in the pouring of *investment casting trees* to produce gold inlays by dentist or to produce bearing caps or small brackets; see also **True Centrifugal Casting** and **Semi-Centrifugal Casting**.

Centripetal Acceleration A *Physics* term; the acceleration acting on a particle (or a body) which undergoes a uniform circular motion; acts along the radius and is directed towards the centre of the circular path (radially inward); the velocity and acceleration are always at right angles; its magnitude is given as:

$$a_c = \frac{v^2}{r} = \omega^2 r = v\omega$$

where

a_c = centripetal acceleration vector
v = velocity of the particle
r = radius of the circular path
ω = angular acceleration vector.

Centripetal Force A *Physics* term; a force required to move a body uniformly in a circular path (circle); acts along the radius and is directed towards the centre of the circular path; for example, the earth exerts the centripetal force on the moon (as a gravitational force) to make it revolving around the earth; the nucleus exerts the centripetal force on an electron (as an electrostatic force) to make it revolving around the nucleus; equals to the product of mass of the body and its centripetal acceleration; given as:

$$F = \frac{mv^2}{r} = m.r.\omega^2$$

where

F = centripetal force
m = mass of the body
v = linear velocity of the body
r = radius of the circular path
ω = angular velocity of the body; see also **Centrifugal Force.**

Centroid A *Strength of Materials* term; also called **Geometric Centre**; the centre of gravity of a plane area; a point where the entire areas of a plane figure can be assumed to be concentrated; see also **Centre of Gravity**.

Centroidal Axis A *Strength of Materials* term; the axis which passes through the centroid of a cross section of a body; classified into *horizontal centroidal axis* and *vertical centroidal axis.*

Centroidal Load A *Strength of Materials* term; a load which passes through the centroid of a cross section of a body.

Ceramic Materials A *Materials Engineering* term; the materials which exhibit *ionic* or *covalent* bonds or both; can withstand *high temperatures.*

Ceramic Mould Casting A *Manufacturing* term related to *casting*; also called **Ceramic Shell Process** as the mould is made of *ceramic shell* or **Shaw Process** named after its British inventors *Clifford and Shaw* or **Cope-and-Drag Investment Casting** as the two parts of the mould are made by using a *ceramic slurry* (the process is called *investment*) consisting of a mixture of *fine-grained zircon* ($ZrSiO_4$), *aluminium oxide* (Al_2O_3) and *fused silica* mixed with bonding agent; used to produce high precision parts made from *ferrous* and other *high-temperature alloys, stainless steels*, and *tool steels;* the typical products made by this method include *turbine impellers, machining cutters, metalworking dies* including *stamping dies* and *extrusion dies*; also used in making *plastic* and *glass moulds.*

Ceramic Shell Process A *Manufacturing* term related to *casting*; see **Ceramic Mould Casting**.

Ceramics A *Materials Engineering* term; the *inorganic crystalline materials* which are *strong* and *hard*, but also *very brittle*; have exceptional *compressive strength*; can withstand *very high temperatures*, and are *good electrical insulators*; *naturally occurring ceramics* include *beach sand* and *rocks*; usually processed by *powder metallurgy technique*; used to make *bricks*, *tableware*, *sanitaryware*, *refractories*, and *abrasives*; also used in *consumer products* such as *paints*, *plastics*, *tyres*, and for industrial applications such as *tiles for the space shuttles*.

Cermets A *Materials Engineering* term; a group of *cutting tool materials* consisting of *ceramics* (TiC, TiN, and TiCN) and *metals* (*nickel* and *molybdenum*), which has the advantages of both; suitable for rough cutting; offer higher cutting speed at increased temperatures.

Chain Drive A *Theory of Machines* term; a mechanism used to transmit *motion* or *power* between two shafts which are *shorter distance apart* as compared to belt or rope drive, using a chain that runs over the shafts and consists of rigid links; a *positive drive* of power transmission which offers a constant velocity ratio to some extent; see also **Belt Drive** and **Rope Drive**.

Chain of Command A *Management* term; the clear and distinct line of authority among the positions in an organization; consists of two components, namely *unity of command* and *scalar principle*.

Chain A *Theory of Machines* term; a series of links which are uniform in length and fitted together to form a continuous band.

Chain-Growth Polymerization A *Chemistry* term; see **Addition Polymerization**.

Chance Error A *Metrology* term; see **Random Error**.

Chance Variation A *Statistics* term; see **Natural Variation**.

Channel Induction Furnace See **Core Induction Furnace**.

Chaplet A *Manufacturing* term related to *casting*; a metallic *supporting device* used to prevent the core from shifting and keeping it in place.

Characteristic Gas Constant A *Chemistry* term; a gas constant (R) appearing in the equation of state for an *ideal gas* ($PV = mRT$, where $P =$ pressure, $V =$ volume, $T = $ absolute temperature, and $m =$ mass of the gas); related to *universal gas constant* (R_u) as $R = R_u/M$, where M is the molar mass (also called molecular weight) of the gas; see also **Universal Gas Constant**.

Charcoal A *Chemistry* term; an *amorphous* allotropic form of carbon (C) which is most reactive of all the allotropic forms and readily burns in air to give carbon dioxide (CO_2); obtained in four forms, namely *wood charcoal, animal charcoal, sugar charcoal*, and *activated charcoal*; used as an excellent *adsorbent* in gas masks and for removing the offensive odour from the air used in air-conditioning processes; used in decolourizing sugar syrup and for *refining* of *oils, fats, glycerine*, etc.; also finds use in the *treatment of drinking water after chlorination to adsorb the excess of chlorine.*

Charge Density A *Physics* term; the number of charge carriers per unit volume.

Charging Dock An *Electric Vehicle* term; see **Electric Vehicle Supply Equipment**.

Charging Point An *Electric Vehicle* term; the location—whether it is at home, at work or in a public space—where electric vehicles can be plugged in and charged.

Charging Station An *Electric Vehicle* term; see **Electric Vehicle Supply Equipment**.

Charging An *Electric Vehicle* term; the process of refilling of electric vehicles with electricity.

Charles Augustin de Coulomb (1736–1806) A famous *French* physicist whose name is inscribed on the *Eiffel Tower*; well known for his *Coulomb's law of electrostatics* and the coulomb (C), the SI unit of electric charge; derived the *torsion formula* in about 1775.

Charpy Test A *Strength of Materials* term; the most commonly used *impact test* used to determine the impact properties of a material like brittleness and impact strength; the Charpy test is conducted using a swinging pendulum which is released from a fixed height to strike on the notched portion of a 55 mm × 10 mm × 10 mm test specimen, and reading the energy absorbed by the specimen from a dial; see also **Izod Test**.

Check Valve A *Fluid Mechanics* term; a type of one-way valve which allows a fluid to flow in one direction only.

Chemical Blanking A *Nonconventional Machining* term; a nonconventional *chemical machining* (CM) process which removes materials from a workpiece by chemical dissolution in the form of metallic salts; a few chemical blanking products include *printed circuit boards, decorative panels*, and *thin parts.*

Chemical Bond A *Chemistry* term; the attractive force which holds together the constituent atoms of a molecule; the base of formation of molecules.

Chemical Compounds A *Chemistry* term; compounds of two or more than two elements chemically formed; for example, *water* (H_2O) is a chemical compound of *hydrogen* and *oxygen.*

Chemical Energy A *Chemistry* term; the energy which results from a chemical change or chemical reaction between molecules of different substances; a stored form of energy; stored in fossil fuels such as coal and wood, and petroleum fuels such as petrol, diesel and kerosene oil; burning of 1 kg of coal produces 3×10^7 J of energy.

Chemical Engineer A technical person who is responsible for the industrial-scale production of chemicals and pharmaceuticals, and for managing the by-products of operations in the *petroleum*, *electronics*, and *biotechnology industries.*

Chemical Equilibrium A *Chemistry* term; a state of equilibrium which occurs when there is no chemical reaction or phase change in the system; see also **Mechanical Equilibrium** and **Thermal Equilibrium**.

Chemical Machining A *Nonconventional Machining* term, abbreviated as **CM**; the oldest nonconventional machining process which uses a strong chemical reagent, called *etchant*, to remove a layer of material either from the entire surface or form selected portions of a workpiece; used to produce very intricate shapes, internal as well as external, which are otherwise difficult to produce; classified into *chemical milling*, *chemical blanking*, and *photochemical blanking*.

Chemical Milling A *Nonconventional Machining* term; a nonconventional *chemical machining* (CM) process which replaces conventional milling by chemical action, hence is so named; a chemical machining process which uses etchant to remove materials from a workpiece, usually to produce shallow cavities on plates, sheets, forgings, and extrusions; extensively used in *aerospace industry* to remove thin layer of materials from many aircraft components to *reduce their weights* and to *improve the stiffness-to-weight ratio.*

Chezy Coefficient A *Fluid Mechanics* term; a coefficient which appears in the *Chezy equation*, named after a *French* physicist and hydraulic engineer *Antoine Chezy* (1718–1798); given as:

$$C = \sqrt{\frac{2g}{C_f}}$$

where

C = Chezy coefficient
C_f = skin friction factor.
g = acceleration due to gravity = 9.81 m/s^2.

Chezy Equation A *Fluid Mechanics* term; one of the oldest equations in the area of fluid mechanics developed in 1768 by a *French* Physicist and hydraulic engineer *Antoine Chezy* (1718–1798), who designed a canal for the *Paris water supply system*, which connects the flow velocity to channel geometry parameters like hydraulic radius and bed slope; given as:

$$V = \sqrt{\frac{2g}{C_f}} \times \sqrt{(R_h S_b)}$$

where

C_f = skin friction factor
R_h = hydraulic radius
S_b = bed slope of the channel
g = acceleration due to gravity = 9.81 m/s^2.

Chief Executive Officer A *Management* term, abbreviated as **CEO**; a high level manager who is responsible for planning the long-term strategic objectives of the organization and managing the organization's relationship with external stakeholders; the most powerful person in a company who controls the allocation of money; reports to the chair of the board of directors (who has the most legal authority); often the same person is both chief executive officer and the chair of the board; selects the key executives to occupy the topmost levels of the managerial hierarchy; his actions and reputation have a major impact on the organization.

Chief Operating Officer A *Management* term, abbreviated as **COO**; a president who is next in line for the job of a chief executive officer (CEO).

Chill A *Manufacturing* term related to *casting*; a *metallic object* of high heat capacity and high thermal conductivity placed in the mould or mould cavity to *increase the cooling rate of castings* or to provide *uniform* or *directional solidification*; usually made of same material as that of casting; classified into *external* and *internal* chills.

Chilled Cast Iron A *Materials Engineering* term; see **White Cast Iron**.

Chilled Water A *Heat Transfer* term; the low-temperature water used in *central air-conditioning plants* to produce cooling effect by the absorption of heat.

Chimney A *Power Plant* term; a *masonry structure* with an opening or outlet through which the hot flue gas can be emitted to the ambient atmosphere.

Chip Thickness Ratio A *Manufacturing* term related to *cutting*; also called **Cutting Ratio**; a ratio of the uncut chip thickness (depth of cut) to the chip thickness; a ratio of the chip length to the length of uncut material; a ratio of the chip velocity to the cutting tool velocity; always less than unity; its reciprocal is called *chip reduction coefficient*.

Chip In *electronics* terms, a thin slice of crystalline semiconductor upon which an electrical circuitry is produced by controlled diffusion; in *metal cutting* terms, the undesired part of a material removed from it during its machining by a cutting tool, and may be classified into *continuous* and *discontinuous* chips.

Chisel A *Manufacturing* term; a workshop tool used for cutting metals or nicking prior to breaking or chipping away pieces of metals from a workpiece; classified into *hot* and *cold* chisels depending upon whether they are used for cutting hot or cold metal respectively; cold chisels have higher cutting angle (60°) compared to hot chisels (30°), as the former is required to transmit more force on the workpiece; usual cross sections of a chisel include *rectangular*, *hexagonal,* or *octagonal*; a few important types of chisels include *flat*, *cross cut*, *half round*, *diamond point,* or *side chisels.*

Chord A *Mathematics* term; a line segment with its end-points lying on a circle; the diameter of a circle is its largest chord.

Chrome Plating A *Chemistry* term; a process of providing a coating cover of *chromium* (Cr) over *iron* (Fe) and other base metals to make them resistant to corrosion; stainless steel uses this process to make it corrosion-resistant.

Chromium A *Chemistry* term; a metallic element having symbol Cr with atomic number 24; a bluish-white lustrous metal which melts at 1875 °C and does not tarnish in air at ordinary temperatures; an important alloying element which is added to steel to increase its corrosion resistance; *stainless steel is spotless because of the presence of chromium*; used extensively for plating metals (called *chrome plating*) as it is very resistant to corrosion; alloys of *nickel* (Ni) and *chromium* (Cr) containing about 20% Cr are used in making *electrical resistances* for *heaters* and *electric furnaces* because of their tendency to resist oxidation effectively even at red heat.

Chromium-Nickel Steel A *Materials Engineering* term; see **Austenitic Stainless Steel.**

CI Engine An *IC Engine* term; see **Diesel Engine.**

CIM A *Manufacturing* term; see **Computer Integrated Manufacturing.**

Circle A *Mathematics* term; the locus of a point which moves in a plane such that its distance from a fixed point in that plane is always constant (the fixed point is called *centre* of the circle and fixed distance its *radius*); a round-shaped closed plane figure; the equation of a circle with its centre (h, k) and radius a, area (A) and circumference (P) are expressed as:

$$(x - h)^2 + (y - k)^2 = a^2$$
$$A = \pi r^2 = \frac{\pi}{4} d^2$$
$$P = 2\pi r = \pi d$$

where

r = radius of the circle
d = diameter of the circle = $2r$

π = a constant = $\frac{22}{7}$.

Circular Frequency A *Physics* term; see **Angular Frequency**.

Circular Pitch A *Gear* term; also called simply **Pitch**; the linear distance traveled in one revolution of a screw; the distance measured parallel to the axis between a point on one tooth to a corresponding point on an adjacent tooth; equals to the sum of the tooth thickness and width of tooth space.

Circular Region A *Mathematics* term; a region circular in shape; the interior part of a circle.

Circular Tube A tube having a circular cross section.

Circulation A *Fluid Mechanics* term; the line integral of the tangential velocity component about any closed curve fixed in a flow; denoted by Γ; expressed as:

$$\Gamma = \oint V.dS$$

where

V = velocity of the fluid
dS = elemental length.

Circumference A *Mathematics* term; the perimeter of a circle; equals to $2\pi r$ or πd, where r is the radius of the circle (equal to one-half of the diameter (d)) and π is a constant which has the value of $\frac{22}{7}$.

Circumferential Strain A *Strength of Materials* term; also called **Hoop Strain**; the strain produced in the circumference of a cylindrical vessel.

Circumferential Stress A *Strength of Materials* term; also called **Hoop Stress**; a tensile stress produced in the circumference of a pressure vessel.

Civil Engineer A technical person who designs and constructs roads, buildings, airports, tunnels, dams, bridges, and water supply systems; the environmental concerns are often at the centre of a civil engineer in order to prevent contamination of ground-water, soil, and the atmosphere.

C-Kanban An *Operations Management* term; see **Withdrawal Kanban**.

Clamp Coupling A *Design* term; also called **Compression Coupling** or **Ribbed Coupling**; a type of *rigid coupling*; a modification and improvement of *sleeve coupling*; consists of two parts to be clamped tightly against the surface of the shaft ends by through bolts, and are machined to fit the shaft, and finished off around the periphery and on both ends.

Clamped Beam A *Strength of Materials* term related to a *beam*; see **Fixed Beam**.

Clapeyron Equation A *Thermodynamics* term; named after a *French* engineer and physicist *Benoit Paul Emile Clapeyron* (1799–1864); relates changes in properties for a change of phase of a substance; used to determine the *enthalpy change* associated with a *phase-change* (h_{fg}) from a knowledge of pressure (p), volume (v), and temperature (T) data alone, by simply measuring the slope of the *saturation curve* on a *P–T* diagram and the specific volume of saturated liquid and saturated vapour at the given temperature; given as:

$$\left(\frac{dP}{dT}\right)_{\text{sat}} = \frac{h_{fg}}{Tv_{fg}}$$

Classical Physics Physics based on a *macroscopic domain* which discusses the matter in its *totality*, includes subjects like *mechanics, thermodynamics, electrodynamics*, and *optics*.

Clausius Inequality A *Thermodynamics* term; a corollary of the second law of thermodynamics; provides the basis of entropy production and entropy balance for a system; for any thermodynamic cycle, the Clausius inequality is given as:

$$\oint\left(\frac{\partial Q}{T}\right) \leq 0$$

where

δQ = heat transfer
T = absolute temperature of heat transfer

the *equality sign* applies when there are *no internal irreversibilities* when the system executes the cycle, and the *inequality sign* applies when *internal irreversibilities are present*.

Clausius Statement A *Thermodynamics* term; the second statement of the second law of thermodynamics, named after a German mathematical physicist *Rudolph Julius Emmanuel Clausius* (1822–1888); states that no device can transfer heat from a cooler body to a hotter body without leaving an effect on the surroundings (without consumption of some energy in the form of work); the statement relates to *refrigerators* or *heat pumps*; see also **Kelvin–Planck Statement**.

Clearance Volume An *IC Engine* term; the minimum volume occupied by the air in a cylinder of an engine when the piston is at top dead centre (TDC).

Clearance A *Design* term; used in the context of mating of cylindrical parts (such as bolt and a hole) in which the internal member is smaller than the external member; the opposite of *interference*; see also **Interference**.

Clockwise Moment A *Physics* term; a moment which can rotate a body in the direction of movement of the hands of a clock; considered as a *negative moment*; see also **Counterclockwise Moment**.

Close-Coiled Helical Spring A *Strength of Materials* term related to a *spring*; a cylindrical helical spring with a small helix angle which can be approximated to zero so that the plane of coil of spring and the axis of spring are normal to each other; see also **Open-Coiled Helical Spring**.

Closed Circuit An *Electrical* term; a path for electrical flow if voltage is applied to it; see also **Open Circuit**.

Closed Curve A *Mathematics* term; a curve whose start and end points coincide; a curve which cuts or crosses itself; see also **Simple Curve**.

Closed System A *Thermodynamics* term; a thermodynamic system of fixed mass; does not allow mass transfer across its boundary, but allows only energy transfer; see also **Open System**.

Closed-Die Forging A *Manufacturing* term; also called **Impression-Die Forging**, as it produces the impression of the die on the workpiece; a hot metal working process used to shape a heated workpiece by placing it between two shaped-dies (that confine metal between them) and then pressed or hammered; see also **Open-Die Forging**.

Closed-Loop System An *Electrical* term; a control system which considers the output of a system and makes adjustments based on that output; see also **Open-Loop System**.

CM A *Nonconventional Machining* term; see **Chemical Machining**.

CMI An *Operations Management* term; see **Co-Managed Inventory**.

CMM A *Manufacturing* term: see **Coordinate Measuring Machine**.

CN A *Chemistry* term; see **Coordination Number**.

CNC Lathe A *Manufacturing* term; a *type of lathe* based on *computer numerical control* (CNC); capable of machining complex part geometries than a conventional lathe; widely used for contour turning operations with precise accuracy.

CNC A *Manufacturing* term; see **Computer Numerical Control**.

CO A *Chemistry* term; see **Carbon Monoxide**.

CO_2 Mould A *Casting* term; see **Carbon Dioxide Mould**.

Coal A *Chemistry* term; a contaminated, *amorphous* form of carbon abundantly available in nature inside the earth's crust and whose purest form is *anthracite* which contains about 95% of carbon; other varieties of coal include *steam coal* (93% carbon), *bituminous coal* (88% carbon), *lignite* (67% carbon), and *peat* (60% carbon); extensively used as a fuel in *boilers*, *heat engines*, and *furnaces*.

Coated Cemented Carbide A *Materials Engineering* term; a carbide material having a thin layer coating of *titanium carbide*, *titanium nitride*, *titanium carbonitride* or *aluminium oxide*; used in making *cutting tools* offering increased wear resistance, which allows increased cutting speeds at elevated temperatures; manufactured by *chemical vapour deposition* or *physical vapour deposition* methods.

Coated Electrode A *Welding* term; see **Covered Electrode**.

Coaxial A *Design* term; having axes in the same straight line.

Cobalt Alloys A *Materials Engineering* term; contains cobalt as the main constituent besides other elements present in various amounts; important cobalt alloys include *cobalt steel*, *stellite*, and *cochrome*.

Cobalt Steel A *Materials Engineering* term; an alloy of *cobalt* (Co) and *iron* (Fe) containing 35% Co; used in making permanent magnets.

Cobalt A *Chemistry* term; a metallic element having symbol Co with atomic number 27 which melts at 1495 °C; a white coloured hard, malleable and ductile metal, which has ferromagnetic nature but much less than iron; used as an important *alloying element* in making many useful alloys such as *cobalt steel*, *stellite*, and *cochrome*; also used for electroplating iron and other base metals using cobalt (II) ammonium sulphate as the bath.

Cochrome A *Materials Engineering* term; an alloy of *cobalt* (Co), *iron* (Fe), and *chromium* (Cr); used in making electrical goods.

Code of Ethics A *Management* term; a formal document which states an organization's primary values and the ethical rules followed by the organizational members; helps to maintain discipline in the organization.

Code A *Metrology* term; a set of specifications for the *analysis*, *design*, *manufacture*, and *construction* of something for achieving a specified degree of *safety*, *efficiency*, and *performance* or *quality*; see also **Standard**.

Coefficient of Contraction A *Fluid Mechanics* term; *a jet coefficient*; a ratio of the area of cross-section at vena contracta to the area of cross-section of the orifice; denoted by C_c; varies between 0.5 and 1.0; related to other jet coefficients as given below:

$$C_c = \frac{C_d}{C_v}$$

where

C_d = coefficient of discharge
C_v = coefficient of velocity

see also **Coefficient of Discharge** and **Coefficient of Velocity**.

Coefficient of Discharge A *Fluid Mechanics* term; *a jet coefficient*; a ratio of the actual discharge to the theoretical discharge of a fluid; denoted by C_d; varies between 0.59 and 0.68; related to other jet coefficients as given below:

$$C_d = C_c \times C_v$$

Where

C_c = coefficient of contraction
C_v = coefficient of velocity

see also **Coefficient of Contraction** and **Coefficient of Velocity**.

Coefficient of Dynamic Friction A *Physics* term; see **Coefficient of Kinetic Friction.**

Coefficient of Elasticity A *Strength of Materials* term; see **Elastic Constants.**

Coefficient of Kinetic Friction Also called **Coefficient of Dynamic Friction**; a ratio of the force of kinetic friction to the normal reaction between two solid bodies in contact, when one body is moving over the surface of another body; denoted by μ_k; always less than the coefficient of limiting friction; see also **Coefficient of Limiting Friction**; the values of coefficient of kinetic friction between some pairs of materials are tabulated below:

Surfaces in Contact	**Coefficient of Kinetic Friction**
Wood on Wood	0.20
Wood on Leather	0.40
Steel on Steel (mild)	0.57
Glass on Glass	0.40
Rubber on Concrete	0.80
Copper on Steel	0.40
Lubricated ball bearings	< 0.01
Metal on ice	0.02

Coefficient of Limiting Friction A *Physics* term; also called **Coefficient of Static Friction**; a ratio of the force of limiting friction to the normal reaction between two solid bodies in contact; depends upon the nature of surfaces in contact, that is, whether they are dry or wet, rough or smooth, polished or unpolished; also depends on the material of the surfaces in contact; denoted by μ_s; equals to the tangent of the angle of friction (θ) between two surfaces in contact ($\mu_s = \tan\theta$); always greater than the coefficient of kinetic friction; see also **Coefficient of Kinetic Friction**; the values of coefficient of limiting friction between some pairs of materials are tabulated below:

Surfaces in Contact	Coefficient of Limiting Friction
Wood on Wood	0.40
Wood on Leather	0.50
Steel on Steel (mild)	0.74
Glass on Glass	0.90
Rubber on Concrete	1.00
Copper on Steel	0.50
Lubricated ball bearings	< 0.01
Metal on ice	0.02

Coefficient of Performance A *Refrigeration* term, abbreviated as **COP**; also called **Energy Ratio**; a ratio of the useful heat to the work input; used to measure the performance of a refrigerating machine or a heat pump; the COP of a heat pump is greater than the COP of a refrigerator by unity; given as:

$$(\mathrm{COP})_{\mathrm{Refrigerator}} = \frac{\text{Heat rejected (useful heat)}}{\text{Workinput}}$$

$$(\mathrm{COP})_{\mathrm{Heat\ pump}} = \frac{\text{Heat added (useful heat)}}{\text{Work input}}$$

Coefficient of Resilience A *Physics* term; see **Coefficient of Restitution**.

Coefficient of Restitution A *Physics* term; also called **Coefficient of Resilience**; a ratio of the relative velocity of separation after collision to the relative velocity of approach before collision; denoted by *e*; for a *perfectly elastic collision*, the relative velocity of separation after collision is equal to the relative velocity of approach before collision, hence the value of *e* is equal to 1; for a *perfectly inelastic collision*, the relative velocity of separation after collision is 0, hence value of *e* for such collision is 0; for all other collisions, the value of *e* lies between 0 and 1.

Coefficient of Rolling Friction A *Physics* term; the amount of length which a rolling object presses on a surface on which it rolls; the harder the rolling object, the less it deforms, and has very little coefficient of rolling friction; usually expressed in *mm*; the values of coefficient of rolling friction between some pairs of materials are tabulated below:

Surfaces in Contact	Coefficient of Rolling Friction (mm)
Steel on Steel	0.03
Steel on Wood	0.20
Tyre on Road	0.10
Hardened steel ball bearings	0.10

Coefficient of Static Friction A *Physics* term; see **Coefficient of Limiting Friction**.

Coefficient of Thermal Expansion A *Physics* term; a property of a material related to the expansion or contraction of the material when it is heated or cooled; the unit is per degree Celsius (1/°C) or per Kelvin (1/K).

Coefficient of Velocity A *Fluid Mechanics* term; *a jet coefficient*; a ratio of the actual jet velocity to the theoretical jet velocity of a fluid; denoted by C_v; has the maximum value among all the three jet coefficients; related to other jet coefficients as given below:

$$C_v = \frac{C_d}{C_c}$$

where

C_c = coefficient of contraction
C_d = coefficient of discharge.

see also **Coefficient of Discharge** and **Coefficient of Contraction**.

Coefficient of Viscosity A *Fluid Mechanics* term; see **Dynamic Viscosity**.

Cogeneration Plant A *Power Plant* term; a plant which produces electrical power along with meeting the process heat requirements of certain industrial processes.

Cogeneration A *Power Plant* term; the production of more than one useful form of energy (such as *process heat* and *electrical power*) from the same energy source.

Cohesive Force A *Physics* term; also called **Force of Cohesion**; the *intermolecular force* acting between *molecules of the same substance*; similar to adhesive force, it also does not obey the inverse square law of forces and increases rapidly with decrease in distance between the molecules (varies inversely as the seventh power of the distance between the molecules); *has the highest value in solids, lesser in liquids and least in gases*; mercury (Hg) does not wet the surface of a glass container because the force of cohesion amongst molecules of mercury is stronger than the force of adhesion between the molecules of mercury and glass; see also **Adhesive Force**.

Coining A *Manufacturing* term; a closed-die cold forging operation used to produce *coins, medals, jewellery,* and other products with exact size and fine details using a shaped die-punch assembly; see also **Embossing**.

Coke A *Chemistry* term; a greyish black hard, brittle substance which consists chiefly of carbon together with small amounts of *hydrogen, oxygen, nitrogen, sulphur,* and *phosphorus*; obtained as a residue left in the *retort* on its roof and sides, when coal is subjected to *destructive distillation* involving strong heating in the absence of air; used in the production of steel.

Cold Extrusion A *Manufacturing* term; an extrusion process carried out at room temperature; requires large extrusion force than hot extrusion; useful for metals and alloys having good ductility like *aluminium, copper, magnesium* and *their alloys*, and *steels* and *stainless steels*; see also **Hot Extrusion**.

Cold Forging A *Manufacturing* term; a cold metal working process used to shape a workpiece by repeatedly hammering it or by applying compressive forces through a ram at *room temperature*; requires greater forces as compared to hot forging; the surface finish and dimensional accuracy of the cold forged products are superior to hot forged products; see also **Hot Forging** and **Warm Forging**.

Cold Shortness A *Manufacturing* term; the ability of a metal to become brittle at lower or room temperature; see also **Hot Shortness**.

Cold Shut A *Casting* term; a type of casting defect and a variation of misrun; occurs *due to lack of molten metal fluidity* (because of low metal temperature) in the centre of a casting having gates at its both sides; see also **Misrun**.

Cold Working A *Manufacturing* term; a metal working process in which the metal is shaped below the recrystallization temperature of metals (usually room temperature) in the plastic region of the stress–strain curve, where metals may be hammered, rolled or drawn; results in a large increase in yield strength, an increase in ultimate strength and a decrease in ductility; the cold-worked parts have *good finish*, are *more accurate*, and *require less machining*; a few examples include *drawing, extrusion*, and *bending*; see also **Hot Working**.

Cold-Chamber Die Casting A *Manufacturing* term related to *casting*; a die casting process in which there is a separate melting furnace to produce molten metal; usually suitable for *high melting point alloys* of *aluminium, magnesium*, and *copper* along with *ferrous metals*; see also **Hot-Chamber Die Casting**.

Collar Bearing A *Design* term; a collar-incorporated bearing generally used to *support a horizontal shaft* under axial load, where the shaft extends through and beyond the bearing; consists of one or more collars which are either fitted to or integral with the shaft; in a single collar bearing, the collar rotate against the solid bearing surface, whereas in multi-collar bearing, more collars rotate against the stationary split bearing surfaces; see also **Foot-Step Bearing**.

Co-Managed Inventory An *Operations Management* term, abbreviated as **CMI**; a system of inventory management in which the seller manages the inventory of *consumable items* at customer's location/premises; see also **Vendor Managed Inventory**.

Combination Layout An *Operations Management* term related to *facilities planning*; see **Hybrid Layout**.

Combined Cycle A *Thermodynamics* term; a thermodynamic cycle which is a combination of the *Brayton cycle* and the *Rankine cycle*; the waste energy of the high-temperature exhaust gases of gas turbine (*Brayton cycle*) is supplied to the boiler of the *Rankine cycle*, thus increasing the power output of the cycle.

Combined Loading A *Strength of Materials* term; a loading condition in which a structural or machine member is subjected to more than one type of loading; a few examples include *combined bending and torsion, combined bending and axial forces,* and *combined torsion and axial forces.*

Combined Stress A *Strength of Materials* term; the stress caused due to several loadings (two or more) applied simultaneously; the effect is determined by using the method of superposition in which the stress caused by each load is determined separately and then added algebraically.

Combustion Chamber An *IC Engine* term; a zone in the engine cylinder where combustion of the fuel–air mixture takes place.

Committee A *Management* term; a group of persons to whom some matter is committed; a tool used to discuss about a matter collectively; headed by a chairperson; the members of the committee must be representative of the interests they are expected to serve and must also possess the required authority to make decisions, make recommendations or give the chairperson some insights into the issue under consideration; the committee should not be too large to avoid the problem of inadequate communication among its members and also should not be too small to avoid the meaningful discussion and promote vested interests.

Common Cause Variation A *Statistics* term; see **Natural Variation.**

Communication Satellite A *Physics* term; see **Geostationary Satellite.**

Competitive Advantage A *Management* term; the ability of one company to outperform another by virtue of its managers' ability to create more value from the resources at its disposal; a tool to increase the market share of a company.

Competitive Benchmarking A *Quality Management* term; the continuous rating of a company's products and practices against the world's best firms, including other organizations in other industries; see also **Functional Benchmarking.**

Complete Combustion An *IC Engine* term; a combustion process which results in products of combustion such as H_2O and CO_2, but no CO, C, H_2, or OH; see also **Incomplete Combustion.**

Component A single handmade or machined part which belongs to a larger whole.

Composite Beam A *Strength of Materials* term; a beam fabricated from more than one material; a few examples include *bimetallic beams* (such as those used in *thermostats*), *plastic coated pipes*, and *wood beams with steel reinforcing plates*; see also **Sandwich Beam.**

Composite Material A *Materials Engineering* term; see **Composite**.

Composite Mould A *Casting* term; a mould made of *two or more different materials* such as *sand*, *plaster*, *graphite*, and *metal*; used in *shell moulding* and other casting processes for producing *complex shapes* such as *impellers for turbines*.

Composite Number A *Mathematics* term; a number which has at least one factor other than 1 and the number itself; a few examples include 4, 6, 8, 9, 10, 12, etc.; see also **Prime Number**.

Composite Shaft A *Strength of Materials* term; a shaft which consists of two or more dissimilar sections of the same or different materials.

Composite A *Materials Engineering* term; also called **Composite Material**; a material composed of two or more materials (*metals*, *ceramics*, *glass* or *polymers*) or *phases*; has a combination of properties not obtainable otherwise like *stiffness*, *strength*, *weight*, *high-temperature performance*, *corrosion resistance*, *hardness*, or *conductivity*.

Compound Cylinder A *Strength of Materials* term related to a *pressure vessel*; a high-pressure bearing capacity cylindrical container which consists of two cylinders that are shrink-fitted after placing one over another.

Compound Gear Train A *Theory of Machines* term related to *gear*; a gear train in which each intermediate gear shaft between the follower and the driver carries two gears that rotate at the same speed; see also **Simple Gear Train**.

Compound Semiconductors An *Electronics* term; the semiconductors available in the compound form; examples include *gallium arsenide* (GaAs) and *cadmium sulphide* (CdS); see also **Elemental Semiconductors**.

Compressibility Factor A *Thermodynamics* term; a dimensionless factor which accounts for the deviation of real gases from ideal-gas behaviour; denoted by Z; given as:

$$\mathrm{Z} = \frac{Pv}{RT}$$

where

P = pressure of the gas
v = volume occupied by the gas
R = universal Gas constant
T = absolute temperature of the gas

$Z = 1$ for ideal gas; for the real gases, Z can be greater than or less than unity; the farther Z is from unity, the more it deviates from the ideal-gas behaviour.

Compressibility A *Fluid Mechanics* term; the volumetric strain per unit compressive stress; the reciprocal of bulk modulus of elasticity (K) of a material; denoted by β; the SI unit is $N^{-1}m^2$; see also **Bulk Modulus of Elasticity**.

Compressible Fluid A *Fluid Mechanics* term; a fluid which can be compressed and whose density varies with its pressure; gases are compressible fluids; see also **Incompressible Fluid**.

Compression Coupling See **Clamp Coupling**.

Compression Ratio An *IC Engine* term; a ratio of the volume of a fluid before compression to the volume of the fluid after compression (clearance volume); a ratio of the volume in the cylinder at BDC to the volume in the cylinder at TDC; used to find the efficiency of an *IC engine* or a *reciprocating pump*; see also **Pressure Ratio**.

Compression Test A *Strength of Materials* term; a *destructive* test conducted on a solid material to know its compressive strength; involves a compressive load to press a solid cylindrical test specimen placed between two flat dies; uses a stress–strain curve to know the behaviour of the material; useful in manufacturing operations such as *forging, rolling,* and *extrusion*; the brittle materials are more stronger in compression than in tension as opposite to the ductile materials which are more stronger in tension than in compression.

Compression Work A *Thermodynamics* term; the work input needed in compressing a working fluid during the compression stroke of a cycle.

Compression A *Physics* term; the act of two forces pushing on an object in opposite directions, causing it to compact; see also **Tension**.

Compressive Strain A *Strength of Materials* term; a type of longitudinal strain produced on account of decrease in length of a body by a compression force; calculated as a ratio of decrease in length of a body to its original length; see also **Tensile Strain**.

Compressive Strength A *Strength of Materials* term; the resistance of a material against a compressive force which prevents its failure by *crushing*.

Compressive Stress A *Strength of Materials* term; a negative normal stress caused by a compression force; results in decrease in length (called contraction) of a body in the direction of force applied; see also **Tensile Stress**.

Compressor A *Fluid Mechanics* term; a *turbomachine* which is basically a *gas pump*; used to move gases with a very high pressure rise and a low to moderate flow rate; a *work-absorbing device* used to compress a gas such as air or a refrigerant vapour to change its state from low pressure to high pressure; a few examples include *air compressors* used to run *pneumatic tools* and *inflate tyres* at automobiles service stations, and *refrigerant compressors* used in *heat pumps, refrigerators,* and *air conditioners*; see also **Fan** and **Blower**.

Computational Fluid Mechanics The study of solution of the equations of a fluid flow through the use of a computer.

Computer Integrated Manufacturing A *Manufacturing* term, abbreviated as **CIM**; a computer-assisted manufacturing methodology which involves the computerized integration of all aspects of *design, planning, manufacturing, distribution*, and *management*.

Computer Numerical Control A *Manufacturing* term, abbreviated as **CNC**; a computer-assisted methodology used to control the movement of machine components through coded instructions in the form of numerical data.

Computer A machine which can be used to solve *complex* and *difficult problems* in a short time, which otherwise may take a very long time or even cannot be solved.

Computer-Aided Design A *Manufacturing* term, abbreviated as **CAD**; a computer-assisted design system which makes use of computers to create *design drawings* and *product models*; allows the development of three-dimensional (3-D) designs from which two-dimensional *orthographic views* with automatic dimensioning can be produced; finds extensive use in *mechanical design* and *geometric modelling* of *products* and *components*; see also **CAM**.

Computer-Aided Manufacturing A *Manufacturing* term, abbreviated as **CAM**; a computer-assisted manufacturing system which makes products using engineering drawings from computer-aided design (CAD) and a numerical control system; eliminates the common problems of conventional manufacturing; see also **CAD**.

Computer-Aided Process Planning An *Operations Management* term, abbreviated as **CAPP**; a computer-assisted planning process which makes use of computers to optimize process plans; helps to *reduce planning costs* and improve the consistency of *product quality* and *reliability*.

Computerized Inventory Control An *Operations Management* term related to inventory; computerized method of inventory management which can be programmed to track the inventory level continuously; takes care of transactions of inventory against the existing stocks and automatically indicates about the time of placing the order.

Concentrated Load A *Physics* and a *Strength of Materials* term; also called a **Point Load**; a load or force which acts at a point; for example, the weight of a body; see also **Distributed Load**.

Conceptual Skill A *Management* term; a management skill proposed by *Robert L. Katz* (1917-) which relates to a manager's mental ability to coordinate all organizational activities; helps a manager to *analyse* and *diagnose* complex situations.

Concrete A *Materials Engineering* term; a *composite* consisting of *aggregate, cement, water*, and *admixtures* (in some cases); finds extensive use in civil structure works.

Condensation Polymerization A *Chemistry* term; also called **Step-Growth Polymerization**; a much slower polymerization process in which two or more chemically different monomers are combined to form a cross-linked or linear polymer along with by-products such as *ammonia, ethanol, methanol* or *water*; a typical example includes the formation of an *ester* by the chemical reaction between *methyl alcohol* (CH_3OH) and *acetic acid* (CH_3COOH) with *water* as a by-product; see also **Addition Polymerization**.

Condensation A *Thermodynamics* term; a process associated with the phase change of a substance from *vapour* to *liquid* as a result of rejecting heat.

Condenser In *refrigeration* terms; a heat exchanger which extracts heat from a vapour passing through it and converts the vapour into a liquid; a component of the refrigeration system which liquefies or condenses the very hot, high-pressure refrigerant vapour from the compressor to a high-pressure warm liquid; in *power plant* terms, a component of the thermal power plants used to condense the used steam into water; classified into *air-cooled*, *water-cooled*, and *evaporative* condensers; also a component of electrical circuit, popularly called *capacitor*.

Condition Monitoring An *Operations Management* term; see **Predictive Maintenance**.

Conductance A *Physics* term; a measure of the ease with which electricity can pass through a material; measured in siemens.

Conduction A *Heat Transfer* term; a basic mode of heat transfer other than convection and radiation in which the more energetic molecules of a solid material at higher temperature transmit heat energy to the less energetic molecules at lower temperature without changing their positions until their temperature levels become same; results due to interaction between molecules; becoming hot of one end of a metal rod while heating it at its other end forms example of heat transfer by conduction; among the metals, *silver is the best conductor*, followed by *copper* and *aluminium*; iron is low in comparison to copper; see also **Convection** and **Radiation**.

Conductivity A *Physics* term; a material's ability to conduct heat or electricity.

Conductor A *Physics* term; see **Electrical Conductor**.

Conglomerate Structure A *Management* term; a type of organizational structure which is typically an assemblage of legally independent firms (subsidiaries) which operate under one corporate umbrella, but controlled through the subsidiaries' board of directors; most appropriate for a large corporation with many product lines in several unrelated industries.

Conical Pin A *Design* term; also called **Taper Pin**; a fastening element having a conical section; used in making a pin joint to connect two circular shafts; placed along the tangent to the shaft or on its diameter; used for light works; see also **Cylindrical Pin**.

Conical Spring A *Design* term; a spring wound in the shape of *a cone* with a round wire.

Conjugate Beam Method A *Strength of Materials* term; an analytical method which uses conjugate beam theorems to find slope and deflection at any section of loaded beams; useful for beams of variable flexural rigidity.

Conjugate Beam A *Strength of Materials* term; an imaginary beam whose length equals to the length of the actual beam but the load is not the actual load rather the elastic weight $\frac{M}{EI}$ that acts at the same point where the actual load acts (M = bending moment produced by the load acting on the beam, E = modulus of elasticity of beam's material, and I = second moment of area of beam's cross section).

Connecting Rod An *IC Engine* term; a metallic rod which connects the piston to the crankshaft of an IC engine.

Conservative Force A *Physics* term; a conservative is conservative, if the particle, after going through a complete round trip, returns to its initial position with the same kinetic energy as it had initially; alternatively, a force is conservative, if the work done by the force in moving a particle from one position to another is independent of the path followed between the two positions; a few examples include *gravitational force*, *elastic force*, *electrostatic force*, and *central force*; see also **Non-Conservative Force**.

Constant Enthalpy Process A *Thermodynamics* term; see **Free Expansion**.

Constant A *Mathematics* term; a quantity whose value does not change during mathematical operations, for example, π, e, *integers*, and *fractions*; see also **Variable**.

Constantan A *Materials Engineering* term; an alloy of *nickel* (Ni) and *copper* (Cu) which consists of 40% Ni and 60% Cu; finds use in *electrical* works.

Consumable Electrode A *Welding* term; an electrode which gets consumed during the welding process; acts as a filler metal; finds use in *shielded metal arc welding*, *MIG welding*, *submerged arc welding*, and others; see also **Nonconsumable Electrode**.

Consumer Goods An *Economics* term; goods bought for personal use or use by a household; classified into *convenience goods*, *shopping goods*, and *speciality goods*.

Consumer Price Index An *Economics* term, abbreviated as **CPI**; a parameter used to measure the retail prices of a fixed 'market basket' of several thousand goods and services purchased by households.

Consumer A *Management* term; see **Customer**.

Consumer's Risk A *Quality Control* term; the probability of accepting a bad lot; see also **Producer's Risk**.

Contact Friction A *Physics* term; see **Friction**.

Contaminant The unwanted material which is usually harmful or is of a nuisance value or both.

Contingency Approach A *Management* term; a managerial approach in which the design of an organization's structure must match the sources of uncertainty faced by an organization.

Continuity Equation A *Fluid Mechanics* term; an equation based on the principle of *conservation of mass*; states that the mass of a fluid flowing through a pipe remains constant at its every section; for an *incompressible* fluid, the continuity equation is given as $AV = \text{constant}$; for a compressible fluid, the continuity equation is given as $\rho AV = \text{constant}$, where $A =$ area of cross section of the fluid flow, $V =$ velocity of the fluid flow, and $\rho =$ density of the fluid.

Continuous Beam A *Strength of Materials* term related to a *beam*; a statically indeterminate beam consisting of more than one span which is continuously lying and supported over more than two simple supports of which one support is essentially a roller support; widely used in *buildings*, *bridges*, and *aircraft*.

Continuous Casting A *Manufacturing* term related to *casting*; also called **Strand Casting** because of the longer continuous length of the castings produced; a casting process earlier used for continuous casting of nonferrous metal strip, but is now used to produce high quality steel with significant cost reduction; an economical method to produce longer continuous lengths; the typical products made by this method include *blooms*, *billets*, *slabs*, and *tubing* which are directly produced from the molten metal.

Continuous Flow Calorimeter A *Chemistry* term; an instrument used to measure the calorific value of *liquid* and *gaseous fuels*; see also **Bomb Calorimeter**.

Continuous Production An *Operations Management* term; also called **Process Production**; a type of a production system which involves a continuous process to produce items continuously round the clock with very little operator intervention; used in *chemical*, *oil*, *pharmaceutical*, and *textile* industries; see also **Mass Production** and **Batch Production**.

Continuous Quality Improvement A *Quality Management* term; see **Total Quality Management**.

Continuous Review System An *Operations Management* term; an inventory management system of *continuously monitoring* the status of inventory in which order for the inventory is placed, when the stock level touches a certain lower limit defined by the *reorder point*; the ordered quantity (called *economic order quantity*) corresponds to the minimum annual total cost of inventory which is the sum of holding cost and ordering cost; the *economic order quantity* (EOQ) model is based on this system; see also **Periodic Review System**.

Continuous Weld A *Welding* term; a weld which is continuously extended for its entire length without any interruption.

Continuum A *Physics* term; the mathematical idealization of continuous distribution of matter which implies that properties of the matter are considered as continuous functions of the space variables.

Contraction A *Strength of Materials* term; the amount of length by which a member has contracted or shrinked, when loaded axially under compression; see also **Elongation**.

Control Charts for Attributes A *Quality Control* term; the most commonly employed control charts used to control the quality of product or service attributes include *p-chart* and *c-chart*; see also **Control Charts for Variables**.

Control Charts for Variables A *Quality Control* term; used to monitor the *mean* and the *variability* of the process distribution; the two control charts used for variables include *X-bar chart* and *R-chart*; an *X-bar chart* and an *R-chart* are frequently *used to monitor the quality of products and services*; the *X-bar chart* monitors the *average value* of the variables being measured, and the *R-chart* monitors the variation among the items within samples; see also **Control Charts for Attributes**.

Control Charts A *Quality Control* term; the most commonly employed quality improving tool; a graphical representation of the status of a conversion process in terms of control parameters such as *UCL*, *LCL*, and *process average*; a control chart is in the form of a *line graph* in which data is plotted over time (the vertical axis represents a measurement and the horizontal axis represents the time); consists of three horizontal lines: the *upper line* represents the *larger value* of measurement, called the *upper control limit* (*UCL*), the *central line* is the *process mean* or *process average* which is the target value to achieve, and the *lower line* represents the *smaller value* of measurement, called the *lower control limit* (*LCL*); a control chart is primarily used for routine monitoring of a production process to determine if its outputs meet quality standards and to identify when the process has gone out of control; classified into *control charts for variables* and *control charts for attributes*; a few control charts include *R-chart*, *C-chart*, and *P-chart*.

Control Limits A *Statistics* term; the limits which define the range (boundary) of variation permitted in a product characteristic such that a conversion process is in control; there are two control limits, namely the *upper control limit* (UCL) and the *lower control limit* (LCL); the higher value of a data represents the UCL and its smaller value represents the LCL; a process data which falls between the UCL and the LCL indicates that the process is under control and exhibits common causes of variation and those falling outside the control limits indicates that the process is not under control and exhibits assignable causes of variation.

Control Surface A *Thermodynamics* and a *Fluid Mechanics* term; the surface which surrounds the control volume; the geometric boundary of a control volume which permits mass and energy transfer across it; may be *real* or *imaginary*, and may be *at rest* or *in motion*.

Control System An *Electrical* term; a set of components working together to perform a given task under the direction of a processor or computer.

Control Volume A *Fluid Mechanics* and a *Thermodynamics* term; an arbitrary volume in space bounded by a control surface through which a fluid flows; a concept that defines a certain volume encircling a system, which can permit mass and energy transfer across its control surface; finds use in the analysis of *an open system.*

Controlling A *Management* term; also called **Monitoring** or **Progressing**; a management function which is concerned with monitoring activities to ensure that the goals are met according to planned schedule.

Convection A *Heat Transfer* term; the second basic mode of heat transfer after conduction in which heat energy is transferred by the circulation of heated portions of a fluid; takes place between a solid surface and the adjacent fluid in motion; cooling of a refrigerator is one of the best examples of convection; explains why the warmest air is usually found near the top of a refrigerator, where it comes in contact with the cold evaporator, and gets cooled, becomes denser and starts to settle down; classified into *free convection* and *forced convection*; see also **Conduction** and **Radiation.**

Convenience Goods An *Economics* term; a type of consumer goods which are purchased without much thought or search time; the most frequently bought items; include household supplies and commonly eaten foods, for example, *detergent, soap, food grains*, and similar other things; usually sold through many retail outlets; heavily dependent on promotions or advertising in order to capture more users; see also **Shopping Goods** and **Speciality Goods**.

Conventional Sources of Energy A *Power Plant* term; the sources of energy used since long times back and are exhaustible; involve high initial costs in extraction and are not environment-friendly; a few examples include *fossil fuels* (such as *coal, petroleum*, and *natural gases*), *water*, and *nuclear fuels*; see also **Nonconventional Sources of Energy** and **Renewable Sources of Energy**.

Conveyance Kanban An *Operations Management* term; see **Withdrawal Kanban.**

COO A *Management* term; see **Chief Operating Officer**.

Cooling System An *IC Engine* term; a system which helps in cooling an engine; a *water-cooling system* consists of radiator and a pump in which water circulates through passages around the cylinders and then travels through the radiator to cool it off; an *air-cooling system* uses air in place of water; the air-cooling system although makes the engine lighter, but keeps it hotter, and generally decreases engine life and the overall performance.

Cooling Tower A *Power Plant* term; a device used to utilize waste heat of the power plants by recovering its heat before being rejected to surroundings and using it to produce cooling atmosphere; also used frequently to provide chilled water for various applications.

Cooling A *Thermodynamics* term; a process which lowers the temperature of a body.

Coordinate Covalent Bond A *Chemistry* term; proposed by *Gilbert Newton Lewis* (1875–1946); also called **Dative Bond**; a *type of covalent bond* formed between two atoms in which one atom has a complete octet and also possesses at least one pair of unshared electrons, while the other atom is short of two electrons; the electrons for sharing between the atoms are contributed by one atom only; for example, the *hydronium ion* (H_3O^+) has a *dative bond*.

Coordinate Covalent Compound A *Chemistry* term; a compound which has a *coordinate covalent bond*; has properties similar to covalent compounds, but does not form ions in water; sparingly soluble in water, but many of them are largely soluble in organic solvents; its volatility lies between those of electrovalent and covalent compounds.

Coordinate Measuring Machine A *Manufacturing* term, abbreviated as **CMM**; a computer numerical controlled (CNC) machine used for the dimensional inspection of parts; a three-dimensional CMM is the most common, which can measure dimensions in X, Y, and Z directions.

Coordinating A *Management* term; see **Integrating**.

Coordination Number A *Chemistry* term, abbreviated as **CN**; the number of adjacent ions (or atoms) which surround a reference ion (or atom).

COP A *Refrigeration* term; see **Coefficient of Performance**.

Cope and Drag Pattern A *Casting* term; a type of pattern whose two split parts are separately mounted on *two match plates*- one match plate contains cope part of the pattern used to produce cope part of the mould and other match plate contains drag part of the pattern used to produce drag part of the mould; the complete mould is produced by bringing together the cope and drag parts of the pattern; see also **Match Plate Pattern**.

Cope-and-Drag Investment Casting A *Manufacturing* term related to *casting*; see **Ceramic Mould Casting**.

Copolymer A *Chemistry* term; a polymer made from more than one type of monomer; for example, *ester*; see also **Homopolymer**.

Copolymerization A *Chemistry* term; a process in which a mixture of more than one monomeric species is allowed to polymerize.

Copper Glance A *Chemistry* term having the chemical formula Cu_2S; an important ore of *copper* (Cu).

Copper Pyrites A *Chemistry* term having the chemical formula $CuFeS_2$; an important ore of *copper* (Cu) from which copper is mostly extracted.

Copper A *Chemistry* term; a *metallic element* having symbol Cu with atomic number 29 and atomic weight 63.54; a reddish brown *nonferrous metal* which melts at 1083 °C; one of the most important technical metals after *iron* and *aluminium*; the native copper, which is 99.9% pure, is found in large quantities in Michigan state of the USA; a *malleable* and *ductile* metal which can be drawn into wires of diameter as small as 0.03 mm; an *excellent conductor of heat* and *electricity* (the best conductor next to silver), but the presence of even small quantities of other metals lowers its conductivity appreciably (presence of just 0.03% of *arsenic* (As) lowers its conductivity by about 15%); the important ores of copper include *copper glance* (Cu_2S), *copper pyrites* ($CuFeS_2$), *cuprites* (Cu_2O), and *malachite* ($Cu(OH)_2.CuCO_3$), but copper is extracted mostly from *copper pyrites*; a *non-reactive* metal which resists the action of most of the common chemical reagents like air, water, non-oxidizing acids, and alkalis; majority of the copper is used in making electric cables and other electric goods due to its high electrical conductivity; finds use in *electroplating* and *electrotyping*; used in making *utensils, containers, kettles, evaporating pans,* and *coins*; also used for alloying with gold and silver to make them hard for the purpose of making ornaments and coins; forms a number of alloys with other metals like zinc, tin, nickel, aluminium, etc.; some important alloys of copper include *brass, bronze, gun metal, bell metal,* and *German silver*.

Core Box A *Manufacturing* term related to *casting*; a *wooden* or *metal* box used for making cores; imparts the desired shape to the core-sand on ramming.

Core Competence a *Management* term; a manager's skill and abilities in value creation activities such as *manufacturing, research & development* (R&D), or *organizational design.*

Core Competency A *Management* term; the organizational strengths which represent its unique skills or resources.

Core Induction Furnace also called **Channel Induction Furnace**; a type of *electric furnace* which uses a *low frequency current* passing through a *water-cooled copper coil*, which surrounds only a small part of the crucible; commonly used in *nonferrous foundries* for *superheating* (to improve the fluidity of the molten metal) and holding the molten metal at a constant temperature for die casting applications; see also **Coreless Induction Furnace.**

Core Print A *Manufacturing* term related to *casting*; the extra projection provided on a pattern to form core seats to support the core against the buoyancy of the molten metal; classified into *cope print*, *drag print*, and *parting line print*.

Core Sand A *Casting* term; a high silica sand used for *making cores.*

Core Shift A *Casting* term; a type of casting defect which occurs in the form of varied thickness of a casting; occurs due to shifting of a core from its designated position when not properly supported by a chaplet.

Core A *Casting* term; a *sand mass* inserted into the mould to produce *holes* (*cavity*) or *passages* for water cooling; also refers to the *interior surface* of the casting.

Coreless Induction Furnace A type of *electric furnace* consisting of a *crucible* made of a refractory material and completely surrounded with a *water-cooled copper coil* carrying a *high frequency current* which produces heat using induction principle; has excellent mixing characteristics for alloying and adding new charge of metal; see also **Core Induction Furnace**.

Corollary A *Physics* term; the consequence or additional inference which logically follows from an established proposition.

Corporate Culture A *Management* term; the collection of *beliefs*, *expectations*, and *values* learned and shared by a corporation's members and transmitted from one generation of employees to another; generally reflects the values of the founder and the mission of the firm; gives a company a sense of identity like '*this is who we are*', '*this is what we do*', and '*this is what we stand for*'.

Corporate Mission A *Management* term; a set of long range goals of an organization; includes statements about the type of business the company wants to be in, who its customers are, its basic beliefs about business, and its goals of survival, growth, and profitability.

Corporate Social Responsibility A *Management* term, abbreviated as **CSR**; the consideration of the impact of a company's actions on society.

Corporate Strategy A *Management* term; also called **Corporate-Level Strategy**; a strategy which describes a company's overall direction in terms of its general attitude towards growth and the management of its various businesses and product lines; focuses on corporation's stability, growth, and retrenchment and helps in protecting and enlarging an organization's existing domain as well as expanding into new domains; for example, *Mercedes-Benz* used its competences in *R&D* and *product development* to enter the household products and aerospace industries, and *Coca-col*a took its marketing skills and applied them globally in the soft-drink industry.

Corporate-Level Strategy A *Management* term; see **Corporate Strategy**.

Corporation A *Management* term; an *intangible* legal business entity and a dominant form of a business organization, a few examples include *Indian Oil Corporation* (OIL), *Oil and Natural Gas Corporation* (ONGC), *Life Insurance Corporation* (LIC), etc.

Corrosion A *Chemistry* term; a chemical process of deterioration of a metal from atmospheric contamination and from oxidation or scaling at elevated temperatures; for example, iron gets rusted due to corrosion.

Corrosive Wear A *Manufacturing* term; a type of wear which occurs when *two smooth surfaces* slide over each other in a *corrosive environment*; see also **Abrasive Wear** and **Adhesive Wear**.

Corthias Casting A *Manufacturing* term related to *casting*; see **Pressed Casting**.

Corundum A *Materials Engineering* term; also called **Alpha-Al_2O_3 (α-Al_2O_3)**; an oxide of *aluminium* (Al_2O_3) which occurs in nature as anhydrous Al_2O_3; can withstand high temperatures; very hard and is not attacked by acids; used as an *abrasive*; see also **Activated Alumina**.

Cost Index An *Operations Management* term; a ratio of the quality cost to the manufacturing cost; see also **Sales Index** and **Labour Index**.

Cost of Poor Quality A *Quality Management* term; the cost associated with poor quality of products or services produced; includes costs relating to *product* (or *service*) *inspection*, *rework*, *duplicate work*, *scrap rejection*, *replacement*, *refunds*, as well as *loss of customers* and *reputation*; a concept first used by *Philip B. Crosby* (1926–2001), one of the famous quality gurus, in his book, '*Quality Is Free*' in 1979.

Cost of Quality A *Quality Management* term; the cost associated with achieving good quality with respect to products or services in order to meet the requirements of a customer.

Cost An *Operations Management* term related to *cost analysis*; the expenditure incurred on an item; there are many types of cost; *fixed cost*, *variable cost*, *incremental cost*, *direct cost*, *indirect cost*, etc.

Cost–Benefit Ratio An *Economics* term; an economic evaluation index which relates to control costs to benefits.

Costing An *Operations Management* term related to *cost analysis*; the process of calculating the cost of an item.

Cotter Joint with a Gib A *Design* term; a cotter joint formed by using a gib; prevents the bending of the lower part of the socket end; also helps to use parallel holes in the spigot and socket; mostly used to connect square rod ends as in the case of big end of a connecting rod; see also **Gib**.

Cotter Joint with Sleeve A *Design* term; the simplest of all the cotter joints; used to fasten two circular rods with a common sleeve (covering) over them.

Cotter Joint with Socket and Spigot Ends A *Design* term; a cotter joint used to fasten (temporarily) two circular rods which does not use a sleeve; modifies the two rod ends in which one end is used as a socket end and other end as a spigot end, where alignment of the joint is made by inserting the spigot end into the socket end and the cotter is fitted in the slots already provided.

Cotter Joint A *Design* term; *a temporary joint formed by using a cotter* for fastening two circular rods subjected to axial forces (tensile or compressive), but not to any rotation; gets locked in position by means of a screw; allows a convenient and rapid assembly and disassembly of the units involved.

Cotter A *Design* term; a *wedge-like fastening element* of rectangular cross section with uniform thickness, but tapered on one side or both sides; finds use in a cotter joint; used to hold two parts together where parts are subjected to axial forces (tensile or compressive) only; usually driven in perpendicular to the axis of the connected parts; resists shear over two transverse sections; generally made of *mild steel.*

Coulomb Friction A *Physics* term; see **Dry Friction**.

Coulomb A *Physics* term, named after *Charles-Augustin de Coulomb* (1736–1806), a *French* physicist who is famous for *Coulomb's law*; the SI unit of *electric charge*; denoted by C; 1 C is the amount of electric charge which is carried by a current of 1 A flowing for 1 s.

Coulomb's Law A *Physics* term; a law used in *electrostatics* to find the magnitude of the electrostatic force F acting between two static point charges q_1 and q_2 separated by a distance r in air or vacuum; given as:

$$F = \frac{1}{4\pi\epsilon_o} \cdot \frac{q_1 q_2}{r^2}$$

where

$\frac{1}{4\pi\epsilon_o}$ $= 9 \times 10^9$ N-m^2/C^2 = a constant
ϵ_0 = absolute electrical permittivity of free space $= 8.85 \times 10^{-12}$ C^{2}N^{-1} m^{-2}

The charge on an electron or a proton is of the smallest magnitude found in nature, represented by $e = 1.6 \times 10^{-19}$ C; the proton has charge + e and the electron has charge – e; all other charges have magnitude ne, where n is an integer.

Counter Flow Heat Exchanger A *Thermodynamics* term; a type of heat exchanger in which hot and cold fluids flow past the heat transfer surface in *opposite direction*; see also **Parallel Flow Heat Exchanger** and **Cross Flow Heat Exchanger**.

Counterclockwise Moment A *Physics* term; also called **Anticlockwise Moment**; a moment which can rotate a body in a direction opposite to the movement of the hands of a clock; considered as a *positive moment*; see also **Clockwise Moment**.

Couple a *Physics* term; two equal and opposite forces having parallel lines of action form a couple, where the sum of the forces is zero, but the sum of the moments of two forces is not zero; effect of a couple acting on a rigid body is to rotate it about an axis perpendicular to the plane of the forces; for example, when we open the lid of a bottle by turning it, our fingers apply a couple on the lid.

Covalency A *Chemistry* term; the number of electrons which an *atom contributes for sharing* in a covalent bond; for example, covalency of *hydrogen* (H), *chlorine* (Cl), *oxygen* (O), and *nitrogen* (N) are respectively 1, 1, 2, and 3.

Covalent Bond A *Chemistry* term; a chemical bond discovered by *Gilbert Newton Lewis* (1875–1946) in 1916; formed by the mutual sharing of electrons between the combining atoms which contribute towards the stability of both the atoms; bonding atoms may share more than one pair of electrons depending upon their requirement of completing the octet; for example, in the formation of oxygen molecule (O_2), every oxygen atom (O) shares *two electrons each* to form a *double bond* (O=O), whereas in the formation of nitrogen molecule (N_2), every nitrogen atom (N) shares *three electrons each* to form a *triple bond* ($N \equiv N$); see also **Ionic Bond**.

Covalent Compound A *Chemistry* term; a compound formed as a result of a *covalent bond*; formed by the mutual sharing of electrons and not by the transfer of electrons; *does not conduct electricity in the fused or dissolved state*; has a *low melting* and *boiling point*; generally *insoluble* or less soluble in *water* or *other polar solvents*, but is *easily soluble in nonpolar solvents*; see also **Ionic Compound**.

Covered Electrode A *Welding* term; also called **Coated Electrode**; an electrode having a cover or coating of *fluxing* or *slagging material* over its *core wire*, which vaporizes in the heat of the arc to form a protective atmosphere (usually of carbon dioxide) that stabilizes the arc and protects the molten metal from contamination; arc welding processes using coated electrodes are more popular; see also **Bare Electrode**.

CPI An *Economics* term; see **Consumer Price Index**.

CPM A *Project Management* term; see **Critical Path Method**.

Cracks A *Forging* term; the *forging defects* which occur *in the form of breaks on the surface of the forged parts* due to *nonuniform distribution of forging pressure*.

Crank an *IC Engine* term: a bar that has one fixed pivot point and is allowed to rotate 360°.

Crankshaft An *IC Engine* term; an engine shaft which converts reciprocating motion of the piston into rotary motion.

Crash Cost a *Project Management* term; the cost associated with crashing of an activity; see also **Normal Cost**.

Crash Time A *Project Management* term; the amount of time by which an activity can be shortened using additional resources; see also **Normal Time**.

Crashing A *Project Management* term; see **Project Crashing**.

Creativity A *Management* term; the thinking of novel and appropriate ideas; the ability to combine ideas in a unique way or to make unusual connections; connected to intellectual thoughts; leads to innovation of new product development.

Creep Curve A *Strength of Materials* term; a graphical plot between linear strain (deformation) and time for a material which undergoes creep deformation; the creep curve has three distinct stages, namely primary, secondary, and tertiary, where primary and tertiary stages are short-lived with higher creep rates, while secondary stage is rather longer with creep rate somewhat linear.

Creep Rate A *Strength of Materials* term; shows how creep changes with respect to time; increases with temperature and applied load; primary and tertiary stages of creep have higher creep rates than its secondary stage; higher creep rate produces more deformation.

Creep Test a *Strength of Materials* term; a *destructive* test used to measure creep strength of a solid material; conducted on a 1 m length and 2–3 mm diameter lead wire, which is subjected to a constant tension load at room temperature and measuring the change in length over a regular interval of time.

Creep A *Strength of Materials* term; a *time-dependent plastic deformation* of a material under a *constant load* acting over a long time period; usually a characteristics of metals, but some non-metallic materials such as thermoplastics and rubber also show this property; a very slow process which is usually carried out at room temperatures, but is accelerated at higher temperatures; *turbine disk* and *blades* of a jet engine of an aircraft, a *ceiling fan*, *reinforced concrete floors*, *high-temperature pressure vessels*, *tools* and *dies* when subjected to hot working operations like *forging* and *extrusion* undergo creep.

Crippling Load A *Strength of Materials* term related to a *column*; see **Buckling Load**.

Cristobalite A *Chemistry* term; one of the three *naturally occurring crystalline forms* of silica (SiO_2), other two being quartz and tridymite; tridymite when heated above 1470 °C changes into cristobalite which melts at 1740 °C.

Criteria A *Design* term; the specific standards against which a design will be judged acceptable or unacceptable.

Critical Activity A *Project Management* term; an important project activity which should not be delayed at any cost and has zero slack; if the critical activity consumes more time than its estimated time, the project will be delayed; see also **Non-Critical Activity**.

Critical Load A *Strength of Materials* term related to a *column*; see **Buckling Load**.

Critical Path Method A *Project Management* term, abbreviated as **CPM**; one of the two project network approaches which was developed in 1956 as a computerized system for scheduling and planning by *James E. Kelly*, a Remington-Rand computer specialist, and *Morgan R. Walker*, a du Pont engineer; a *deterministic approach* of project management which uses definite time estimates for every activity; uses four time estimates, namely the *earliest start time* (EST), the *earliest finish time* (EFT), the *latest start time* (LST) and the *latest finish time* (LFT); especially useful for repetitive and large projects consisting of many activities such as construction projects, for example constructing a major hospital, building a ship and constructing an oil pipeline; see also **Program Evaluation and Review Technique**.

Critical Reynolds Number A *Fluid Mechanics* term; the Reynolds number at which the flow becomes turbulent; varies with different geometries and flow conditions; the value of the critical Reynolds number for a flow in a circular pipe is 2300.

Critical Speed A *Design* term; the speed at which a shaft becomes unstable while rotating and vibrates loudly; causes a large deflection in the shaft and occurs due to a large centrifugal force.

Croning Process A *Manufacturing* term related to *casting*; see **Shell Mould Casting**.

Crookes Glass A *Chemistry* term; a *special type of glass* which contains *cerium oxide*; used for *optical purposes.*

Cross Flow Heat Exchanger A *Thermodynamics* term; a type of heat exchanger in which hot and cold fluids flow past the heat transfer surface *at right angles* to each other; see also **Parallel Flow Heat Exchanger** and **Counter Flow Heat Exchanger**.

Cross Product of Vectors A *Physics* term; see **Vector Product of Vectors.**

Cross Section A *Strength of Materials* term; the section of a body normal to its longitudinal axis.

Cross Slide A *Lathe* term; a principal part of a lathe which allows movement of the cutting tool at right angle to the lathe centre line, and can also support *compound slide.*

Crossed Belt Drive A *Theory of Machines* term; a belt drive in which two parallel shafts move in opposite direction, and the total length of the crossed belt is given as:

$$L = \pi(r_1 + r_2) + \frac{(r_1 + r_2)^2}{x} + 2x$$

where

r_1 = radius of the bigger pulley
r_2 = radius of the smaller pulley
X = centre-to-centre distance between two pulleys; see also **Open Belt Drive**.

Cross-Linking A *Chemistry* term; a process of joining of adjacent linear molecules by a chemical bonding; for example, *vulcanization of rubber*.

Cross Section A *Strength of Materials* term; a section perpendicular to the longitudinal axis of a member.

CRP An *Operations Management* term; **Capacity Requirement Planning**.

Crucible Furnace A furnace in the form of a *crucible* heated from outside by *gas flames, fuel oil, fossil fuel,* or *electricity*; the oldest furnace earlier used for making steel castings, but is now mainly used for nonferrous castings.

Cryogenics A *Thermodynamics* term; the science of producing and using temperatures below − 250 °F.

Crystalline Materials A *Materials Engineering* term; the materials in which the constituent atoms are arranged in a *regular, repeating pattern*; for example, *all metals* are crystalline materials; see also **Amorphous Materials**.

Crystalline A *Materials Engineering* term; the property of having the constituent atoms of a material being arranged together in a regular, repeating pattern.

Crystallization A *Chemistry* term; a chemical process which makes the materials attain regularity in their atomic or molecular arrangements by forming new strain-free grains; results in completely new properties in the materials.

CSR A *Management* term; see **Corporate Social Responsibility**.

Cubic Boron Nitride A *Materials Engineering* term; a cutting tool material, abbreviated as **CBN**; sometimes also called *artificial diamond* due to its artificial preparation and extreme hardness next to natural diamond; also known as **Borazon**, when used as abrasives; finds use in making cutting tools offering many advantages over the cemented carbide tools such as *higher cutting speeds, higher metal removal*, and *better surface finish* (cutting speed and metal removal rate are approximately five times higher than cemented carbide tools); used with *negative rake angle cutting tools* which can withstand high temperatures, and can be subjected to interrupted cutting; also finds use in the machining of *difficult-to-machine materials* such as *hardened steel, chilled cast iron*, and *super alloys*.

Cup Drawing A *Manufacturing* term; see **Deep Drawing**.

Cupola Furnace The simplest furnace used in casting to convert *pig iron* into *grey cast iron*; its main advantages include its *simple construction, high melting rate*, less maintenance cost and *wide range of capacity*; the charging of the cupola furnace consists of alternate layers of coke, pig iron mixed with scrap castings and a flux, usually limestone ($CaCO_3$).

Cuprites A *Chemistry* term having the chemical formula Cu_2O; an important ore of *copper* (Cu).

Cupronickel A *Materials Engineering* term; an alloy of *copper* (Cu) and *nickel* (Ni) consisting of 20% Ni; used in making *bullet sheaths*.

Curie A *Chemistry* term related to *radioactivity*; the *unit of radioactivity* named after *Marie Curie* (1867–1934) and her husband *Pierre Curie* (1859–1906) who together discovered two radioactive elements, namely *polonium* (Po) (atomic number 84 and atomic weight 210) and *radium* (Ra) (atomic number 88 and atomic weight 226); the amount of a radioactive substance which has a decay rate of 3.7×10^{10} dis/s, which is equivalent to 1 g of radium which disintegrates at the rate of 3.7×10^{10} dis/s; *millicurie* (mci) and *microcurie* (μci) are the other units of radioactivity, where 1 mci = 3.7×10^7 dis/s and 1 μci = 3.7×10^4 dis/s.

Curing A *Materials Engineering* term; a process by which a material transforms from a liquid or plastic state to a permanent solid state.

Curl V A *Mathematics* term; represents the curl of the *velocity vector* (**V**); expressed as: $\nabla \times \mathbf{V}$.

Current Assets A *Financial Management* term; also called **Working Capital** or **Revenue Assets**; the short-term assets which are expected to be converted into cash within one year or less; represents the portion of investment which circulates from one form to another in the ordinary conduct of business; include *stocks* and *work-in-progress* (WIP) which are used up fairly quickly; see also **Current Liabilities**.

Current Liabilities A *Financial Management* term; the liabilities of short-term which need to be met within the coming year (within one year or less) such as *money owed to suppliers* or *bank overdrafts*; see also **Current Assets**.

Curved Bar A *Strength of Materials* term; a bar curved to a definite curvature.

Curved Beam A *Strength of Materials* term; a beam which is curved in shape and whose neutral axis does not coincide with the centroidal axis of the cross section; the flexural stress distribution for a curved beam is nonlinear with the distance from the neutral axis; used in frames of machines like *planers, shapers, drilling machines* as well as *crane hooks*.

Customer A *Management* term; also called **Consumer** or **External Customer**; the people who buy a firm's products or services; usually the end-user of the products or services; the largest outside stakeholder group of an organization.

Cut-off Ratio An *IC Engine* term; a ratio of the volume in the cylinder at the end of combustion to the clearance volume in a diesel cycle.

Cutting Ratio A *Manufacturing* term; see **Chip Thickness Ratio**.

Cutting Torch A *Welding* term; a device used to mix *oxygen* and *acetylene* or any other fuel gas in the required proportion for *cutting operation*; see also **Welding Torch**.

Cyaniding A *Manufacturing* term related to *heat treatment*; a *surface-hardening process* in which *carbon* and *nitrogen* are absorbed in the outer layer of *low carbon steel* parts, when they are heated at a temperature of 1550–1600 °F in a cyanide atmosphere; results in surface hardness to a depth of 0.003–0.020 in.

Cycle Inventory An *Operations Management* term; also called **Lot Size Inventory**; the inventory ordered or purchased in lots of some specified number in order to reduce the production costs, rather than procuring them as and when required; the *EOQ formula* is based on cycle inventory; see also **Buffer Inventory, Movement Inventory**, and **Anticipation Inventory**.

Cyclic Engine A *Thermodynamics* term; an engine which operates in a thermodynamic cycle.

Cyclic Frequency A *Physics* term; the frequency with which a process repeats itself; denoted by f, n or υ; measured in cycles per second which is equivalent to *hertz* (Hz) named after the German physicist *Heinrich Rudolf Hertz* (1857–1894); 1 Hz = 1 cycle/s; see also **Angular Frequency**.

Cyclic Heat Engine A *Thermodynamics* term; a thermodynamic device which works on the principle of a thermodynamic cycle where there is a net heat transfer to the system and a network transfer from the system.

Cyclic Motion A *Physics* term; see **Periodic Motion**.

Cycloid A *Theory of Machines* term related to *gear*; a parameter used to define the geometrical shape of a gear tooth; the locus of a point on the circumference of a circle which rolls without slipping on a fixed straight line; classified into *epicycloid* and *hypocycloid*; see also **Involute**.

Cyclotron A *Chemistry* term related to *radioactivity*; a device invented by *Ernest Orlando Lawrence* (1901–1958), for which he was awarded the 1939 Physics Nobel Prize; used to increase the velocity of *alpha* (α) *particles*, *protons*, and *deuterons* in order to make them more effective for nuclear reactions; uses very high frequency A.C. voltage of the order of 10,000–20,000 V to produce the particle's velocity of about 40,000 km/s.

Cylinder A cylindrical container which may be used to contain high pressure gas or liquid; may be thin- or thick- walled; an important component of an *IC engine* in which fuel is burnt to produce power; the piston reciprocates in the cylinder of an engine; usually made of grey cast iron.

Cylindrical Pin A *Design* term; also called **Straight Pin**; a fastening element having a uniform cylindrical section; used in making a pin joint to connect two circular shafts; also used as a key; most suitable for low power drives; see also **Conical Pin**.

D

DA A *Chemistry* term; see **Dissolved Acetylene**.

Dam A *Civil Engineering* term; an engineering structure constructed across a river, mostly for the purpose of *irrigation* and *electricity generation.*

Damped Oscillation A *Physics* term; see **Free Oscillation**.

Daniel Bernoulli (1700–1782) A *Swiss* scientist and mathematician who presented the famous *Bernoulli's equation* in 1738, used as a very powerful tool in fluid mechanics.

Darcy Friction Factor A *Fluid Mechanics* term; see **Darcy–Weisbach Friction Factor**.

Darcy-Weisbach Equation A *Fluid Mechanics* term; also called **Darcy-Weisbach Formula**; an equation named after a French hydraulic engineer *Henri Philibert Gaspard Darcy* (1803–1858) and a German mathematician *Julius Weisbach* (1806–1871); used to find the major frictional head loss (h_f) for any fully developed, steady and incompressible pipe flow; given as:

$$h_f = \frac{flv^2}{2gD}$$

where

f = friction factor
l = length of the pipe
D = diameter of the pipe
v = velocity of the flow in the pipe
g = acceleration due to gravity = 9.81 m/s^2.

Darcy–Weisbach Formula A *Fluid Mechanics* term; see **Darcy–Weisbach Equation**.

D. K. Singh, *Dictionary of Mechanical Engineering*,
https://doi.org/10.1007/978-981-99-1722-8_4

Darcy–Weisbach Friction Factor A *Fluid Mechanics* term; also called simply **Darcy Friction Factor**, named after the French engineer *Henry Darcy* (1803–1858) and the German engineer *Julius Weisbach* (1806–1871); a dimensionless quantity associated with frictional head loss of a fluid flow in a pipe; related to Reynolds number (Re); denoted by f, given as:

$$f = \frac{\Delta p(D/l)}{\left(\rho v^2/2\right)}$$

$$= \frac{64}{\text{Re}} \text{(for a fully developed pipe laminar flow)}$$

$$f = \frac{0.316}{\text{Re}^{0.25}} \left(\text{for turbulent flow, where } 4 \times 10^3 < \text{Re} < 1 \times 10^5\right)$$

where

Δp = pressure drop in the pipe
l = length of the pipe
D = diameter of the pipe
v = velocity of the flow in the pipe
ρ = density of the fluid.

Dative Bond A *Chemistry* term; see **Coordinate Covalent Bond**.

Datum Energy A *Physics* term; see **Potential Energy**.

Datum Head A *Fluid Mechanics* term; see **Elevation Head**.

DBT A *Thermodynamics* term related to *psychrometry*; see **Dry Bulb Temperature**.

DC Charging An *Electric Vehicle* term; see **Level 3 Charging**.

DC Motor An *Electrical* term; an electric motor which runs on *direct current* (DC) electricity; finds use in *electric razors*, *electric car windows*, and *remote control cars*; classified into *brushed dc motor* and *brushless dc motor*.

DC An *Electrical* term; see **Direct Current**.

Debt Capital A *Financial Management* term; all long-term borrowing which is incurred by the firm including bonds; see also **Equity Capital**.

Decagon A *Mathematics* term; a polygon having ten (10) sides; see also **Polygon**.

Decarburization A *Welding* term; the loss of carbon from a *ferrous alloy* in a reactive atmosphere at high temperatures; see also **Carburization**.

Decay Constant A *Chemistry* term; see **Disintegration Constant**.

Deceleration A *Physics* term; see **Retardation**.

Decentralization A *Management* term; a process of delegating the authority of making important decisions about the organizational resources and initiating new projects to managers at all levels in the hierarchy; *democracy* is based on decentralization; see also **Centralization**.

Decision-Making A *Management* term; the selection of a course of action from among alternatives; a major part of *planning*.

Dedendum Circle A *Theory of Machines* term related to *gear*; a circle passing through the *roots* of all the gear teeth; its standard diameter equals to the pitch circle diameter minus (2 × 1.157) module; see also **Addendum Circle**.

Dedendum A *Theory of Machines* term related to *gear*; the radial distance from the bottom land to the pitch circle of the gear or the depth of a gear tooth *below the pitch circle*; has the standard value of 1.157 times of one module; see also A**ddendum**.

Deep Drawing A *Manufacturing* term; also called **Cup Drawing** or **Radial Drawing**; a metal working process used to produce *cup-shaped objects* by forcing a sheet metal blank into a die cavity using a punch.

Defects Per Million Opportunities A *Quality Control* term abbreviated as **DPMO**; a *six sigma* approach which does not allow more than 3.4 defects per million of products or services produced.

Deflection Curve A *Strength of Materials* term; see **Elastic Curve**.

Deformation A *Strength of Materials* term; a process of producing change in shape or size of a body caused by the application of force, pressure, heat or any other physical phenomenon; classified into *linear*, *lateral*, *angular*, *volumetric*, *elastic*, or *plastic* deformations.

Deforming Force A *Physics* term; a force which tends to deform or change the configuration (shape or size) of a body.

Degree of Freedom A *Materials Engineering* term; the number of *independent variables* required to specify an equilibrium microstructure.

Degree of Polymerization A *Chemistry* term; the number of *repetitive units* (or mers) present in the chain of a linear polymer; the average length of the chains in a linear polymer.

Degree of Reaction A *Fluid Mechanics* term; a ratio of the energy transfer by means of static pressure to the total energy transfer for an *incompressible flow turbomachine*; also defined as a ratio of the energy transfer by means of enthalpy change to the total energy transfer for *compressible flow machines*; a parameter used to classify a turbomachine based on its value, which varies between 0 and 1; for *pure impulse machine* (Pelton turbine), the degree of reaction is zero, whereas a degree of reaction of 0.4–0.6 are most frequently employed for reaction machines (Francis turbine); the unit value of degree of reaction is not possible in practice.

Degree of Saturation A *Thermodynamics* term related to *psychrometry*; a psychrometric property; a ratio of the actual specific humidity (of moist air) to the specific humidity of saturated air at a given temperature; indicates the capacity of air to absorb moisture; denoted by μ; given as:

$$\mu = \frac{\omega}{\omega_s} = \phi\left[\frac{1 - \frac{p_s}{p}}{1 - \frac{p_v}{p}}\right]$$

where

ω = actual specific humidity
ω_s = specific humidity of saturated air
ϕ = relative humidity = $\frac{p_v}{p_s}$
p_s = partial pressure of saturated air
p_v = partial pressure of water vapour
p = total atmospheric pressure; see also **Specific Humidity**.

Dehumidification A *Thermodynamics* term related to *psychrometry*; a process which removes the moisture content of the air through heating; see also **Humidification**.

Del Operator A *Mathematics* term; see **Gradient Operator**.

Delphi Method An *Operations Management* term related to *forecasting*; also called **Delphi Technique**; a *qualitative* forecasting technique developed by *Olaf Helmer* (1910–2011), a *German-American* logician and futurologist and his colleagues at the *Rand Corporation*; involves making a consensus at the forecast value using a *questionnaire* shared among 6 to 8 executives *anonymously* in *five-six rounds*; a highly *unbiased* method which has wide acceptance, and eliminates the problem of personality conflicts; useful for *long-range* forecasts of product demand and new-product sales projections.

Delphi Technique An *Operations Management* term related to *forecasting*; see **Delphi Method**.

Demand Pattern An *Operations Management* term; the general behaviour of the pattern of a product with respect to time; may be of constant, seasonal, trend type, or combination of these types.

Demand An *Economics* term; the amount of a product which people are willing to purchase at every possible price during a given period of time; a measure of the relationship between the price and quantity of a particular good or service demanded.

Deming Award A *Quality Management* term; also called **Deming Prize**; a quality award/prize established by the *Japanese* government in 1951 in honour of an *American* quality expert *W. Edwards Deming* (1900–1993) to honour his contributions in the field of quality management; awarded annually to a company which distinguishes itself in quality management programs.

Deming Cycle An *Operations Management* term; see **Deming Wheel**.

Deming Prize A *Quality Management* term; see **Deming Award**.

Deming Wheel An *Operations Management* term; also called **Plan-Do-Check-Act Cycle** or **PDCA Cycle** or **Shewhart Cycle** or **Deming Cycle**; a four-step quality improvement process originally formulated by *Walter A. Shewhart* (1891–1967), but renamed after American quality expert *W. Edwards Deming* (1900–1993); represents the concept of continuous improvement as an endless cycle of any action; involves four steps of continuous improvement such as *plan*, *do*, *check* and *act*, where the first step involves planning for the necessary improvement, the second step is the implementation of the plan, the third step is to check the results of the plan, and the last step is to act upon the results of the plan.

Deming's 14 Points of Management A *Management* term; 14 guidelines for improving quality proposed as part of total quality control by *Willium Edwards Deming* (1900–1993) who is known as the '*father of quality control*' in Japan.

Democratic Leader A *Management* term; also called **Participative Leader**; a person who consults with subordinates on proposed actions and decisions, and encourages participation from them; see also **Autocratic Leader**.

Density A *Physics* and a *Fluid Mechanics* term; also called **Mass Density**; the mass contained in unit volume of a body; an *intensive* property of a substance and a scalar quantity which depends on *temperature* and *pressure*; the density of solids, liquids and gases decreases with rise in temperature, but the density of solids and liquids does not change with pressure, as they are incompressible substances; the density of gases increases with pressure, as they are compressible in nature; the SI unit is Kg/m^3, has the dimension of $[ML^{-3}]$; the density of water at 4 °C (277 K) is 1000 kg/m^3, which is considered as its standard value.

Departmentalization A *Management* term; a process of grouping jobs or tasks according to some logical arrangement; the five most commonly used groupings include *business function*, *process*, *product or service*, *customer*, and *geography*.

Dependent Demand An *Operations Management* term; the demand for an item which depends on the demand for some other items; for example, the demand of *component parts*, *raw materials*, and *subassemblies*; see also **Independent Demand**.

Dependent Variable A *Mathematics* term; a variable quantity whose value depends upon the numerical value assigned to the independent variable; see also **Independent Variable**.

Deposition Rate A *Welding* term; the speed at which a filler metal is added to a weld joint; measured in terms of volume of metal deposited per minute.

Deriaz Turbine A *Fluid Mechanics* term; *a diagonal-flow propeller turbine* having *adjustable blades*, where the flow is directed inwards as it passes through the blades.

Derived Units A *Physics* term; the units of measurement expressed in terms of fundamental units of *meter* (m), *second* (s) and *kilogram* (kg); a few examples in SI unit include *newton* (the unit of force), *joule* (the unit of work and energy), m^2 (the unit of area) and m^3 (the unit of volume); see also **Fundamental Units** and **Supplementary Units.**

Desiccant A *Refrigeration* term; a solid material capable of removing moisture from a gas, liquid, or solid; finds use in the drier to trap moisture from the refrigerant.

Design of Experiment A *Statistics* term abbreviated as **DOE**; the science of designing sets of experiments which will generate enough useful data to make sound decisions without involving too much costs or taking too much time; a very useful tool for conducting research.

Design Stress A *Strength of Materials* term; see **Working Stress.**

Destructive Inspection A *Materials Engineering* term; see **Destructive Testing.**

Destructive Testing A *Strength of Materials* term; also called **Destructive Inspection**; a mechanical test in which a test material gets damaged during the test and can't be reused; the destructive test is conducted to know the important properties of a material like its soundness, strength, toughness, etc.; a few examples include *tension test, compression test, impact test, fatigue test,* and *bend test*; see also **Nondestructive Testing.**

Determinants of Demand An *Economics* term; the factors other than the price of the good which influence demand; include *income, tastes, prices* of related goods and services, *expectations,* and *number of buyers.*

Detonation An *IC Engine* term; an undesirable phenomenon associated with a *petrol engine*; *analogous to knocking* in a diesel engine; see also **Knocking.**

Deviation A *Metrology* term; the difference between a single value and the mean value.

Dew Point Temperature A *Thermodynamics* term related to *psychrometry*; abbreviated as **DPT**; the temperature to which the moist air must be cooled at *constant pressure* to *start condensation.*

Diagonal A *Mathematics* term; a straight line which connects the opposite corners of a *rectangle, square,* or *parallelogram.*

Dial Caliper A *Metrology* term; a type of vernier calliper which contains a dial to show the measurements; a precision measurement device which is accurate to at least 1/1000th of an inch; used to measure inside and outside dimensions and depths of materials with the output read from a dial.

Diameter A *Mathematics* term; a line segment which passes through the centre of a circle and forms the end-points on the circle; the largest chord of a circle which is equal to two times the radius of the circle; expressed in the units of length such as mm, cm, inch, or m.

Diametral Pitch A *Theory of Machines* term related to *gear*; the number of gear teeth per inch of pitch circle diameter.

Diamond Pyramid Hardness Test A *Strength of Materials* term; see **Vickers Test**.

Diamond A *Chemistry* term; the *purest* and *crystalline* form of *carbon* (C) which contains 100% *carbon*; the hardest naturally occurring material whose hardness varies between 7000 and 8000 HK where HK represents the Knoop hardness number; a *colourless* and *transparent* substance which has a three-dimensional *tetrahedral structure* consisting of a large number of carbon atoms arranged tetrahedrally by utilising their sp^3 hybrid orbitals; its increased hardness, high density (3.51 g/cm^3 at 25 °C) and unusually *high melting point* (being close to 3500 °C) are due to strong covalent bonds holding numerous carbon atoms together; a bad conductor of electricity but has the highest thermal conductivity of any known substance (about 5 times that of copper), that is why, *diamond tipped tools* do not overheat, and hence they are extensively used for *drilling* and *cutting purposes*; chemically resistant towards almost all chemical reagents, but burns in air when heated at 900 °C giving carbon dioxide; stable in vacuum up to 1500 °C, but changes into graphite at 1800 °C and more rapidly at 2000 °C; finds use in *cutting glass*, *making cutting tools* for *grinding* and *polishing* of *hard materials*, and making *abrasives*; its *high refractive index* (2.417) helps light to refract from its surfaces, making it looking shining when it is cut and polished, and hence finds use in making precious *gems* and *jewellery*; also used for making dies for drawing thin wires from metals; the value of the diamond is expressed in terms of its *weight* and *purity*, and 1 carat = 200 mg; the famous '*Kohinoor*' diamond which decorates the crown of the British queen at present weighs 108.93 carats; see also **Artificial Diamond**.

Diathermal Wall A *Thermodynamics* term; a wall which permits the flow of heat across it; see also **Adiabatic Wall**.

Dichlorodifluoromethane A *Refrigeration* term; a refrigerant widely known as Freon-12; finds wide use for domestic purposes.

Die Casting A *Manufacturing* term related to *casting*; also called **Pressure Die Casting** as it involves high pressure (0.6 to 275 MPa) to force-fed the molten metal into the mould cavity: does not use any riser; can produce thin-walled parts (0.5 mm thickness) with complex shapes, good dimensional accuracy and surface details and good strength (not obtainable by other methods); castings usually does not require any further machining but suffer from gas porosity and chilling effect; commercially used to produce castings made from *nonferrous metals and alloys* such as *zinc*, *aluminium*, *magnesium*, and *copper-based alloys*; the typical products made by this method include *carburettors*, *motors*, *hand tools*, and *toys*; classified into *hot-chamber die casting* and *cold-chamber die casting*.

Die In *casting* terms, a *metallic mould* (also called *permanent mould*) which can be used repeatedly many times to produce castings of better surface finish and high-dimensional accuracy; offers high rate of production; in *gearing* terms, a tool to produce *external threads* on a round bar such as a *bolt*; see also **Tap**.

Dieing A *Manufacturing* term; also called **Threading**; a process of making *external threads* on a solid bar by means of a *die* (e.g. on a *bolt*); see also **Tapping**.

Dielectric An *Electrical* term; an electrically insulating material; for example, a plastic.

Diesel Cycle A *Thermodynamics* term related to *air standard cycle*; an air standard thermodynamic cycle used in *diesel engine*; named after its *German* inventor *Rudolf Diesel* (1858–1913); consists of *four* reversible processes, namely reversible adiabatic (isentropic) compression, constant pressure heat addition, reversible adiabatic (isentropic) expansion, and constant volume heat rejection; for the same compression ratio and same heat rejection, the Diesel cycle has lower thermal efficiency than the Otto cycle; see also **Otto Cycle**.

Diesel Engine An *IC Engine* term; an internal combustion engine named after its *German* inventor *Rudolph Diesel* (1858–1913); works on the principle of *Diesel cycle*; also called **Compression Ignition Engine** abbreviated as **CI Engine** in which ignition is initiated by the high temperature of air compressed alone in the cylinder (instead of compressing fuel–air mixture in case of petrol engine) and the fuel is injected in the cylinder in the spray form that catches fire by the high temperature of compressed air; uses a very high compression ratio; used in heavy and transport vehicles such as *bus*, *trucks*, *railways*, and *tractors*; also used for small power generation in a captive power plant; produces more air pollution than petrol engine and also suffers from noise and vibration; see also **Petrol Engine**.

Differential Control Volume A *Fluid Mechanics* and a *Thermodynamics* term; an infinitesimal control volume.

Differential Cost An *Operations Management* term related to cost *analysis*; see **Marginal Cost**.

Differential Equation of Elastic Curve A *Strength of Materials* term; see **Differential Equation of Flexure**.

Differential Equation of Flexure A *Strength of Materials* term; also called **Differential Equation of Elastic Curve**; a differential equation which incorporates the flexural rigidity (EI) of a beam, where $E =$ modulus of elasticity of the beam's material and $I =$ second moment of area of the beam's cross section; used to find slope and deflection at various sections of a loaded beam.

Differential Gear A *Theory of Machines* term related to a *gear*; a gear set which permits variable motion to the inner and outer wheels of an automobile while taking a turn.

Differential Manometer A *Fluid Mechanics* term; a metering device used to measure the difference in pressure between two points of a fluid flow, using Bernoulli's principle.

Differential Pressure A *Fluid Mechanics* term; the difference between two pressures.

Diffuser A *Fluid Mechanics* term; a flow passage device of varying cross-sectional area in which pressure of a fluid (gas or liquid) increases in the direction of flow at the expense of its velocity drop; see also **Nozzle**.

Diffusion A *Chemistry* term; the movement and mixing of atoms or molecules from one material (or from high concentration zone) to another (or low concentration zone).

Digital Signal An *Electrical* term; an electrical signal which has an integral number of discrete levels or values within a given range.

Digital Thermometer A device which uses a sensor called a *thermoresistor* or *thermistor* to measure the temperature.

Digital-to-Analog Converter An *Electrical* term, abbreviated as *DAC*; the conversion of digital signal to an analog equivalent such as a voltage; see also **Analog-to-Digital Converter**.

Dilatant Fluid A *Fluid Mechanics* term; also called **Shear Thickening Fluid**; a fluid whose viscosity increases with increase in applied shear stress; example includes *quicksand* (a mixture of sand and water); see also **Pseudoplastic Fluid**.

Dilatation A *Physics* and a *Strength of Materials* term; see **Volumetric Strain**.

Dimension An *Engineering Drawing* and a *Machine Drawing* term; a numerical value usually expressed in mm; dimension is shown on the drawings of an object using *lines*, *symbols*, *holes*, etc. so that all its features are completely defined.

Dimensional Analysis A *Fluid Mechanics* term; a tool as well as the foundation of the theories of similitude and modelling based on the notion of *dimensional homogeneity*; used for checking the dimensional consistency of a physical equation and to understand the behaviour of a physical system without the need for complex mathematics; used to determine the number and form of the dimensionless groups describing any fluid system using the *Buckingham Pi theorem*; provides a means to design an efficient experimental program.

Dimensional Constants A *Physics* term; the *dimensional physical quantities* whose values are constant; a few examples include *velocity of light in vacuum*, *universal gas constant*, *Boltzmann constant*, and *Planck's constant*; see also **Dimensional Variables**.

Dimensional Homogeneity A *Fluid Mechanics* term; a condition which ensures that all the terms of a physical equation must have the same dimensions; see also **Dimensional Analysis**.

Dimensional Variables A *Physics* term; the *dimensional physical quantities* whose values are variable; a few examples include *area, volume, density, velocity,* and *acceleration*; see also **Dimensional Constants.**

Dimensioning An *Engineering Drawing* and a *Machine Drawing* term; a process of producing dimension using *lines, symbols, figures,* and *notes* for furnishing information about the size of a drawing of an object; includes information such as the *distance between the surfaces*, the *location of holes*, the *nature of surface finish*, and the *type of object material.*

Dimensionless Constants A *Physics* term; the *nondimensional physical quantities* whose values are constant; a few examples include pure numbers like 1, 2, 3 ..., and mathematical constants such as π and *e*; see also **Dimensionless Variables.**

Dimensionless Group A *Fluid Mechanics* term; a *nondimensional group* of physical quantities having a dimensional representation of unity, when multiplied together; for example, the term '$\rho VL/\mu$', called the *Reynolds number* (ρ = flow density, V = flow velocity, μ = dynamic viscosity of the fluid, and L = characteristic length) is a dimensionless group, since.

$$\frac{\rho VL}{\mu} = \frac{[\frac{M}{L^3}][\frac{L}{T}][L]}{[M]/[LT]} = 1$$

Other dimensionless groups include *Euler number, Mach number, Froude number,* and *Weber number.*

Dimensionless Variables A *Physics* term; the *nondimensional physical quantities* whose values are variable; a few examples include *angle, strain,* and *specific gravity*; see also **Dimensionless Constants.**

Dipole Moment A *Chemistry* term; the product of charge and the separation distance between centres of positive and negative charges in a dipole; see also **Dipole.**

Dipole A *Chemistry* term; the asymmetrical distribution of positive and negative charge associated with a secondary bonding; see also **Dipole Moment.**

Direct Cost An *Operations Management* term related to *cost analysis*; the direct costs are directly attributable to a specific output or work activity; for example, *material* and *labour* costs associated with a *product, service* or *construction activity* are categorized as *direct costs*; direct cost increases in proportion to volume of product or service; see also **Indirect Cost.**

Direct Current An *Electrical* term, abbreviated as **DC**; a type of electrical current whose polarity is always *fixed* and always flows in the *same direction*; *more dangerous* than the alternating current; see also **Alternating Current.**

Direct Extrusion A *Manufacturing* term; also called **Forward Extrusion**; *the most basic extrusion process* in which metal confined in a chamber is forced to come out through a stationary shaped-die opening in the form of required product in the direction of force applied; see also **Indirect Extrusion**.

Direct Strain A *Strength of Materials* term; see **Longitudinal Strain**.

Direct Stress A *Strength of Materials* term; see **Normal Stress**.

Dirt A *Casting* term; a type of casting defect which occurs in the form of *small angular holes* (called dirts) left out on the surface of a casting; occurs when the sand particles are removed that have fallen and embedded on the surface of the casting from the cope part of the moulding flask.

Discharge A *Fluid Mechanics* term; see **Volume Flow Rate**.

Disintegration Constant A *Chemistry* term related to *radioactivity*; also called **Decay Constant**; the fraction of the total number of atoms of a radioactive substance at any time which disintegrates per second; a characteristic of a radioactive element; denoted by λ; the relationship between the original number of atoms (N_o) of a radioactive element and the number of atoms (N) at any time t is given as:

$$N = N_o e^{-\lambda t} \tag{1}$$

The relationship between decay constant (λ) and half-life period ($t_{1/2}$) of a radioactive element is given as:

$$t_{1/2} = \frac{0.693}{\lambda} = \text{constant} \tag{2}$$

Dislocation A *Materials Engineering* term; also called **Linear Defect**; a one-dimensional (linear) imperfection (defect or disorder) in a material primarily associated with mechanical deformation; classified into *edge dislocation*, *screw dislocation*, and *mixed dislocation*; see also **Point Defect**.

Dispatching An *Operations Management* term; the process of initiating the scheduled work on the shop floor; dispatching includes giving out *order tickets*, *route sheets*, *part drawings*, and *job instructions*.

Displacement Volume An *IC Engine* term; see **Swept Volume**.

Disposable Pattern A *Manufacturing* term related to *casting*; a pattern made from *polystyrene* (a type of *plastic*) which gets consumed and is not required to be removed from the mould; offers advantages such as *better surface finish* of castings, *less time consuming*, and *no need for the pattern allowances*; see also **Removable Pattern**.

Dissolved Acetylene A *Chemistry* term, abbreviated as **DA**; the *trade name of acetylene*; acetylene kept dissolved in *acetone* to contain its increased volume in a small space.

Distributed Force System A *Physics* term; a force system in which forces may act continuously along a line or on a surface or on a volume; a few examples include *linear*, *surface*, and *body* forces.

Distributed Force A *Strength of Materials* term; see **Distributed Load**.

Distributed Load A *Physics* and a *Strength of Materials* term; a load which is not concentrated at a point but is exerted along a line, over an area, or throughout an entire solid body; see also **Point Load** and **Uniform Load**.

Divergence Theorem A *Mathematics* term; also called **Gauss's Theorem**, named after the German mathematician *Johann Carl Friedrich Gauss* (1777–1855); used to transform a *volume integral* of the divergence of a vector into an *area integral* over the surface which defines the volume.

Dividend A *Financial Management* term; the periodic distribution of earnings to the stockholders of a firm or the gains which are realized through increase in share price.

Division of Labour An *Operations Management* term, coined by a Scottish economist *Adam Smith* (1723–1790) in his book 'The Wealth of Nations' in 1776; a process of dividing a job into a series of small tasks to be separately performed efficiently by different workers and to extract the economic benefits out of it; helps in increasing the efficiency of a worker because of expertness and dexterity of works.

Divisional Manager A *Management* term; a manager who sets the strategy not for the organization as a whole, but for the division which he heads.

Divisional Structure A *Management* term; a type of organizational structure in which functions are grouped together according to the specific demands of products, markets or customers; helps to create smaller, more manageable subunits within an organization; suitable for a large corporation with many product lines in several related industries; for example, General Motors groups its various auto lines into the separate divisions of *Chevrolet*, *Pontiac*, *Saturn*, *Oldsmobile*, *Buick*, and *Cadillac*.

DOE A *Statistics* term; see **Design of Experiment**.

Dot Product of Vectors A *Physics* term; see **Scalar Product of Vectors**.

Double Integration Method A *Strength of Materials* term related to *deflection*; an analytical method used to find the slope and deflection at various sections of a loaded beam; involves two times integration of the differential equation of flexure, where the first integration gives slope and the second integration gives deflection at any section of a loaded beam.

Double Sampling A *Quality Control* term; the acceptance sampling carried out on a *large size sample*, when results are inconclusive with a small sample.

Double-Acting Cylinder A *Mechatronics* term; a common type of *linear actuator* controlled by fluid pressure in both directions; see also **Single-Acting Cylinder**.

Double-Acting Engine A *Thermodynamics* term; an engine in which the working fluid is made to act on both sides of the piston; most of the reciprocating steam engines and marine diesel engines are double-acting; theoretically develops twice the power developed by a single-acting engine; see also **Single-Acting Engine**.

Double-Helical Gear A *Gear* term; see **Herringbone Gear**.

Douglas McGregor (1906–1964) An *American* motivation expert who is best known for his formulation of two sets of behavioural assumptions, called *Theory X* and *Theory Y*, which are used as important theories of motivation.

Downsizing A *Management* term; an activity in an organization which aims to reduce the size of the organization's workforce in order to create greater efficiency by eliminating certain jobs.

Downsprue A *Manufacturing* term related to *casting*; see **Sprue**.

DPMO A *Quality Control* term; see **Defects Per Million Opportunities**.

DPT A *Thermodynamics* term related to *psychrometry*; see **Dew Point Temperature**.

Draft Allowance A *Casting* term; see **Draft**.

Draft Tube A *Fluid Mechanics* term; a gradually expanded closed passage provided at the exit of a hydraulic reaction turbine; enables the turbine to be set above the tailrace water level (downstream water) without losing any head, that is, negative (suction) head can be established at runner exit of the turbine; also acts as a *recuperator of energy* as it converts a large proportions of kinetic energy of the fluid rejected from the runner by reducing its flow speed into useful pressure energy.

Draft In *casting* terms, a type of pattern allowance (also called **Draft Allowance**) provided in *sand-mould patterns* in the form of small taper to facilitate their withdrawal from the mould without damaging the latter; in *forming* terms, the difference between initial thickness and final thickness of a blank in a *rolling operation.*

Drag Force A *Fluid Mechanics* term; see **Drag**.

Drag A *Fluid Mechanics* term; also called **Drag Force**; a force that a fluid flow exerts on a body in the direction of the flow; see also **Lift**.

Drawing Sheet An *Engineering Drawing* and a *Machine Drawing* term; a standard size paper on which the drawings of objects are prepared; the preferred sizes for the drawing sheets are indicated below:

Designation	Dimensions (mm)
A0	841 × 1189
A1	594 × 841
A2	420 × 594
A3	297 × 420
A4	210 × 297

Drawing A *Manufacturing* term; a metal working process used to reduce the cross section and increase the length of *a metal bar*, *wire* or *tube* by pulling it through a series of *conical, tapering holes* in a *die plate*; can be performed in both *cold* and *hot* states; also a picturization process of the imagination of a scene (e.g. an object) in its totality by an individual; see also **Tube Drawing**.

Drikold A *Refrigeration* term; the trade name of the commercial refrigerant *solid carbon dioxide.*

Drill press A *Manufacturing* term; a standard machine tool used for *drilling*; the upright drill press is the most basic one.

Drilling A *Manufacturing* term; a process of making a circular hole in a workpiece by means of a drill; see also **Boring** and **Reaming**.

Driving Dynamometer A *Theory of Machines* term; a type of dynamometer used to measure *torque* or *power* as well as to supply energy to operate the device under test; useful in determining performance characteristics of pumps and compressors.

Drop Forging A *Manufacturing* term; a method of applying forces in a forging operation in which a hated workpiece is shaped by an *impact force* using closed matching dies; most of the impact energy is wasted unutilized as it is absorbed by the machine and foundation; see also **Press Forging**.

Drop A *Casting* term; a type of casting defect which occurs in the form of *some projection*; produced when a cope part of a casting falls on the drag surface due to poor cohesiveness of the moulding sand.

Drum Gate A *Fluid Mechanics* term; another common type of *underflow gate* which has a *circular cross section* with a streamlined surface; used to control the liquid flow rate in *open channels*; see also **Sluice Gate**.

Dry Bulb Temperature A *Thermodynamics* term related to *psychrometry*; abbreviated as **DBT**; the *actual temperature* of the *moist air*; see also **Wet Bulb Temperature**.

Dry Friction A *Physics* term; also called **Unlubricated Friction** or **Solid Friction** or **Coulomb Friction**, named after a *French* physicist and engineer *Charles-Augustin de Coulomb* (1736–1806); occurs between bodies having dry or unlubricated contact surfaces due to their microscopic roughness; causes high rate of wear.

Dry Ice A *Chemistry* term; see **Solid Carbon Dioxide**.

Dry Sand Mould A *Casting* term; a mould made of *dry sand* which does not contain any moisture.

Dry Sand A *Casting* term; the moulding sand which does not contain any *moisture*, and is dry.

Dryness Fraction A *Thermodynamics* term related to *steam*; a measure of quality of the wet steam; calculated as a ratio of the mass of dry steam (m_g) to the mass of total wet steam ($m_g + m_f$), where m_f is the mass of water vapour; denoted by x; see also **Wetness Fraction**.

Dual Combustion Cycle A *Thermodynamics* term related to *air standard cycle*; see **Dual Cycle**.

Dual Cycle A *Thermodynamics* term related to *air standard cycle*; also called **Dual Combustion Cycle** or **Mixed Cycle** or **Solid Injection Cycle**; a thermodynamic cycle of an engine in which combustion takes place both at *constant volume* and at *constant pressure*; consists of *five* reversible processes, namely isentropic compression, constant volume heat addition (that forms one part of heat addition), constant pressure heat addition (that forms another part of heat addition), isentropic expansion, and constant volume heat rejection.

Duct Flow A *Fluid Mechanics* term; also called **Internal Flow**; the flow of liquids or gases in a *duct* (completely bounded by solid surfaces); satisfies the steady-flow conditions, and thus can be analysed as a *steady-flow process;* for example, water flow in a pipe; see also **External Flow**.

Duct A *Fluid Mechanics* term; a closed conduit of non-circular cross section.

Ductile Cast Iron A *Materials Engineering* term; see **Nodular Cast Iron**.

Ductile Iron A *Materials Engineering* term; see **Nodular Cast Iron**.

Ductile Material A *Materials Engineering* term; a material which shows large plastic range beyond elastic limit, and can withstand a significant amount of stretching before it fractures; the breaking point of ductile materials is widely separated from elastic limit on the stress–strain curve; can be easily drawn into *wires*; well suited for use in *structures* and *machines*; undergoes *neck formation* before it finally breaks; used in making *springs* and *sheets*; a few examples include *copper*, *silver*, *iron*, and *aluminium*; see also **Brittle Material**.

Ductility A *Strength of Materials* term; a mechanical property, usually of a ductile material, which enables the material to be drawn into thin wires or to be stretched significantly before it fractures; a few high ductility materials include *copper*, *aluminium*, and *mild steel.*

Dummy Activity A *Project Management* term; a fictitious activity which does not consume any time, labour or material resources; inserted into a project network just to show a precedence relationship.

Duralumin A *Materials Engineering* term; an important alloy of *aluminium* (Al) which consists of 95% *aluminium* (Al), 4% *copper* (Cu), 0.5% *magnesium* (Mg), and 0.5% *manganese* (Mn); due to light weight, it finds use in making parts of *aircrafts* and *automobiles*.

Durometer A *Strength of Materials* term; an instrument used to measure the hardness of *soft materials* such as *rubber* and *plastic*.

Dye A *Chemistry* term; a soluble, organic colorant used for polymers.

Dynamic Equilibrium A *Physics* term; also called **Translational Dynamic Equilibrium**; the equilibrium connected to a rigid body in uniform motion (constant velocity) along a particular direction; the body under dynamic equilibrium has *constant linear momentum* and *zero linear acceleration;* see also **Static Equilibrium**.

Dynamic Friction A *Physics* term; see **Kinetic Friction**.

Dynamic Load A *Physics* term; a load which changes its magnitude with time, and tends to set oscillations in a body; classified into *impact* and *fluctuating* loads. See also **Static Load**.

Dynamic Pressure A *Fluid Mechanics* term; the pressure linked to motion of a fluid; represents the pressure rise when a fluid in motion is brought to a rest isentropically; see also **Static Pressure, Hydrostatic Pressure,** and **Stagnation Pressure**.

Dynamic Programming An *Operations Research* term; named after *Richard Bellman* (1920–1984), an *American* applied mathematician who developed it in the 1950s; a method based on *Bellman's principle of optimality*; used to solve a variety of problems such *as finding the shortest path from one station to another, cargo loading problem, reliability problem, capital budgeting problem,* and *inventory problem.*

Dynamic Pump A *Fluid Mechanics* term; see **Rotodynamic Pump**.

Dynamic Similarity A *Fluid Mechanics* term; the similarity of forces acting on a model and its prototype; see also **Geometric Similarity** and **Kinematic Similarity**.

Dynamic Viscosity Coefficient A *Fluid Mechanics* term; **Dynamic Viscosity**.

Dynamic Viscosity A *Fluid Mechanics* term; also called **Coefficient of Viscosity** or **Absolute Viscosity** or **Absolute Coefficient of Viscosity** or **Dynamic Viscosity Coefficient** or **Kinetic Viscosity**; a viscous force which acts tangentially on unit area of a liquid layer having a unit velocity gradient perpendicular to the direction of liquid flow; denoted by η (eta) or μ (mu); strongly depends on temperature; its CGS unit is *poise*, named after a *French* physician and scientist *Jean Poiseuille* (1797–1869), which is equivalent to dyne s cm^{-2} or gm cm^{-1} s^{-1}; the SI unit is N s m^{-2} or kg m^{-1} s^{-1}; the relationship between its CGS and SI unit is: 1 *poise* = 0.1 N s m^{-2}; has dimension of $[ML^{-1}\ T^{-1}]$; see also **Kinematic Viscosity**.

Dynamics A *Physics* term; a branch of mechanics which deals with the action of forces on a body in motion; classified into *kinematics* and *kinetics*; see also **Statics**.

Dynamometer A *Theory of Machines* term; a brake incorporating mechanical device used to measure *frictional resistance* offered by a *brake* or to measure *energy*, *torque* or *power* developed by a prime mover at its rated speed; classified into *absorption*, *driving*, and *transmission* dynamometers.

Dyne A *Physics* term; the CGS unit of force; exerted on a body of mass 1 g and produces an acceleration of 1 $\frac{cm}{s^2}$; newton and dyne are related as: 1 dyne = 10^{-5} N; see also **Newton**.

E

Earliest Finish Time (EFT) A *Project Management* term; the earliest time an activity can be completed; finds use in *critical path method* (CPM); equals to the earliest start time (EST) of the immediate predecessor activity plus activity time (t); EFT = EST + t; when more than one activity immediately precedes the activity, the maximum of the EST of all the preceding activities is considered as the EFT of the activity; see also **Latest Finish Time (LFT)**.

Earliest Start Time (EST) A *Project Management* term; the earliest time an activity can start subject to preceding activities; finds use in *critical path method* (CPM); see also **Latest Start Time (LST)**.

Earning Assets A *Financial Management* term; see **Fixed Assets**.

EBM A *Nonconventional Machining* term; see **Electron Beam Machining**.

E-Business A *Management* term; see **Electronic Commerce**.

Eccentric Axial load A *Strength of Materials* term; an axial load which does not pass through the centroid of the cross section of a body.

ECG A *Nonconventional Machining* term; see **Electrochemical Grinding**.

ECM A *Nonconventional Machining* term; see **Electrochemical Machining**.

E-Commerce A *Management* term; see **Electronic Commerce**.

Economic Good An *Economics* term; an item which is scarce or anything for which a condition exists that the amount of the good which people want is greater than the amount that is available; see also **Free Good**.

Economic Growth An *Economics* term; the increase in the total output of an economy which occurs when a society acquires new resources or when it learns to produce more using *existing resources*.

D. K. Singh, *Dictionary of Mechanical Engineering*,
https://doi.org/10.1007/978-981-99-1722-8_5

Economic Life A *Financial Management* term; the length of time an asset is economically useful.

Economic Order Quantity An *Operations Management* term, abbreviated as **EOQ**; an optimal order quantity of inventory which results in minimum total inventory cost which consists of annual ordering cost and inventory holding cost; proposed by *F. W. Harris* in 1915; denoted by Q; calculated as:

$$Q = \sqrt{\left(\frac{2D\,S}{H}\right)}$$

where

D = annual demand of items
S = setup cost or ordering cost (cost per order)
H = annual holding (carrying) cost per item.

Economic Production Quantity Model An *Operations Management* term; see **Production Quantity Model.**

Economics The study of how individuals and societies choose to use the scarce resources which nature and previous generations have provided so as to satisfy their unlimited wants; classified into *macroeconomics* and *microeconomics.*

Economizer A *Power Plant* term related to a *boiler*; an important *boiler accessory* used to extract (absorb) heat from the hot flue gases in the furnace and use it to heat the *feed water* being supplied to the boiler; reduces heat input to the boiler and increases its efficiency; fitted near the chimney of the boiler; see also **Superheater**.

EDD An *Operations Management* term; one of the priority rules for sequencing jobs which stands for *earliest due date*; earlier due date jobs are assigned first; minimizes maximum tardiness, which may be necessary for jobs which have a very heavy penalty after a certain date.

Edgar Buckingham (1867–1940) An *American* scientist who used the symbol *π (pi)* to represent a dimensional product; well known for the *Buckingham pi theorem* used in dimensional analysis.

Edge Dislocation A *Materials Engineering* term; a type of *linear defect* in a material which runs along the edge of the extra row of atoms; the *Burgers vector* is perpendicular to the dislocation line; see also **Screw Dislocation** and **Mixed Dislocation.**

Edge Preparation A *Welding* term; a process of preparing the edges of a workpiece (making them straight) to be joined to ensure a strong weld joint; a butt joint requires edge preparation, but a lap joint does not.

EDI An *Operations Management* term; see **Electronic Data Interchange**.

EDM A *Nonconventional Machining* term; see **Electrical Discharge Machining**.

Effective Length A *Strength of Materials* term related to a *column*; also called **Equivalent Length**; the distance between the adjacent points of inflexion on the elastic curve of a column; the effective length varies according to end conditions of the column; for actual length L of the column, the values of effective length for various end conditions are tabulated below:

End conditions	Effective length
Both ends of the column are hinged/pinned	L
Both ends of the column are fixed	$L/2$
One end of the column is fixed, while its other end hinged	$L/\sqrt{2}$
One end of the column is fixed, while its other end free	$2L$

Efficiency Ratio An *IC Engine* term; a ratio of the thermal efficiency of an actual heat engine to the thermal efficiency of the corresponding ideal heat engine in which all the processes are reversible; used as an *index of performance* for a heat engine; see also **Work Ratio**.

Effort Force A *Physics* term; an input force generated by a person, motor, engine, magnetic field, spring, moving water, wind, or by any other phenomenon which serves to impart energy into a machine or system.

EGL A *Fluid Mechanics* term; see **Energy Gradient Line**.

Ego Need A *Management* term related to *motivation*; see **Esteem Need**.

Elastic Body A *Physics* and a *Strength of Materials* term; a body which regains its original shape and size after the removal of external load applied on it; a perfectly elastic body is an ideal concept and does not exist in practice as some deformation always stays with the material; see also **Rigid Body**.

Elastic Collision A *Physics* term; the collision involving no loss of kinetic energy, that is, the kinetic energy remains conserved; the linear momentum and the total energy of the system are also conserved; the forces involved during elastic collisions are conservative forces; the collisions between *atomic*, *nuclea*r, and *fundamental particles* are usually elastic; also the collisions between two *ivory balls* or two *glass balls* are approximately elastic; see also **Inelastic Collision**.

Elastic Constants A *Strength of Materials* term; also called **Coefficient of Elasticity**; a ratio of the stress to the strain produced within elastic limit; three elastic constants include *Young's modulus of elasticity* (E), *modulus of rigidity* (G), and *bulk modulus of elasticity* (K); their relationships are expressed as: $E = 2G(1 + \upsilon)$; $E = 3K(1 - 2\upsilon)$; $E = 9\text{ kg}/(3K + G)$, where υ is the Poisson's ratio.

Elastic Curve A *Strength of Materials* term; also called **Deflection Curve**; a curve obtained as a result of deflection of a loaded beam.

Elastic Deformation A *Strength of Materials* term; a process of deforming a body with the application of an external force which vanishes immediately after the removal of the force, and the body recovers its original shape and size; the civil structures undergo elastic deformation; see also **Plastic Deformation**.

Elastic Demand An *Economics* term; the demand for a product or service is said to be elastic, if a price reduction gives more than a proportionate increase in demand, or if a price increase gives more than a proportionate decrease in demand; for example, demand of *luxury goods*; see also **Inelastic Demand**.

Elastic Fatigue A *Strength of Materials* term; a property of an elastic body by virtue of which its behaviour becomes less elastic under the action of repeated alternating deforming forces, and the material regains its original degree of elasticity, when allowed to rest for some time.

Elastic Limit A *Strength of Materials* term; the maximum limit of elasticity of a material, when stress is linearly proportional to strain and the body recovers its original conditions fully on the removal of the deforming force, but at a slightly slower rate as compared to the limit of proportionality; beyond this point, the material no longer remains elastic, and it reaches into plastic state; see also **Limit of Proportionality**.

Elastic Material A *Materials Engineering* term; a material which fully recovers its unstrained (original) state on the removal of the load applied on it, when stressed within the elastic limit; the unloading path coincides with the loading path, and the process is *reversible*; for example, *steel* is an elastic material; see also **Plastic Material**.

Elastic Modulus A *Strength of Materials* term; see **Young's Modulus of Elasticity**.

Elastic Potential Energy A *Physics* term; the energy stored in an *elastic body* such as a *spring* or a *bar* on being stretched or bent and has always positive value; for a spring, it is expressed as:

$$U = \frac{1}{2} \cdot K \cdot \Delta L^2$$

where

U = elastic potential energy stored in the spring
K = stiffness of the spring
ΔL = elongation or compression in the spring.

Elastic Strain Energy A *Strength of Materials* term; the strain energy which can be recovered during unloading of a member; see also **Inelastic Strain Energy**.

Elastic-After-Effect A *Physics* term; a temporary delay in regaining the original configuration by an elastic body after the removal of a deforming force that usually occurs when an elastic body is stressed beyond its limit of proportionality till it reaches the elastic limit; negligibly small for *quartz* and *phosphor bronze*, which makes them suitable as suspension elements in *galvanometers* and *electrometers*, but is very large for *glass fibre*.

Elasticity of Demand An *Economics* term; also called **Price Elasticity of Demand**; a phenomenon of change in demand of a product with rise or fall in its price.

Elasticity A *Physics* term; a property of a metallic body which enables it to resist a deforming force and helps to regain its original conditions (shape and size) after the removal of the deforming force; an important property used in *structural members*; see also **Plasticity**.

Elastomer A *Materials Engineering* term; a material for which the stress–strain relationship is not linear within elastic limit; the elastic region is very large; does not obey *Hooke's law* over most of the elastic region; has no plastic range; the breaking point lies just close to the elastic limit; a few examples include *rubber* and *veins* which carry blood from the heart.

Elbow Meter A *Fluid Mechanics* term; one of the simplest types of flow measuring device used in a pipeline; uses the pressure difference created due to the centrifugal effects at the bend to measure the flow; consists of piezometer taps installed at the inner and outer walls of a 90° elbow in the pipeline.

Electric Arc Welding A *Welding* term; a type of welding process which uses heat of an arc to join metals; one of the most extensively used welding processes.

Electric Arc A *Welding* term; also called simply **Arc**; a flash produced between two current carrying conductors when they are touched momentarily; discovered by *Sir Humphry Davy* (1778–1829) in 1801 while he was conducting experiments in electricity.

Electric Furnace A furnace which uses *electricity to produce heat* using a continuous arc between two or three graphite electrodes and the charged metal; used extensively to produce steels because of its *high production rate*, *non-polluting nature*, and the ability to hold the molten metal for prolonged duration.

Electric Vehicle Battery An *Electric Vehicle* term, abbreviated as *EVB*; a battery used to power an electric vehicle (EV); can be recharged by plugging the EV into an electrical output.

Electric Vehicle Supply Equipment An *Electric Vehicle* term, abbreviated as **EVSE;** also called the **Charging Station** or the **Charging Dock**; a commercial station having infrastructural facilities for charging the batteries of electric vehicles safely in large numbers; analogous to petrol/diesel pump stations.

Electric Vehicle A relatively new *automobile* term, abbreviated as **EV**; a vehicle completely powered by electricity and driven by on-board motors.

Electrical Circuit An *Electrical* term; a closed path through which electricity can flow between the terminals of a power source and through one or more electrical components.

Electrical Conductivity An *Electrical* term; a *physical property* of a substance which indicates its ability to conduct electrical current; chiefly depends on the number of *free electrons* in the materials; an important design consideration while selecting a material for electrical appliances; *metals* are the *good conductor* of electricity, and *insulators* are the *bad conductor* of electricity.

Electrical Conductor A *Physics* term; often called simply **Conductor**; a material which offers very low or no resistance to current flow and is able to conduct the electrical current; for example, *all metals*; see also **Insulator**.

Electrical Discharge Machining A *Nonconventional Machining* term, abbreviated as **EDM**; also called **Electro Discharge Machining**; one of the most widely used *nonconventional machining* processes which uses *electrical spark* or thermal energy to erode materials from a workpiece to produce the desired shape; capable of machining *hard* and *difficult-to-machine materials* (like *thin* and *fragile*) or *brittle materials* without producing any distortion; used to produce parts with complex, precise, and irregular shapes for forgings, press tools, extrusion dies, and difficult internal shapes for aerospace and medical applications; used only for *electrically conducting materials*; see also **Wire-EDM**.

Electrical Energy A *Physics* term; the energy associated with the movement of electrical charges, (electrons); the most common form of energy which has vast domestic and commercial applications; electrical energy causes a bulb to glow, a fan and a turbine to rotate, and a bell to ring.

Electrical Engineer A technical person who is involved in the design and production of integrated circuits, wireless communication networks, aviation electronics, robot control systems, and the transmission and distribution of electrical power.

Electrical Resistance An *Electrical* term; the opposition to electrical current as it flows through a conductor; measured in *ohm* (Ω); lead has 10 times electrical resistance as compared to copper.

Electrical Torsionmeter A *Metrology* term; a device used to measure torque by measuring the angle of twist over a fixed length of a shaft using electrical means such as transducers which are placed on two identical toothed wheels attached to the shaft at a fixed distance apart.

Electrical Work A *Thermodynamics* term; the work produced by an electric current; for example, lifting of a weight by a pulley attached to the shaft of a motor driven by an electric current or moving of an electric fan.

Electricity A *Physics* term; the flow of electrons through a conductor.

Electro Discharge Machining A *Nonconventional Machining* term; see **Electrical Discharge Machining.**

Electrochemical Grinding A *Nonconventional Machining* term, abbreviated as **ECG**; also called **Electrolytic Grinding**; a nonconventional machining process in which material from a workpiece is removed by the combined action of *electrochemical machining* (ECM) and *conventional grinding*, the former contributes about 90%; a fast process in which tool wear is less; suitable only for *electrically conducting materials.*

Electrochemical Machining A *Nonconventional Machining* term, abbreviated as **ECM**; a nonconventional machining process which uses the principle of *reverse electroplating* to remove material from a workpiece; capable of producing complex shapes in hard and difficult-to-machine materials.

Electrode Holder A *Welding* term; a device used to mechanically hold an electrode and allow the current to flow through it.

Electrolytic Grinding A *Nonconventional Machining* term; see **Electrochemical Grinding.**

Electromagnet A *Physics* term; a device consisting of a coil of conductive wire wrapped around an iron core which generates a magnetic field when current passes through the coil.

Electromagnetic Casting A *Manufacturing* term related to *casting*; also called **Levitation Casting**; a casting process which uses an induction coil to produce and confine the molten metal through magnetic suspension and then allowed to flow into an *investment casting mould* placed directly below the coil; the process combines the advantages of both *induction heating* and *investment casting*; produces smooth surface castings having homogeneous and fine-grained structure free from refractory inclusions and gas porosity.

Electromagnetic Force A *Physics* term; a force which acts between charged particles; called an *electrostatic force*, when charges are at rest.

Electromagnetic Induction A *Physics* term; the generation of voltage within a conductor as a result of passing the conductor through a magnetic field.

Electromagnetic Radiation A *Physics* term; the radiation which has the ability to transfer energy through empty space, unlike conduction and convection, which need matter to transfer energy; a few examples of electromagnetic radiation include *X-rays*, *gamma rays*, *microwaves*, *radio waves*, and *television waves*; see also **Thermal Radiation.**

Electromotive Force An *Electrical* term, abbreviated as **emf**; an electrical pressure which moves through the electrical circuit of an instrument and pushes electrons through a conductor; measured in volts.

Electron Affinity A *Chemistry* term; the amount of *energy released when an electron is added* to a gaseous isolated atom of an element, producing a *negatively charged ion* (*anion*); shows the affinity of a neutral atom to attract electron towards itself; the greater the energy released in the process of taking extra electron, the greater is the electron affinity; noble gases have zero electron affinities, because of their stable electronic configuration which does not allow any addition of extra electron; generally decreases in going from top to bottom in a group and increases in going from left to right across a period; see also **Ionization Energy**.

Electron Beam Machining A *Nonconventional Machining* term, abbreviated as **EBM**; a nonconventional machining process which removes material with the help of a high-energy *electron beam.*

Electron Volt A *Physics* term, abbreviated as **eV**; *a unit of energy of an electron*; the energy associated with an electron having a charge of 1.602×10^{-19} C moving under a potential difference of 1 V; 1 eV = 1.602×10^{-19} Joule (J).

Electron A *Chemistry* term; one of the three fundamental particles of an atom of a chemical element discovered by *J. J. Thomson* (1856–1940), a British physicist in 1897; a *negatively charged* particle which moves around the nucleus of an atom in different circular orbits; has an absolute charge equal to 1.60206×10^{-19} C and a mass equal to 9.1091×10^{-31} kg; see also **Proton** and **Neutron**.

Electronegativity A *Chemistry* term; the tendency of an atom to attract electrons towards itself; atoms with almost completely filled up outermost orbits are strongly electronegative and readily accept electrons; for example, *chlorine* (Cl), *bromine* (Br), *iodine* (I), and *fluorine* (F) have the highest electronegativity in the periodic table; atoms with nearly empty outermost orbits have low electronegativity as they readily give up electrons, for example, *sodium* (Na) and *potassium* (K); atoms having high atomic number have low electronegativity.

Electronic Commerce A *Management* term which is more commonly called **E-Commerce** or **E-Business**; a trade which makes use of electronic means such as *internet*, *intranets*, and *extranets* to run the business activities between businesses; has made a business process *hassle-free*, *transparent*, *responding*, *accountable*, and *quicker*; the two widely used forms of e-commerce include business-to-business (B2B) and business-to-consumer (B2C).

Electronic Data Interchange An *Operations Management* term, abbreviated as **EDI**; a paperless communication method which uses a computer-based system to electronically transfer business data within and between organizations; offers faster transmission of data at low cost and with fewer errors.

Electronic Materials A *Materials Engineering* term; the materials used for making *electrical* and *electronic* devices for various applications; a few examples include *silicon* (Si), *copper* (Cu), *aluminium* (Al), *superconductors*, and *dielectric materials*; *silicon* is used to make *integrated circuits* for computer chips, and *superconductors* are used to make *powerful magnets*; *copper* and *aluminium* are used as *conductors* in *power transmission* and in *microelectronics*; dielectric materials such as *barium titanate* ($BaTiO_3$) and *tantalum oxide* (Ta_2O_5) are used to make *ceramic capacitors* and other devices.

Electrostatic Force A *Physics* term; a force which acts between charged particles at rest; governed by Coulomb's law; force between like charges is repulsive and between unlike charges is attractive; directly proportional to the product of charges and inversely proportional to the square of distance between the charges.

Electrostatic Precipitation An *Air Pollution* term; an air pollution control equipment used to remove the particulates under the influence of an *electrostatic field*.

Electrovalent Bond A *Chemistry* term; see **Ionic Bond**.

Electrovalent Compound A *Chemistry* term; see **Ionic Compound**.

Element A *Chemistry* and a *Theory of Machines* term; in *chemistry* terms, a pure substance which contains only one kind of particles like atoms or molecules; a few examples include *carbon*, *sulphur*, *iron*, *lead*, *mercury*, *hydrogen*, *oxygen*, and *nitrogen*; in *theory of machines* terms, a machine part more commonly called a *link*.

Elemental Semiconductors An *Electronics* term; the semiconductors available in the elemental form; for example, *germanium* (Ge), *silicon* (Si), and *tin* (Sn); see also **Compound Semiconductors**.

Elevation Head A *Fluid Mechanics* term; also called **Potential Head** or **Datum Head**; an energy parameter which appears in the *Bernoulli equation*; represents the potential energy of the fluid; denoted by z; see also **Pressure Head** and **Velocity Head**.

Elongation A *Strength of Materials* term; also called **Extension**; the amount of length by which a member has increased, when loaded axially under tension; see also **Contraction**.

Embossing A *Manufacturing* term; a cold metal working process in which impressions of the mating die and punch are produced on a thin metal blank, and one side of the product is the reverse of its other side; used in making *nameplates*, *medals*, or *aesthetic design* on a thin sheet metal or foil; see also **Coining**.

emf An *Electrical* term; see **Electromotive Force**.

Emissive Power A *Heat Transfer* term; the thermal radiation emitted by a body per unit area of its surface; proportional to the fourth power of absolute temperature of the body; for a black body, the emissive power is given as:

$$E_b = \sigma \cdot T^4$$

where

E_b = emissive power
σ = Stefan-Boltzmann Constant
= 5.67×10^{-8} Wm^{-2} K^{-4}
T = absolute surface temperature; see also **Emissivity**.

Emissivity A *Heat Transfer* term; a radiation property of the surface of a body; a dimensionless property which indicates how efficiently the surface emits radiative energy; a ratio of the emissive power of a body to the emissive power of a *black body*; depends on the temperature and the wavelength of the radiation; denoted by ε (epsilon); its value lies between 0 and 1; for a black body $\varepsilon = 1$; see also **Emissive Power**.

Employee Empowerment An *Operations Management* term; an approach of assigning *authority* and *responsibility* to an employee to take ownership of his job; helps in improving the employee performance.

Employee A *Management* term; the people who are working in an organization for which they are paid *wages* or *salary* to satisfy their worldly needs.

Empowerment A *Management* term; a process of giving employees the responsibility and authority to make decisions without asking their superiors for permission.

Emulsion A *Fluid Mechanics* term; a mixture of two immiscible liquids along with other additives.

End Condition A *Strength of Materials* term related to *column*; the positioning of the two ends of a column; used to find the *effective length* of columns; the four end conditions of a column include *both ends hinged*, *both ends fixed*, *one end fixed while other end hinged*, and *one end fixed while other end free*.

End measurement A *Metrology* term; the measurement of distance between *surfaces* by using end faces of *gauge blocks* or *slip gauges*; *the most common method of measurement in industrial practice*, as it can produce accuracy of higher order; see also **Line Measurement**.

Endurance Limit A *Strength of Materials* term; also called **Endurance Strength** or **Fatigue Limit**; the maximum stress which results from cyclic or periodic loading of a material before its fatigue failure; depends on the ultimate tensile strength of the metals; for example, the endurance limit of steel is about one-half of its tensile strength.

Endurance Strength A *Strength of Materials* term; see **Endurance Limit**.

Energy Balance A *Fluid Mechanics* term; a condition in which energy of the components on input and output sides of a device is balanced; the concept is used in *Bernoulli's equation*; see also **Mass Balance**.

Energy Gradient Line A *Fluid Mechanics* term, abbreviated as **EGL**; also called **Total Energy Line**, abbreviated as **TEL**; a line joining the points having the *total head* (the sum of *pressure head* ($P/\rho g$), *velocity head* ($v^2/2\ g$), and *elevation head* (z)) at the various cross section of a fluid flow in a pipe; always a distance $v^2/2\ g$ (*velocity head*) above the *hydraulic gradient line* (HGL); the slope of EGL is equal to the rate of energy loss; see also **Hydraulic Grade Line**.

Energy in Transition A *Thermodynamics* term; the energy which crosses the boundary of a system; a few examples include *heat*, *work*, and *electricity*; see also **Stored Energy**.

Energy Level A *Chemistry* term; see **Orbit**.

Energy Ratio A *Refrigeration* term; see **Coefficient of Performance**.

Energy Shell A *Chemistry* term; see **Orbit**.

Energy A *Physics* term; the capacity of a body to do the work which causes it to move or transform; a body capable of doing more work is said to possess more energy; like work, energy is also a scalar quantity; *the CGS unit is erg,* and *the SI unit is joule* ($1\ J = 10^7$ erg); its other units include *kilowatt-hour* (kWh) and *electron volt* (eV); classified into *kinetic energy*, *potential energy*, chemical energy, *nuclear energy*, *solar energy*, etc.

Engine Lathe A *Manufacturing* term; a basic *machine tool* previously driven by a steam engine and was mainly used for turning operations.

Engineer A technical person who applies the theories and principles of *science* and *mathematics* in the design and development of technological solutions to human problems; turns concepts into functioning hardware; often described as being *can-do people with excellent problem-solving skills* can be classified into *mechanical*, *civil*, *electrical*, *computer*, *electronics* and *chemical* engineers.

Engineering Design An important subject of *engineering discipline*; a process of applying science and engineering methods to define a structure or system in detail to permit its realization.

Engineering Drawing A basic subject of all engineering disciplines; a visual representation of an object carried out scientifically according to certain national and international standards of practice; *a graphic language for an engineer*.

Engineering Graphics The engineering drawing which deals with *plane* and *solid geometry*.

Engineering Strain A *Strength of Materials* term; see **Nominal Strain**.

Engineering Stress A *Strength of Materials* term; also called **Cauchy Stress**, named after the French mathematician *Augustin-Louis Cauchy* (1789–1857); a direct or normal stress obtained by dividing load by the original cross-sectional area of a metal test specimen during tension or compression test; see also **True Stress**.

Engineering A technical course which uses the tools of *mathematics* and *science* to develop hardware that can produce cost-effective solutions to the technological problems facing our society; acts as a bridge between *scientific discovery* and *product applications*; a driver of *social* and *economic* growth; generally classified into five traditional fields, namely *mechanical, electrical, civil, materials*, and *chemical* engineering, which are further subdivided into other fields.

Enterprise Resource Planning An *Operations Management* term, abbreviated as **ERP**; a *computerized technique* used to integrate various functions of an organization (enterprise) by consolidating its supply chain processes; reduces non-productive activities and ensures smooth flow of information between different functions of the organization.

Enthalpy of Combustion An *IC Engine* term; the energy released when a compound undergoes *complete combustion* at constant temperature and pressure.

Enthalpy of Formation An *IC Engine* term; the energy absorbed or released when a compound is formed from its components.

Enthalpy A *Thermodynamics* term; a *composite thermodynamic property* which consists of *internal energy* (U) and *flow work* (pV); denoted by H; the SI unit is J or kJ; expressed as:

$$H = U + pV;$$ see also **Specific Enthalpy**.

Enthalpy–Entropy Diagram A *Thermodynamics* term; see **Mollier Diagram**.

Entrepreneur A *Management* term; a person who *organizes, manages*, and *assumes* the risks of a firm and who takes a new idea or a new product and turns it into a successful business.

Entrepreneurship A *Management* term; a process by which people recognize opportunities to satisfy needs and then gather and use resources to meet those needs.

Entropy Change A *Thermodynamics* term; the difference in the values of entropy of a working fluid between two equilibrium states; denoted by dS or ΔS.

Entropy A *Thermodynamics* term; a *thermodynamic property* which measures the disorder of a system; a measurement of the amount of unusable energy in a system; always increases in real processes so that available energy decreases; the existence of entropy is inferred as a corollary of the second law of thermodynamics; denoted by S; always defined in terms of change between equilibrium states; expressed as: $dS = \frac{\delta Q}{T}$, where δQ is the incremental heat addition to the system maintained at temperature T; the SI unit is J/K; see also **Specific Entropy**.

EOQ An *Operations Management* term; see **Economic Order Quantity**.

Epicyclic Gear Train A *Theory of Machines* term related to *gear*; the gear train in which the axes of the gear shafts move relative to a fixed axis; gives *higher velocity ratio* for a given number of gears; used in differential gears of *automobiles*, *wrist watches*, *hoists*, and *pulley blocks*.

Epicycloid A *Theory of Machines* term related to *gear*; the locus of a point on the circumference of a circle, which rolls without slipping outside another circle of a definite radius; *used in making tooth profile of a gear*; see also **Hypocycloid**.

Equal Vectors A *Physics* term; two or more vectors having the same magnitude, direction, and sense; produce the same effect.

Equation of Continuity A *Fluid Mechanics* term; see **Principle of Continuity**.

Equation of State A *Chemistry* term; also called **Ideal Gas Equation**; an equation which relates the state properties of a gas such as *pressure* (p), *volume* (v), and *temperature* (T); given as:

$$pv = nRT$$

where

n = number of moles of a gas
R = universal gas constant.

Equations of Rotational Motion A *Physics* term; the equations of motion for a body which undergoes rotation; the following equations are used for a rotating body:

$$\omega = \omega_0 + \alpha t \quad (1)$$

$$\theta = \omega_0 t + \frac{1}{2}\alpha t^2 \quad (2)$$

$$\omega^2 = \omega_0^2 + 2\alpha\theta \quad (3)$$

where

ω_0 = initial angular velocity of the rotating body
ω = final angular velocity of the rotating body
α = angular acceleration (constant)
θ = angle traced by the body after time t
t = time of rotation.

Equatorial Orbit A *Physics* term; the orbit which encircles the equator of the earth.

Equatorial Plane A plane which contains the equator of the earth.

Equilateral Triangle A *Mathematics* term; a triangle whose all the three sides are of equal lengths and whose three angles are also equal (60°); an acute-angled triangle; see also **Scalene Triangle**.

Equipotential Lines A *Fluid Mechanics* term; the fluid lines along which velocity potentials have the same value.

Equity Capital A *Financial Management* term; the long-term funds which are provided by a firm's owners, that is, the stockholders; see also **Debt Capital**.

Equity A *Management* term; one of the 14 principles of management proposed by *Henri Fayol* (1841–1925); indicates *kindness* and *fairness* to subordinates.

Equity-Holder A *Financial Management* term; a person who owns a company and is next to receive payment left over after being paid to debtors.

Equivalent Fractions A *Mathematics* term; the fractions having same value; a few examples include $\frac{2}{4}$, $\frac{3}{6}$, $\frac{4}{8}$, etc. (value of each fraction is $\frac{1}{2}$).

Equivalent Length A *Strength of Materials* term related to a *column*; see **Effective Length**.

E-REV An *Electric Vehicle* term; see **Extended-Range Electric Vehicle**.

Erg A *Physics* term; the CGS unit of work and energy; equivalent to dyne-cm; 1 J $= 10^7$ erg or 1 erg $= 10^{-7}$ J; see also **joule**.

Ergonomics The science which deals with the application of human sciences like *anatomy*, *physiology*, and *psychology* to the design of work environment and jobs; fitting the task to a person.

Ericsson Cycle A *Thermodynamics* term related to *air standard cycle*; named after *John Ericsson* (1803–1889), a *Swedish-American* engineer who invented it; a gas power thermodynamic cycle which uses gas (hot air) as the working fluid; a closed cycle which consists of *two reversible isotherms* (constant temperature process) and *two reversible isobars* (constant pressure process); heat addition and heat rejection take place at *constant pressures*, but compression and expansion processes take place at constant temperatures; a regenerative Ericsson cycle has the same thermal efficiency as the Carnot cycle; see also **Stirling Cycle**.

Ernst Mach (1838–1916) An *Austrian* physicist, philosopher, and one of the pioneers in the field of supersonic aerodynamics who investigated the shock waves of supersonic projectiles in the 1880s; best known for the Mach number (a nondimensional group).

ERP An *Operations Management* term; see **Enterprise Resource Planning**.

Error Due to External Causes A *Metrology* term; a type of *systematic error* affected by external conditions such as changes in *temperature, pressure, humidity,* and *wind velocity.*

Error A *Metrology* term; the difference between the measured value and the true value of a physical quantity; the true value indicates the actual magnitude of an input signal to a measuring instrument, and the measured value is the value indicated by the instrument; classified into *systematic error*, *random error*, and *gross error*.

Escape Velocity A *Physics* term; the minimum velocity with which if a body is projected upward from the surface of the earth (or any other planet), it goes out of the gravitational field of the earth (or other planet), and never returns back on its own; does not depend upon the mass of the body and its angle of projection from the surface of the earth or planet, but depends on the mass and radius of the earth (or planet); has different values for different planets; for earth, the escape velocity is 11.2 km/s, and for sun, it is 618 km/s; given as:

$$V_e = \sqrt{\frac{2GM}{R}} = \sqrt{2gR} = \sqrt{\frac{8\pi\rho G R^2}{3}}$$

where

V_e = escape velocity of the body
G = universal gravitational constant = 6.67×10^{-11} N m^2 kg^{-2}
M = mass of the earth (or any other planet)
R = radius of the earth (or any other planet)
g = acceleration due to gravity = 9.81 ms^{-2}
ρ = mean density of the material of the earth (or any other planet)
π = A constant = $\frac{22}{7}$ = 3.14.

Esprit De Corps A *Management* term; one of the 14 principles of management proposed by *Henri Fayol* (1841–1925); indicates the *team spirit, harmony*, and *unity* within an organization.

Esteem Need A *Management* term related to *motivation*; also called **Ego Need**; one of the five levels of human needs proposed by *Abraham Harold Maslow* (1908–1970) in his '*Need-Hierarchy Theory*' of motivation; represents self-esteem need and the esteem need of others; self-esteem need relates to the development of *self-confidence*, whereas the esteem need of others represents the need for status and recognition from others.

Etching A *Chemistry* term; a process of *chemical attack* on a material; *chemical machining* uses this principle for material removal from a workpiece.

Ethics A *Management* term; the discipline which deals with what is good and what is bad; focuses on *moral duty* and *obligation*; see also **Business Ethics**.

Euler Number A *Fluid Mechanics* term; a dimensionless (nondimensional) group, named after *Leonhard Euler* (1707–1783), a great *Swiss* scientist who is credited with being the first to recognize the role of pressure in fluid motion; a ratio of the *pressure force* to the *inertia force*; denoted by Eu, given as:

$$\mathrm{Eu} = \frac{\Delta p}{\rho V^2}$$

where

Δp = pressure drop
V = velocity of fluid flow
ρ = density of the fluid.

Euler's Formula A *Strength of Materials* term related to *column*; named after *Leonhard Euler* (1707–1783), a great *Swiss* scientist; a mathematical formula most commonly used to find the *crippling* (*critical*) *load* for *long columns*; expressed as:

$$P = \frac{\pi^2 EI}{l_e^2}$$

where

P = Euler's crippling load
E = modulus of elasticity of the column material
I = second moment of area of the column's cross section
l_e = Equivalent/effective length of the column
π = (22/7) a constant.

Eulerian Method A *Fluid Mechanics* term; named after *Leonhard Euler* (1707–1783), a *Swiss* mathematician and physicist who proposed it; a mathematical method used to describe the motion of various fluid particles at some fixed points in space; the most widely used method because of its simplicity; see also **Lagrangian Method**.

Eutectic Reaction A *Materials Engineering* term related to *phase diagram*; a chemical reaction which transforms a *eutectic liquid* (*L*) to a relatively fine-grained microstructure of two solid phases (α and β) upon cooling; expressed as:

$$L(\text{eutectic}) \rightarrow \alpha + \beta$$

Eutectoid Reaction A *Materials Engineering* term related to *phase diagram*; a conversion process in *iron-carbon phase diagram* which occurs at 723 °C to convert *austenite* to *ferrite* and *cementite*; a solid-state reaction which transforms a *eutectoid* (a solid-state microstructure means eutectic-like) into microstructure of two solid phases (α and β) upon cooling; expressed as:

$$\gamma\,(\text{eutectoid}) \rightarrow \alpha + \beta$$

Eutectoid Steel A *Materials Engineering* term; a state of phase of steel in *iron-carbon phase diagram* which contains 0.77% carbon; classified into *hypo-eutectoid* and *hyper-eutectoid steels*.

eV A *Physics* term; see **Electron Volt**.

EV See **Electric Vehicle**.

Evangelista Torricelli (1608–1647) An *Italian* physicist who is known for the *Torricelli equation* used to obtain free jet velocity and related barometric height to weight of atmosphere; invented the *barometer*.

Evaporating Cooling A *Thermodynamics* term; the cooling which results from the evaporation; effective in hot and relatively dry climates; involves either spraying liquid water into air or forcing air through a water-soaked pad; for example, cooling of a space using an air cooler.

Evaporation A *Thermodynamics* term; also called **Vaporization**; a process associated with a phase change from liquid to vapour as a result of absorbing heat; results when the vapour pressure is less than the saturation pressure of the liquid at a given temperature; a few examples include *drying of clothes, fruits, and vegetables*, and the *evaporation of sweat to cool the human body*.

Evaporative Condenser A *Power Plant* term; a condenser which uses both air and water as the condensing medium; see also **Air-Cooled Condenser** and **Water-Cooled Condenser**.

Evaporator A *Refrigeration* term; a component of the refrigeration system used to take away the heat from the refrigerator to maintain low temperature.

EVB An *Electric Vehicle* term; see **Electric Vehicle Battery**.

Even Number A *Mathematics* term; any number divisible by 2; a few examples include 4, 6, 20, etc.; the sum of any two even numbers is always an *even number*; see also **Odd Number**.

Event A *Project Management* term; indicates the beginning or completion point of an activity; does not consume any time or involve any labour or material resources; see also **Activity**.

Exact Differential A *Thermodynamics* term; the differential of a *point function* whose integral depends only on the end points; for example, differential of *temperature*, symbolized as dT is expressed as:

$$\int_{T_1}^{T_2} dT = T_2 - T_1$$

where

T_1 = temperature at state 1
T_2 = temperature at state 2; see also **Inexact Differential**.

Exergy A *Thermodynamics* term; see **Available Energy**.

Exhaust Valve An *IC Engine* term; a valve through which the combustion products are expelled from the cylinder of an engine; see also **Intake Valve**.

Expanded Polystyrene Casting A *Manufacturing* term related to *casting*; see **Full-Mould Casting**.

Expansion Stroke A *Thermodynamics* term related to an *IC engine*; see **Power Stroke**.

Expectancy Theory of Motivation A *Management* term related to *motivation*; see **Vroom's Expectancy Theory**.

Expected Time A *Project Management* term; the time duration (t_e) in which a project activity is expected to be completed; represents the mean of a *beta distribution curve,* and its value is more close to the most likely time, given as:

$$t_e = \frac{a + 4m + b}{6}$$

where

a = optimistic time of the activity
b = pessimistic time of the activity
m = most likely time of the activity.

Expenses An *Operations Management* term related to *cost analysis*; the expenses are the cost other than material and labour costs; classified into *direct* and *indirect* expenses and also *fixed* and *variable* expenses; *direct expenses* include cost of designs or drawings produced for a specific job or transportation cost or consultation fees; a few examples of *indirec*t expenses include *rent of the building, insurance premium, telephone bill, electricity bill,* etc.; *fixed expenses* remain relatively constant, and its example include *taxes on land and building, rent,* etc.; *variable expenses* vary with volume of production, and its example includes royalties paid by a publisher to an author.

Experience Curve A *Management* term; see **Learning Curve**.

Expert System A complex computer program capable of performing and solving difficult real-life problems in a way that human minds do.

Explosive Welding A *Welding* term; a *solid-state welding* in which a detonator is exploded to produce a high-velocity impact to bring workpieces together to make a weld.

Exponential Smoothing Method An *Operations Management* term related to *forecasting*; also called **Exponentially Weighted Moving Average Method** as it gives the *highest weightage* to the *most recent data,* and the *weightage reduces exponentially with older data*; also called **Adaptive Forecasting** as it corrects for the past errors in demand forecasting; the most frequently used formal forecasting method because of its simplicity and the small amount of data needed in this method; requires only *three* data: the last period's forecast, the demand of this period, and a smoothing constant (α) whose value lies between 0 and 1.0; useful for *short-range* forecasts; uses the following formula to find the forecast value:

$$F_{\mathrm{t}+1} = \alpha\, D_{\mathrm{t}} + (1 - \alpha) F_{\mathrm{t}}$$

where

$F_{\mathrm{t}+1}$ = forecast for the current period
D_{t} = actual demand for the past period
F_{t} = forecast for the past period
α = smoothing constant.

The zero value of α indicates that the current period's forecast becomes equal to the past's period forecast, that is, the forecast does not change; when α is 1, then the current period's forecast becomes equal to the past's period actual demand, that is, demand once set never changes; larger value of α points towards recent data.

Exponentially Weighted Moving Average Method An *Operations Management* term related to *forecasting*; see **Exponential Smoothing Method**.

Extended-Range Electric Vehicle An *Electric Vehicle* term, abbreviated as E-REV; an electric vehicle which uses a small gasoline engine to extend the range.

Extension Spring A *Design* term; a spring subjected to a *tensile loading.*

Extension A *Strength of Materials* term; see **Elongation.**

Extensive Property A *Thermodynamics* term; a property which depends on mass of a system and varies proportional to the mass of a body; a few examples include *volume, surface area,* and *energy* of all kinds; see also **Intensive Property**.

Extensometer A *Strength of Materials* term; a mechanical gauge or electromechanical sensor used to measure the elongation of a bar during its tensile loading.

External chill A *Manufacturing* term related to *casting*; a metallic object placed *externally* in the mould adjacent to the casting; often used to reduce the number of risers (by retarding the solidification process) required for a casting; see also **Internal chill.**

External Combustion Engine An engine in which combustion of the fuel takes place outside the cylinder of the engine; for example, *steam engine.*

External Customer A *Management* term; see **Customer**.

External Failure Cost A *Quality Management* term; a part of the cost of quality related to the defects which pass through the system; includes *costs for customer warranty replacement, loss of customers or goodwill, handling complaints,* and *products repair*; see also **Internal Failure Cost**.

External Flow A *Fluid Mechanics* term; a flow that occurs on an immersed body in an unbounded fluid, for example, water flow over a submarine or air flow over a ball; see also **Internal Flow**.

External Force A *Physics* term; a force which acts on a body, including the reactive forces caused by the supports; classified into *surface* and *body* forces; see also **Internal Force**.

External Friction A *Physics* term; see **Friction.**

Extrapolation A *Mathematics* term; the calculation of a parameter outside the range of observations; see also **Interpolation**.

Extrinsic Factors A *Management* term related to *motivation*; also called **Hygiene Factors**; one of the two factors of *Frederick Herzberg* (1923–2000)'s '*Two-Factor Theory*' of motivation; include *lower level needs* which form *external factors of motivation* such as *company policies, supervision, working conditions, interpersonal relations, salary, status,* and *security*; their absence creates dissatisfaction, and their presence does not create satisfaction (may be because of greedy nature of human beings); see also **Intrinsic Factors**.

Extrinsic Semiconductor An *Electronics* term; the semiconductor in which impurities are added (the process is called *doping*); its conductivity is higher because of doping and is dependent on doping concentration, but is not affected by the temperature; classified into *n-type semiconductor* and *p-type semiconductor*; see also **Intrinsic Semiconductor**.

Extrusion A *Manufacturing* term; a manufacturing operation carried out to shape a product by applying a large pressure on a metal billet or blank (the raw material) which causes it to flow through a restricted orifice (shaped-die opening); capable of producing any solid or hollow cross section; the most widely used technique for processing *thermoplastics*; classified into *hot extrusion* and *cold extrusion*.

F

F. W. Harris (1877–1962) An *American* production engineer who formulated the basic *economic order quantity* (EOQ) *inventory model* in 1914.

FA An *IC Engine* term; see **Fuel–Air Ratio**.

Face A *Theory of Machines* term related to *gear*; the surface of a gear tooth along its width *above the pitch surface*; see also **Flank**.

Face-Centred Cubic Structure A *Chemistry* term, abbreviated as **FCC Structure**; one of the three basic types of atomic arrangements found in metals in which the unit cell consists of *eight corner atoms* and *one atom at the centre of each face*; the atomic packing factor (APF) equals to 0.74; materials having fcc structure have very high formability (the ability to be plastically deformed without fracture); a few examples include *calcium* (Ca), *aluminium* (Al), *copper* (Cu), *nickel* (Ni), *silver* (Ag), *lead* (Pb), and *iron in* γ*-phase* (that occurs between 912 and 1394 °C); see also **Body-Centred Cubic Structure** and **Hexagonal Close-Packed Structure**.

Facility Layout An *Operations Management* term related to *facilities planning*; also called **Plant Layout**; a process of physically arranging the resources (material and human) such as *space*, *raw materials*, *machines* and *equipment*, and *workforce* in a plant to facilitate smooth flow of materials and information; can be suitably modified as per the requirements of the plant; includes planning for the location of all machines, utilities, employee workstations, customer service areas, material storage areas, offices and customer rooms, and other facilities; a bad plant layout produces production bottlenecks, increases level of in-process inventory, delays production schedule, increases production costs, and renders unpleasant working environment.

Facing Sand A *Casting* term; a *freshly prepared sand* which contains *a small amount of carbonaceous material*; used to form the *face* (*inner surface*) of a mould; that is, this sand is rammed around the pattern; provides *better surface finish* to the castings.

Factor Earnings An *Economics* term; the incomes earned from the factors of production, namely *land*, *labour*, and *capital*.

D. K. Singh, *Dictionary of Mechanical Engineering*,
https://doi.org/10.1007/978-981-99-1722-8_6

Factors of Production An *Economics* term; also called **Resources**; the inputs into a production process or goods used to produce other goods; for example, *land*, *labour*, and *capital* are the three key factors of production.

Factory Cost An *Operations Management* term related to *cost analysis*; the sum of *prime cost* and *factory overhead*, which includes all the overheads needed to produce a good.

Fahrenheit Scale A *Thermodynamics* term; a *temperature scale* used in the SI system and now considered in the English system; named after the *German* instrument maker *Daniel Gabriel Fahrenheit* (1686–1736); uses the unit *degree Fahrenheit* (°F) for temperature measurement; the ice and steam points on this scale are assigned the values of 32 °F and 212 °F, respectively; see also **Celsius Scale**, **Kelvin Scale**, and **Rankine Scale**.

Failure Criteria A *Strength of Materials* term; also called **Failure Theories** or **Theories of Failure**; the stress conditions useful in predicting the failure of materials; expressed in terms of various theories such as *Rankine's theory*, *Saint Venant's criterion*, *Haigh's criterion*, *Tresca's yield criterion*, and *von Mises yield criterion*.

Failure Theories A *Strength of Materials* term; see **Failure Criteria**.

Fan Draught A *Power Plant* term related to a *boiler*; a *boiler accessory* used to produce *air* to be supplied to the grate to increase heat supply to the boiler; fitted near the grate of a boiler; see also **Jet Draught**.

Fan A *Fluid Mechanics* term; a *turbomachine* which is basically a *gas pump*; used to move gases (mostly air) with relatively low pressure rise and high flow rate; a few examples include *ceiling fans*, *house fans*, and *propellers*; see also **Blower** and **Compressor**.

Fanning Friction Factor A *Fluid Mechanics* term; see **Friction Coefficient**.

Fastener A *Design* term; also called **Fastening Element**; a device used to connect two parts together; classified into *temporary* and *permanent* fastener; a few examples include *nail*, *screw*, *key*, *cotter*, *nut* and *bolt*, *rivet*, etc.

Fastening Element A *Design* term; see **Fastener**.

Fatigue Limit A *Strength of Materials* term; see **Endurance Limit**.

Fatigue Test A *Strength of Materials* term; a *destructive* test used to measure the fatigue strength of a solid body; the test is conducted by subjecting a test specimen under cyclic loading (usually in alternate tension and compression mode) or torsion; the stress amplitudes (S) and the number of stress cycles (N) required to produce the failure are recorded and drawn into a curve, called *S–N curve*.

Fatigue A *Strength of Materials* term; a deformation phenomenon which results from repeated loading and unloading or when a solid material is subjected to a fluctuating (cyclic or periodic) load which causes it to fail earlier than its expected life; *wings* of an aircraft, *crankshaft* of an automobile engine, and *gear teeth* in a machinery are subjected to fatigue.

FBD A *Physics* term; see **Free Body Diagram.**

FCC Structure A *Chemistry* term; **Face-Centred Cubic Structure.**

FCFS An *Operations Management* term; one of the priority rules for sequencing jobs which stands for *first come first served*; jobs are completed in the order they arrived; has the advantage of being fair to customers.

Feather Key A *Design* term related to a *key*; a *variation of parallel sunk key* fitted to one member of the pair and *permits relative axial movement*; usually fitted into the hub of the mounting, but not on to the shaft; classified into *peg*, *single-headed*, *double-headed*, and *splined* keys (splined shaft); see also **Parallel Sunk Key.**

Feed Check Valve A *Power Plant* term related to a *boiler*; a *boiler mounting* used to continuously feed and regulate water supply to the boiler; fitted in the feed water supply line very close to the furnace area.

Feed Pump A *Power Plant* term related to a *boiler*; a *boiler accessory* used to feed (supply) *water* to a boiler; feed pump is usually a reciprocating pump (single stage) or a rotary pump.

Feed Water A *Power Plant* term; the water which enters a boiler from a pump.

Ferro-Chrome Alloy A *Materials Engineering* term; an alloy of *iron* (Fe) and *chromium* (Cr).

Ferro-Molybdenum Alloys A *Materials Engineering* term; the alloys of *iron* (Fe) and *molybdenum* (Mo); a steel containing just 2% of molybdenum does not soften on heating (increased hot hardness) and hence finds use in making *cutting tools* for high-speed lathes; a steel containing molybdenum along with nickel (called nickel-molybdenum steel) is used in making *gun barrels*; a steel containing molybdenum (2–3%) and chromium (about 60%) is acid-resistant; an alloy containing molybdenum (about 4%), nickel (79%), and iron (17%) has high magnetic permeability and is used in making *magnetic cores.*

Ferrous Metals A *Materials Engineering* term; metals having high *iron* (Fe) content along with some alloying elements; include many types of *steel and its alloys*, *cast iron*, and *wrought iron*; offer good mechanical properties, especially high strength and hardness, which make them suitable for extensive industrial applications; see also **Nonferrous Metals.**

Ferro-Vanadium Alloy A *Materials Engineering* term; an alloy of *iron* (Fe) and *vanadium* (V) containing 35–45% of *vanadium*; finds use in steel industry as a *scavenger* for oxygen and nitrogen, and also as a strengthening agent; for example, when added to molten steel, vanadium takes up oxygen and nitrogen and at the same time dissolves in the molten iron to improve its ductility, strength, and shock-resistance.

Fick's Law of Diffusion A *Mass Transfer* term; named after a *German* physician and physiologist *Adolf Eugen Fick* (1829–1901) who proposed it in 1855; a law connected to the diffusion of two gases; states that the mass flux of an element per unit area (also called the diffusion rate) is proportional to the concentration gradient; expressed as:

$$\frac{m}{A} = -D\frac{\partial C}{\partial x}$$

where

m = mass flow per unit time (kg/h)
A = area through which mass flows (m^2)
D = diffusion coefficient (m^2/h)
C = mass concentration of the component (kg/m^3).

File A *Manufacturing* term; the most commonly used fitting tool used to remove undesired material from a workpiece in small quantity (usually in the form of dust particles) while shaping the workpiece; made of *high carbon steel*; classified into *flat*, *hand*, *square*, *triangular*, *pillar*, *round*, *needle*, and *warding* files.

Filler Metal A *Welding* term; a metal used to fill up the gap between two thicker parts being joined so as to make a stronger joint; has composition similar to that of parent metals that are being joined.

Film Coefficient A *Heat Transfer* term; see **Heat Transfer Coefficient**.

Finance A *Management* term; the art and science of managing money; deals with the process, institutions, markets, and instruments which are involved in the transfer of money among and between individuals, businesses, and governments; the study of finance helps to make better financial decisions.

Financial Capital A *Financial Management* term; includes *stocks* and *bonds*; see also **Capital**.

Financial Management An important discipline of management which is connected with the *acquisition*, *financing*, and *management of assets*.

Financial Manager A *Management* term; a manager who manages the financial affairs of any type of business, whether financial or nonfinancial, private or public, large or small, profit-seeking or not-for-profit.

Financial Statement: A *Financial Management* term; a statement which includes three components, namely the *balance sheet*, the *profit and loss statement*, and the *source and application of the fund statement*.

Finite Element Analysis A *Design* term abbreviated as **FEA**; a computer-based design analysis tool which allows the user to apply virtual forces and pressures to a 3-D CAD model in order to determine deflections, stress concentrations, and other effects.

Fins and Rags A *Forging* term; the *forging defects* which occur *in the form of projections on the surface of the forged parts* due to *attachment of loose metal.*

Fire Tube Boiler A *Power Plant* term related to a *boiler*; a type of boiler in which products of combustion (hot flue gases) pass through one or more tubes surrounded by water; important features of a fire tube boiler include slow steam generation, low evaporative capacity, and low steam pressure range; free from the limitations of using only soft water; suitable for small power plants and locomotives; a few examples include *Lancashire boiler*, *locomotive boiler*, and *Cornish boiler*; Lancashire boiler is the most commonly used fire tube boiler; see also **Water Tube Boiler**.

Firm An *Economics* term; an organization which transforms resources (inputs) into products (outputs); the primary producing unit in an economy.

First Angle Projection An *Engineering Drawing* and a *Machine Drawing* term; a projection method in which an *object is imagined to be placed in the first quadrant* for the purpose of its projection; as the observer normally looks from the right side of the quadrant to obtain the front view, here the *object lies between the observer and the vertical plane of projection*; the *object in this case has to be transparent,* and the projectors (the projection lines) are imagined to be extended from various points of the object to meet the projection plane, and the meeting points on being joined together (in order) form the image of the object on the plane of projection; the Bureau of Indian Standards (BIS) [SP-46:1988] prefers this method of projection; see also **Third Angle Projection**.

First Conjugate Beam Theorem A *Strength of Materials* term; a conjugate beam-based theorem used to find slope at any section of a loaded beam; states that the slope at any section of a loaded beam is equal to the shear force at the corresponding section of the conjugate beam; see also **Second Conjugate Beam Theorem**.

First Law of Thermodynamics A *Thermodynamics* term; a statement of the *conservation of energy* for a system.

First Line Manager A *Management* term; see **Low Level Manager**.

First-Class Lever A *Physics* term; a lever in which the fulcrum is positioned between the load and the effort; only class of lever in which the effort force and the load force act in the same direction; also the only class of lever which can be modified to amplify an effort force or the distance that is moved by the load; moving the fulcrum towards the load will decrease the effort force needed to lift the load; moving the fulcrum towards the effort will increase the distance travelled by the load but require an effort force that is greater than the load; a few common examples of first-class lever include a *pry-bar* (used to remove nails from furniture) and a *see-saw* (a playing device for children).

Fishbone Diagram A *Quality Management* term; see **Cause-and-Effect Diagram.**

Fitting Shop A *Manufacturing* term; a part of the central workshop where an object is given the desired shape through cutting or removing materials from a given raw metal stock and secures proper fit, using various fitting tools.

Fitting Tools A *Manufacturing* term; tools used in the fitting shop to make a right size job; a few examples include *scriber, chisel, try square, hacksaw,* and *files.*

Five S An *Operations Management* term, abbreviated as **5S**; a management tool used to effectively organize the workplace in an organization; consists of five Japanese words starting with English alphabet *S*, namely *Seiri, Seiton, Seison, Seiketsu,* and *Shitsuke* which mean *housekeeping, workplace organization, clean up, cleanliness,* and *discipline,* respectively.

Fixed Assets A *Financial Management* term; also called **Earning Assets** or **Capital Assets**; the long-term assets used in the *long-term* production of goods and not to be converted to cash in the near future; a few examples include *land, buildings, plant, machinery,* and *equipment* which have a much longer life; entered in the *balance sheet* at their original cost value; generally provide the basis for the firm's earning power and value.

Fixed Beam A *Strength of Materials* term related to a *beam*; also called **Restrained Beam** or **Fixed-End Beam** or **Fixed–Fixed Beam** or **Clamped Beam**; a *statically indeterminate* beam that is confined or fixed between the two supports; can resist a force in any direction, as well as a moment or a couple at both support points.

Fixed Costs An *Operations Management* term related to *cost analysis*; the fixed costs remain constant or unchanged with the volume of output produced; they are considered even if no units are produced; a few examples of fixed costs include the *annual cost of renting* or *buying new equipment* and *facilities* (including *depreciation, interest, taxes, insurance, debt,* and *mortgage payments, salaries,* etc.; see also **Variable Costs**.

Fixed Error A *Metrology* term; see **Systematic Error**.

Fixed Order Quantity An *Operations Management* term; a lot sizing rule by which orders are placed always for a fixed order quantity irrespective of the nature of demand; see also **Periodic Order Quantity** and **Lot-for-Lot**.

Fixed Support A *Strength of Materials* term; a support which neither translates nor rotates, and hence can resist both a moment and a force in any direction.

Fixed Vector A *Physics* term; a vector which has a *fixed point of application.*

Fixed-End Beam A *Strength of Materials* term related to a *beam*; see **Fixed Beam**.

Fixed–Fixed Beam A *Strength of Materials* term related to a *beam*; see **Fixed Beam**.

Fixed-Position Layout An *Operations Management* term related to *facilities planning*; a type of plant layout in which the product being manufactured remains in a fixed position and the machines, materials, and workers are transported to and from the product; used when a product is *very bulky, large, heavy,* or *fragile*; a few examples include *missile assembly, ship construction, dam,* or *bridge construction, large aircraft assembly,* and *making of large pressure vessels.*

Fixed-Simple Beam A *Strength of Materials* term related to a *beam*; see **Propped Cantilever Beam**.

Fixture A *Manufacturing* term; a device used to locate and hold a workpiece in a definite position while working on it during a process; used in conjunction with a jig; usually secured to the table of a machine, but does not guide the cutting tool; the cutting tool is either moved into position for the operation, or the table is moved under the cutting tool; helps in rapid and accurate manufacturing; see also **Jig**.

Flame Hardening A *Manufacturing* term related to *heat treatment*; a *surface-hardening process* which involves heating of *low carbon* or *low alloy steel* parts in an *oxyacetylene gas flame* and then cooled quickly by *water*; suitable for parts which are too bulky to put into a furnace; typical applications include *ways of lathes, spindles, shafts, pulleys,* and *gear teeth.*

Flange Coupling A *Design* term; a *temporary fastening device* and *a type of rigid coupling*; consists of *two cast iron flanges* keyed to the shaft ends and bolted together, but the best practice is to press or shrink the flanges on the shafts; for more accurate connection, the ends of the flange couplings and the centring surfaces are machined after assembly on the shaft; *ensures the most accurate, rigid,* and *strong connection of shafts*; the *shafts of steam turbo generators* in three bearings, the *shafts of vertical hydroturbines,* and the *marine propeller shafts* are usually connected by *flange coupling.*

Flank A *Theory of Machines* term related to *gear*; the surface of a gear tooth along its width *below the pitch surface*; see also **Face**.

Flash Butt Welding A *Welding* term; see **Flash Welding**.

Flash Welding A *Welding* term; also called **Flash Butt Welding**; a type of *resistance welding*; the joint is produced when two abutting work surfaces are pressed after being heated by a *flash* (arc) produced by a high voltage in the circuit; see also **Upset Welding**.

Flash In *casting* terms, the metal squeezed out in space near the parting line or into clearances around the core due to high pressure in the die casting, which may be machined off as being undesirable; in *welding* terms, an arc to produce heat for joining.

Flat File A *Manufacturing* term; one of the types of a file used in the fitting shop which has a *rectangular cross section* and is parallel for about two-third of its length and then tapered in width and thickness; consists of double cut on its both faces and single cut on edges; used for filing *flat surfaces*; see also **Hand File**.

Flat Organization A *Management* term; an organization which has few levels in its hierarchy relative to its size; experiences less communication, motivation, and cost problems than a tall organization; see also **Tall Organization**.

Flat Saddle Key A *Design* term related to a *key*; a saddle key which is similar to hollow saddle key, except that its underside surface is flat which fits on the flat surface provided on the shaft; see also **Hollow Saddle Key**.

Flat Spring A *Strength of Materials* term related to a *spring*; a spring made of a flat strip of usually rectangular section and may be in the form of a cantilever or a simple beam; acts as a structural member as well as an energy absorbing device; used as a device for diving into a swimming pool.

Flexible Automation An *Automation* term; see **Soft Automation**.

Flexible Coupling A *Design* term; a *temporary fastening device* or a coupling that provides flexibility in connecting two machine shafts so as to take care of any *misalignment* between them; helps to absorb the impact from the fluctuation of torsional moment or of angular speed; a few examples include *Oldham coupling*, *universal joint coupling*, and *pin-type coupling*.

Flexible Manufacturing System A *Manufacturing* term related to *automation*; abbreviated as **FMS**; a computer-assisted, highly efficient, modern manufacturing system which consists of numerous programmable machine tools connected by an automated material handling system and controlled by a common computer network; offers high degree of flexibility which helps to produce an enormous variety of items including customized products.

Flexible Plant An *Operations Management* term; a plant with equipment easy to install and dismantle as per the requirements; can quickly adapt to changes.

Flexible Workers An *Operations Management* term; the workers having *multiple skills* and *abilities* which make them switch over easily from one kind of task to another; helps in attaining the goals of JIT.

Flexural Centre A *Strength of Materials* term; see **Shear Centre**.

Flexural Rigidity A *Strength of Materials* term related to *bending*; also called **Bending Modulus**; the product of Young's modulus (E) and the second moment of area (I); the external bending moment which can produce unit radius of curvature in a loaded beam; used as a measure of resistance of a beam against bending; a beam having higher value of flexural rigidity is difficult to bend; hence, it is a requirement of a strong beam; the SI unit is N-m^2.

Flexure Equation A *Strength of Materials* term; see **Flexure Formula**.

Flexure Formula A *Strength of Materials* term; also called **Bending Formula** or **Flexure Equation** or **Bending Equation**; valid for pure bending of a beam; used to determine the bending stresses; given as:

$$\frac{\sigma}{y} = \frac{M}{I} = \frac{E}{R}$$

where

σ = stress due to bending
y = distance from the neutral axis
M = bending moment due to applied load
I = moment of inertia of beam's cross-sectional area
E = modulus of elasticity of the beam material
R = radius of curvature of the bent beam.

Flexure Stress A *Strength of Materials* term related to *bending*; see **Bending Stress**.

Flint Glass A *Chemistry* term; a glass of high refracting power prepared by fusing litharge, potassium carbonate, and silica in proper proportions; has higher density and transparency than ordinary glass; approximately includes 45% silica (SiO_2), 4% sodium oxide (Na_2O), 4% potassium oxide (K_2O), 3% calcium oxide (CaO), and 44% lead oxide (PbO); used for making *optical instruments* and also for *ornamental purposes.*

Floor Moulding A *Casting* term; the process of making a mould on the floor of the foundry shop; useful for *large size castings* which are difficult to handle.

Floor Rammer A *Manufacturing* term; a *foundry tool* and a *type of rammer* used to ram the sand mass in *floor moulding* for making *large size moulds*; see also **Peen Rammer** and **Hand Rammer**.

Flow Area A *Fluid Mechanics* term; the cross section of a fluid flow perpendicular to the flow direction.

Flow Control Valve A *Fluid Mechanics* term; a type of valve used to control the volume of a fluid as it flows in one direction only; often used to control the speed of an actuator.

Flow Energy A *Thermodynamics* and a *Fluid Mechanics* term; see **Flow Work**.

Flow Meter A *Fluid Mechanics* term; a device used to measure the flow rate of a fluid.

Flow Process A *Thermodynamics* term; a process which involves flow of mass across the boundary of an open system or control volume.

Flow Rate A *Fluid Mechanics* term; see **Volume Flow Rate**.

Flow Separation A *Fluid Mechanics* term; a process in which a fluid stream detaches itself from the surface of a solid body; occurs at sufficiently high velocities at sharp corners of a body or where there are abrupt changes in the shape of the solid surface.

Flow Velocity A *Fluid Mechanics* term; a ratio of the distance travelled by a drop of fluid to the amount of time that it takes to travel that distance.

Flow Work A *Thermodynamics* and a *Fluid Mechanics* term; also called **Flow Energy**; the work produced by the *pressure force* of a fluid, which can move the fluid; analogous to displacement work; significant only in a flow process or an open system; calculated as the product of pressure (of the working fluid) and change in its volume during expansion or compression; see also **Flow Process**.

Flowchart An *Operations Management* term; see **Process Flowchart**.

Fluctuating Load A *Physics* term; a *dynamic load* which persists for long periods of time and continuously varies in intensity; produced by *rotating machinery*, *traffic*, *water waves*, *earthquakes*, and *manufacturing processes*; see also **Impact Load**.

Flue Gas A *Power Plant* term; the gaseous emissions discharged through a flue or stack; see also **Fly Ash**.

Fluid Dynamics A *Fluid Mechanics* term; also called **Hydrodynamics**; a branch of fluid mechanics which deals with the properties and behaviour of a fluid in motion; see also **Fluid Statics**.

Fluid Friction A *Fluid Mechanics* term; see **Viscosity**.

Fluid Machine A *Fluid Mechanics* term; a device which either performs work or extracts work (or power) from a fluid; a few examples include *hydraulic turbine*, *steam turbine*, and *pump*.

Fluid Mechanics A core subject of *mechanical* as well as *civil* engineering; the study of the properties of gases and liquids that are at rest or in motion; classified into *fluid statics* and *fluid dynamics*.

Fluid Particle A *Fluid Mechanics* term; the particle of a fluid.

Fluid Power A *Fluid Mechanics* term; the use of a confined fluid flowing under pressure to transmit power from one location to another.

Fluid Statics A *Fluid Mechanics* term; also called **Hydrostatics**; a branch of fluid mechanics which deals with the properties and behaviour of a fluid at rest, or there is no relative motion between fluid particles; that is, no shearing stresses exist, and the only stress that exists is a normal stress, called *pressure*; see also **Fluid Dynamics**.

Fluid A *Fluid Mechanics* term; a substance which begins to flow when an external force is applied on it; a substance which deforms continuously when subjected to shear forces, however small the shear force may be; characterized by the presence of a *free surface*; examples include *liquid* and *gas*.

Fluidity A *Fluid Mechanics* term; the tendency of a substance to undergo continuous deformation when subjected to shear forces.

Flux A *Chemistry* and a *Welding* term; a substance which chemically combines with *gangue* (earthy impurities) present in an ore to form *slag*; helps to remove the impurities from the ore; classified into *acidic* and *basic* fluxes.

Flux-Cored Arc Welding A *Welding* term; an arc welding process which uses a continuous length consumable tubular electrode that contains flux in its central part (core) along with some alloying elements to produce shielding atmosphere around the weld pool; mainly used for joining *steels* and *stainless steels*.

Fly Ash A *Power Plant* term; the *unburned particulate matter* discharged from a *stack*; see also **Flue Gas**.

Flywheel Vortex A *Fluid Mechanics* term; see **Forced Vortex**.

Flywheel A *Theory of Machines* term; a heavy rotating body used to store and deliver energy in intermittent operations such as in *punch presses* and *bulldozers*; *used to control the fluctuation of speed* caused by the non-uniform flow of power during each cycle of operation as in case of *internal combustion engines*, *reciprocating engines*, and *steam engines*; see also **Governor**.

FMS A *Manufacturing* term related to *automation*; see **Flexible Manufacturing System**.

Focused Factory A factory which produces customized products for particular markets to fulfil the specific requirements of a customer.

Foot The FPS unit of length; 1 foot = 12 in. = 30.5 cm; its plural is feet; see also **Inch**.

Foot-Step Bearing also called **Pivot Bearing**; a type of *thrust bearing* used to *support a vertical shaft* under axial load which terminates at the bearing on its disc-shaped surface; the bush fitted in the main body supports the shaft in position and takes care of possible radial loads coming on the shaft; see also **Collar Bearing**.

Force of Adhesion A *Physics* term; see **Adhesive Force.**

Force of Buoyancy A *Fluid Mechanics* term; see **Buoyant Force.**

Force of Cohesion A *Physics* term; see **Cohesive Force.**

Force of Friction A *Physics* term; see **Frictional Force.**

Force of Gravity A *Physics* term; see **Gravity.**

Force A *Mechanics* term; an external action on a body which changes or tends to change the state, orientation, shape, or size of the body; specified by the magnitude, direction, and sense as well as the line of action; classified into *point*, *line*, *surface*, and *body forces*; quantitatively, it is equal to the product of *mass* and *acceleration*; the SI unit is newton (N); see also **Load.**

Forced Convection A *Heat Transfer* term; a mode of heat transfer by convection in which heat is transferred by a fluid whose flow is produced by an external source such as a *pump* or a *fan*; a few examples include *the flow of water maintained by a pump in shell and tube condenser of a refrigeration plant* and *the flow of air maintained by a fan in the air-cooled condenser of an air-conditioner*; see also **Free Convection.**

Forced Oscillation A *Physics* term; the oscillation of a body which occurs under the action of an *external periodic force*, where the body oscillates with the frequency of the applied force; the amplitude of the forced oscillation depends on the relationship between the frequency of the applied force and the natural frequency of the body; for example, a stretched string carrying an alternative current kept between two poles of a magnet oscillates with the frequency of the current; see also **Free Oscillation.**

Forced Vortex A *Fluid Mechanics* term; also called **Flywheel Vortex**; a fluid flow in which the fluid particles spin about their own axes, while moving along circular streamlines; see also **Free Vortex.**

Forecast Accuracy An *Operations Management* term related to *forecasting*; indicates how close forecast values come to actual data; if forecast values are very close to the actual data, it implies that forecasting has high accuracy, and the forecast error is low.

Forecast Error An *Operations Management* term related to *forecasting*; the difference between the actual demand and the forecast value.

Forecast An *Operations Management* term; see **Forecasting.**

Forecasting: An *Operations Management* term; also called **Forecast**; a process which aims to determine what is likely to happen in future using past data, if no major changes occur; uses the scientific tools and techniques to arrive at an informed guess.

Forgeability A *Manufacturing* term; the ability (property) of a material which enables it to be easily *forged* (hammering a heated workpiece) into useful products; *mild steel* and medium carbon steel have good forgeability.

Forging Defects A *Forging* term; the defects produced in *forged parts*; may arise due to *initial defects in the work material*, *incorrect design of forging dies*, *poor forging design*, *improper forging method*, *non-uniform heating and cooling of the work material*, or *microstructural changes in the work material because of phase transformation*; some important forging defects include *laps*, *scale pits*, *cracks*, and *fins* and *rags*.

Forging A *Manufacturing* term; also called **Hot Forging**; a *hot metal working process* used to shape a heated workpiece using a die by repeatedly hammering it or by applying compressive forces through a ram; produces grain alignment in a particular direction and results in increased strength and toughness of the workpiece; the typical forged products include *connecting rods*, *bolts*, *rivets*, and *shafts for turbines*; see also **Cold Forging** and **Warm Forging**.

Formability A *Strength of Materials* term; the ability of a material to be plastically deformed without fracture so as to give the desired shape to materials; *high ductility materials* and materials having *face-centred cubic* (*fcc*) structure show *high degree of formability*, whereas materials having *hexagonal close-packed* (*hcp*) structure have *poor formability*.

Formal Organization A *Management* term; the intentional structure of roles in a formally organized enterprise; see also **Informal Organization**.

Formalization A *Management* term; a process of making use of written rules and procedures to standardize the operations.

Forming A *Manufacturing* and a *Theory of Machines* term; in *manufacturing* terms, a manufacturing process which involves permanently changing the shape of the material by applying pressure; a few forming operations include *drawing*, *bending*, *spinning*, etc; in *theory of machines* term, a slow *method of gear manufacturing* (*copying*); used to produce gear teeth on a gear blank using a former tool; best suited for *limited manufacturing*; see also **Generating**.

Forward Extrusion A *Manufacturing* term; see **Direct Extrusion**.

Forward Scheduling An *Operations Management* term; one of the two methods of *job scheduling* in which activities are scheduled once the job requirements are known; used in organizations such as *hospitals*, *clinics*, *restaurants*, and *machine tool manufacturers*, where jobs are performed to customer order, and delivery is typically scheduled at the earliest possible date; see also **Backward Scheduling**.

Fossil Fuels A *Power Plant* term; the naturally occurring fuels preserved in the earth's crust and formed from the remains of tiny plants and animals that died millions of years ago; consist of hydrocarbons and produce heat on combustion; a few examples include *coal*, *petroleum*, and *natural gases*; mostly used in industrial applications for producing power and for transportation.

Foundry Shop A *Manufacturing* term; a part of the central workshop where casting-related activities are carried out.

Foundry Tools A *Manufacturing* term; tools used in the foundry shop to shape a job in the molten state; a few examples include *moulding box*, *rammer*, *swab*, *trowel*, *draw spike*, *shovel*, and *riddle*.

Four-Stroke Engine A *Thermodynamics* term related to an *IC engine*; an internal combustion engine which works on a *four-stroke cycle* consisting of *suction*, *compression*, *expansion*, and *exhaust* strokes; the cyclic operation is completed in every two revolutions of the crankshaft; widely employed for large power requirements, for example, heavy vehicles for transportation such as *bus*, *trucks*, and *cars*; see also **Two-Stroke Engine**.

Four-Bar Linkage A *Theory of Machines* term; a common type of linkage system consisting of three movable bars, two fixed pivot joints, two pin joints, and a fixed bar which is often a part of a machine's frame.

Fourier's Law of Dimensional Homogeneity See **Principle of Dimensional Homogeneity**.

Fourier's Law A *Heat Transfer* term; named after the *French* scientist *Jean Baptist Joseph Fourier* (1768–1830); used to find the heat flow by conduction along the length of a metal rod; given as:

$$Q = \frac{k\,A\,\Delta t}{L}(T_{\mathrm{h}} - T_l)$$

where

Q = quantity of heat that flows along the rod
k = thermal conductivity of the material of the rod
A = cross-sectional area of the rod
Δt = time interval of heat flow
L = length of the rod
T_{h} = high temperature of one end of the rod
T_l = low temperature of the other end of the rod.

Four-Stroke Cycle A *Thermodynamics* term related to an *IC engine*; a thermodynamic cycle which is completed in *four strokes* of a piston or in two revolutions of the crankshaft of an internal combustion (*IC*) engine; consists of *suction*, *compression*, *expansion*, and *exhaust* strokes; finds use in petrol or diesel engines; see also **Two-Stroke Cycle**.

Fourth-Party Logistics An *Operations Management* term, abbreviated as **4PL**; the non-asset-owning third-party logistics firms, which only provide coordination services; see also **Third-Party Logistics**.

Fraction Time A *Time Study* term; see **Allowance Time.**

Fraction A *Mathematics* term; a number which represents a part of a whole single object or a group of objects; expressed as a ratio of two numbers or in decimal form; classified into *proper*, *improper*, and *mixed* fractions.

Fractional Error A *Metrology* term; see **Relative Error.**

Fracture The act of breaking apart a component; represents total structural failure.

Frame of Reference A *Physics* term; a geometric reference with respect to which measurements are made; a coordinate system which is fixed at the earth's surface.

Frame A *Strength of Materials* term; a stationary and fully constrained civil structure used to support loads; consists of multi-force members, and its members are acted upon by three or more forces, unlike a truss where each member is a two-force member: see also **Truss.**

Francis Turbine A *Fluid Mechanics* term; a hydraulic turbine, named after *James B. Francis* (1815–1892), an eminent *American* hydraulic engineer who designed, built, and tested the first efficient inward-flow hydraulic turbine in 1849; a *mixed-flow, single-stage, inward-flow reaction hydraulic turbine*; suitable for *medium head* installations (15 m $\leq$ H $\leq$ 300 m); has self-governing property against speed instability.

Frank Gilbreth (1868–1934) and Lillian Gilbreth (1878–1972) The *two American pioneers* (*husband* and *wife*) of *scientific management* who proposed *therbligs*; *Frank Gilbreth* is probably best known for his experiments in reducing the number of motions in bricklaying; *Lillian Gilbreth* was a psychologist; the two Gilbreths used motion picture films to study work arrangements in order to eliminate wasteful hand-and-body motions; also experimented with design and use of the proper tools and equipment for optimizing work performance; devised an instrument, called *microchronometer* which recorded time to 1/2000 of a minute.

Frederick Herzberg (1923–2000) An *American* psychologist who introduced the '*Motivator-Hygiene theory*', also called '*Two-Factor Theory*' of motivation. His 1968 publication, '*One More Time, How Do You Motivate Employees?*' has sold 1.2 million reprints by 1987 and was the most requested article from the Harvard Business Review.

Frederick Winslow Taylor (1856–1915) An *American* mechanical engineer who originated the scientific management principles and is called *the 'father of scientific management'*; introduced scientific methods to improve the production efficiency; authored the famous book '*Principles of Scientific Management*' in 1911; emphasized on scientific way of doing the job and using tools such as flow diagrams and process charts to find the most economic way to perform a task; invented *high-speed steel* (HSS) cutting tools and developed the famous *Taylor's tool life equation* $V\,T^n$

$= constant$, which finds use in metal cutting operations (V = velocity of cutting tool, T = time of cutting, and n = an exponent which depends on material being cut); his ideas were embraced and further extended by *Frank* and *Lillian Gilbreth, Henry Gantt*, and *Henry Ford*.

Free Body Diagram A *Physics* term, abbreviated as **FBD**; the sketch or diagram of a body which is completely isolated (free) from all other bodies or its assembly and shows all the external forces (both known and unknown) acting on the body including their *location, magnitude,* and *direction* along with its own *weight*; a useful concept in the analysis of an engineering problem.

Free Convection A *Heat Transfer* term; also called **Natural Convection**; a mode of heat transfer by convection in which heat is transferred by a fluid motion caused by buoyancy forces which are induced by the difference in its density due to the variation of temperature; an example includes the flow of air in the air-cooled condenser of a *domestic refrigerator*; see also **Forced Convection**.

Free Expansion A *Thermodynamics* term; also called **Constant Enthalpy Process**; an *irreversible process* which involves expansion of a gas without any change in its internal energy or enthalpy and without doing any external work; see also **Throttling**.

Free Fall A *Physics* term; the falling down of a body under the action of gravity towards the centre of the earth; an ideal motion of a body in which air resistance and the small variations in acceleration with latitude and altitude are ignored.

Free Good An *Economics* term; an item for which there is no scarcity, is abundantly available, and does not require any price to pay for (zero price); for example, *air, sunlight,* and *river water*; see also **Economic Good**.

Free Market System An *Economics* term; a market system left to operate on its own with no outside interference; there is no government involvement and is not affected by the government's plan or directives; the individuals pursuing their own self-interest will go into business and produce the products and services which people want.

Free Oscillation A *Physics* term; also called **Damped Oscillation** as its amplitude decreases to zero due to frictional resistance (damping); the oscillation of a body which occurs by itself and is not influenced by an external periodic force; produced when a body is displaced from its equilibrium position and then released; the frequency of free oscillation of a body is called *natural frequency* of the body.

Free Slack A *Project Management* term; the amount of spare (slack) time for an activity, when all the succeeding activities in the network can be started as early as possible; that is, the earliest start times (EST) of the succeeding activities are not affected; an activity may have total slack, but not a free slack; the free slack is always less than or equal to the total slack; see also **Total Slack**.

Free Surface A *Fluid Mechanics* term; the upper layer of water, where atmospheric pressure acts.

Free Vector A *Physics* term; a vector which can be moved freely anywhere in space, provided its magnitude, direction, and sense remain fixed; examples include moment of a force vector and area vector.

Free Vortex A *Fluid Mechanics* term; also called **Potential Vortex** or **Simple Vortex**; a fluid flow in which the fluid particles move along circular streamlines without rotating about their own axes; formed due to flow separation in the wake region; has zero vorticity; resembles a common whirlpool that is found while paddling a boat or while emptying water from a bathtub; see also **Forced Vortex**.

Freehand Sketching An *Engineering Drawing* and a *Machine Drawing* term; one of the most effective ways to communicate a pictorial or verbal idea to a workman; it helps in visualizing the components.

Freely Falling Body A *Physics* term; a body falling towards the centre of the earth under the action of acceleration due to gravity; the following equations are used for a freely falling body:

$$v = u + gt \tag{1}$$

$$y = ut + \frac{1}{2}gt^2 \tag{2}$$

$$v^2 = u^2 + 2gy \tag{3}$$

In case of bodies moving vertically upward, the acceleration due to gravity acts in the opposite direction, that is, it is a case of retardation, and the equations are accordingly modified as:

$$v = u - gt \tag{4}$$

$$y = ut - \frac{1}{2}gt^2 \tag{5}$$

$$v^2 = u^2 - 2gy \tag{6}$$

where

u = initial velocity of the body (when $t = 0$)
v = final velocity of the body (at time t)
t = time of motion
g = acceleration due to gravity of the body = 9.81 m/s^2
y = vertical displacement of the body.

Free-Stream Velocity A *Fluid Mechanics* term; the velocity of the fluid approaching a body.

Freezing Point A *Physics* term; a temperature at which a liquid substance is converted into a solid; for example, the freezing point of water is 0 °C.

Freon-11 A *Refrigeration* term; also called **R11**; a *methane-based fluorocarbon refrigerant* having chemical formula CCl_3F.

Freon-12 A *Refrigeration* term; also called **R12**; a *methane-based fluorocarbon refrigerant* having chemical formula CCl_2F_2; widely used for domestic purposes.

Freon-22 A *Refrigeration* term; also called **R22**; a *methane-based fluorocarbon refrigerant* having chemical formula $CHClF_2$.

Freon A *Refrigeration* term; the trade name for *fluorocarbon refrigerants* derived from *hydrocarbons* such as *methane*, *ethane*, and others; a few examples include Freon-11 or R11 (CCl_3F), Freon-12 or R12 (CCl_2F_2), and Freon-22 or R22 ($CHClF_2$).

Frequency A *Physics* term; the number of oscillations or periodic motions executed by a body in one second; *the reciprocal of the periodic time*, that is, ($\frac{1}{T}$); usually denoted by n; the SI unit is *hertz* (Hz) named after the German physicist *Heinrich Rudolf Hertz* (1857–1894).

Friction Coefficient A *Fluid Mechanics* term; also called **Fanning Friction Factor**, named after the *American* engineer *John Fanning* (1837–1911); a *friction factor* denoted by C_f; equals to *one-fourth* of the *Darcy-Weisbach friction factor*; see also **Darcy-Weisbach Friction Factor**.

Friction Stir Welding A *Welding* term; see **Friction Welding**.

Friction Welding A *Welding* term; also called **Friction Stir Welding**; a *solid-state welding* process which uses frictional heat to join metals without fusion or filler metals produced by a *rotating tool* when pressed against the parts being joined; particularly suitable for joining *aluminium alloyed parts* in the form of *plates, sheets*, and *hollow pipes*; can also join dissimilar metals; finds application in *ship building*, *aerospace*, *railway*, and *automotive industry*.

Friction A *Physics* term; also called **External Friction** or **Contact Friction**; the resistance which occurs, when two solid bodies in contact with each other try to move, or there is an actual relative motion between the two; see also **Internal Friction**.

Frictional Force A *Physics* term; also called **Force of Friction**; an opposing force which comes into play when one solid body moves (slides or rolls) or even tries to move over the surface of another solid body; occurs due to interaction between molecules of the contacting surfaces; acts always in the direction opposite to the direction of motion; disadvantageous in some machines and processes such as bearings, gears, power screws, and fluid flow in pipes where it is desirable to reduce its effect, whereas in other machines such as brakes, clutches, and belt drives, it is advantageous.

Front View An *Engineering Drawing* and a *Machine Drawing* term; the view of an object obtained as a projection on a *vertical plane* of projection by looking at the object normal to its *front surface*; the object is usually positioned in such a way that its view from the front reveals most of the important features; see also **Top View**.

Frosting A *Physics* term; the reverse process of a vapour turning directly into a solid.

Froude Number A *Fluid Mechanics* term; a *dimensionless* (*nondimensional*) number, named after *William Froude* (1810–1879), a *British* civil engineer, mathematician, and naval architect who pioneered the use of towing tanks for the study of ship design; a ratio of the *inertia force* to the *gravity force*; useful for the flows influenced by gravity such as wave action set up by a ship (ship hydrodynamics), the flow of water in open channels, the forces of stream on a bridge pier, the flow over a *spillway* and a *weir*, and the flow of a jet from an orifice; denoted by Fr, given as:

$$Fr = \frac{V}{\sqrt{gL}}$$

where

V = fluid velocity
L = length (liquid depth in case of an open-channel flow)
g = acceleration due to gravity.

FSN Analysis An *Operations Management* term related to *inventory*; a selective inventory control technique which classifies the inventories based on the consumption pattern of components into three categories: *F* (*fast moving*) items, *S* (*slow moving*) items, and *N* (*no moving*) items; *F items* are regularly used items (e.g. *eggs*, *milk*, *fruits and vegetables*, etc.); *S items* are time-specific items (*old cars*, *old furniture*, *housing units*, etc.); *N items* are *overstocked goods*.

Fuel Cell Electric Vehicle An *Electric Vehicle* term; a vehicle which employs fuel cells, usually hydrogen-based, to produce electricity that powers an on-board motor to drive the vehicle.

Fuel–Air Ratio An *IC Engine* term, abbreviated as **FA**; a ratio of the mass of fuel to the mass of air; the reciprocal of air–fuel ratio; see also **Air–Fuel Ratio**.

Fulcrum A *Physics* term; the point around which a lever, wheel, or linkage pivots.

Full-Mould Casting A *Manufacturing* term related to *casting*; also called **Lost-Foam Casting** as the mould made from *foamed polystyrene* is finally consumed in the process or **Expanded Polystyrene Casting** as the hard beads of polystyrene are first steam expanded and dried; a casting process which uses an *expanded polystyrene pattern* that gets melted by the heat of the molten metal being poured, thus creating a space to act as a full-mould cavity; offers advantages such as reusable mould sand, its simplified handling, no need for the separate cope and drag parts of the mould, and the reasonable precision of cast parts which do not require many machining and finishing operations; the typical products made by this method include *cylinder heads, crankshafts, brake components*, and *manifolds* for automobiles and *machine bases* which are commonly cast from *aluminium, iron, steel, nickel alloys, copper*, and *stainless steel.*

Full Proofing An *Operations Management* term; see **Poka-Yoke**.

Fullerene A *Chemistry* term; a recently discovered (1985) *third crystalline allotrope of carbon* (C), other than two previously well-known allotropes, *diamond* and *graphite*; discovered collectively by three scientists, namely *Richard Smalley* (1943–2005) and *Robert Floyd Curl* (1933-) of Rice University, Houston (USA), and *Harry Walter Kroto* (1939–2016) of the University of Sussex, Brighton (UK), for which they were awarded Nobel Prize of Chemistry in 1996; made by heating graphite in an electric furnace in the presence of inert gases such as *helium* (He) or *argon* (Ar); being covalent, fullerenes are soluble in organic solvents but diamond and graphite are not.

Functional Benchmarking A *Quality Management* term; a process of comparing functions such as *administration, customer service,* and *sales operations* with those of outstanding firms in any industry; for example, the *Xerox* benchmarked its distribution function against *L. L. Bean*'s, as *Bean* is a leading retailer in distribution efficiency and customer service.

Functional Layout An *Operations Management* term related to *facilities planning*; see **Process Layout**.

Functional Manager A *Management* term; a manager who is responsible for developing the core competences in the organization's various departments; trains and develops the subordinates to ensure that the organization has skills which match or exceed the skills of its competitors.

Functional Organization A *Management* term; the organizational structure suggested by *F. W. Taylor* (1856–1915), called *the 'father of scientific management'*; based on the functional requirements (works speciality) in an organization such as *manufacturing, marketing, finance,* and *human resources*; the work quality and productivity improve because of works speciality; suitable for *medium-sized firm* with several related product lines in one industry.

Functional Strategy A *Management* term; also called **Function-Level Strategy**; the approach taken by a functional area to achieve corporate and business unit objectives and strategies by maximizing resource productivity; relates to developing and nurturing a distinctive competence; for example, *Mercedes-Benz* invests heavily to improve its skills in *R&D* and *product design*, and *Coca-cola* invests heavily to devise *innovative approaches* to *marketing*.

Functional Structure A *Management* term; an organizational design which groups similar or related occupational specialities together.

Function-Level Strategy A *Management* term; see **Functional Strategy**.

Fundamental Laws of Motion A *Physics* term; see **Laws of Motion**.

Fundamental Particles A *Chemistry* term; the basic particles of an atom of a chemical element; include *electron, proton,* and *neutron.*

Fundamental Quantities A *Physics* term; the most basic physical quantities which cannot be obtained from one another; all other physical quantities can be obtained from them; *seven* fundamental quantities include *mass, length, time, temperature, electric current, luminous intensity,* and *amount of a substance.*

Fundamental Units A *Physics* term; the units of measurement of *seven fundamental physical quantities* in SI system of units; include *kilogram* (kg) for mass, *metre* (m) for length, *second* (s) for time, *Kelvin* (K) for temperature, *ampere* (A) for electric current, *candela* (cd) for luminous intensity, and *mole* (mol) for amount of a substance; see also **Supplementary Units** and **Derived Units**.

Fusible Plug A *Power Plant* term related to a *boiler*; a *boiler mounting* which melts in case of low water level in the boiler and thus preventing boiler's overheating by allowing water and steam to escape and extinguish the fire in the furnace of the boiler.

G

Gabriel Lame (1795–1870) A famous *French* mathematician whose name is inscribed on the *Eiffel Tower*; proposed *Lame curves* and introduced *Curvilinear curves* in mathematics; also known for *Lame's constants* used in the design of *pressure vessels* in strength of materials.

Galena A *Chemistry* term having the chemical formula of PbS; the chief ore of *lead* (Pb) which contains 6–8% of lead along with small quantities of silver (up to 0.1%).

Gallium Arsenide A *Materials Engineering* term; a *semiconducting compound* consisting of *gallium* (Ga) and *arsenic* (As) having the chemical formula GaAs; used as a *high-temperature rectifier* and as a *laser material*; see also **Cadmium Sulphide**.

Galvanization A *Chemistry* term; see **Galvanizing**.

Galvanized Steel A *Materials Engineering* term; the *zinc-coated mild steel* used for *corrosion* protection.

Galvanizing A *Chemistry* term; also called **Galvanization**; a method of providing a coating (a thin film) of *zinc* over *iron* components, such as iron rods or sheet to prevent them from *rusting* and *corrosion*, popularly called the *GI pipes* or *GI sheet*.

Game Theory An *Operations Research* term; proposed by *John von Neumann* (1903–1957), who is known as the '*Father of Game Theory*'; the *mathematical analysis* of *game* that is fundamentally based upon the *minimax* criterion; has only *theoretical* aspects rather than its application to solving the *real problems*.

Game An *Operations Research* term; an activity between two or more people governed by a set of rules, at the end of which each person receives some benefits or suffers loss; a few examples include *bidding*, *advertisements*, *elections*, and *sports*.

D. K. Singh, *Dictionary of Mechanical Engineering*,
https://doi.org/10.1007/978-981-99-1722-8_7

Gamma Ray A *Physics* term; an *electromagnetic radiation* discovered by *Ernest Rutherford* (1871–1937), a *New Zealand* physicist who is known as the '*father of nuclear physics*'; emitted by radioactive substances such as *uranium* (U), *plutonium* (Pu), and *thorium* (Th); consists of *neutral particles*; not frequently emitted radiation as compared to alpha or beta ray emission; arises from energy rearrangements within the nucleus after the emission of an alpha or beta particles; cannot be deflected even in the strongest electric or magnetic field; has an extremely high penetrating power, even greater than that of *X*-ray, and can pass through 0.15 mm of steel; harmful to living tissues; see also **Alpha Ray** and **Beta Ray**.

Gamma A *Mathematics* term; a *mathematical function* which is denoted by Υ; used as a symbol to denote an *angle*; also a ratio of the two *specific heats* (C_p/C_v); air has the Υ value of 1.4.

Gamma-Al_2O_3 (γ-Al_2O_3) A *Materials Engineering* term; see **Activated Alumina**.

Gamma-Alloy A *Materials Engineering* term, abbreviated as **γ-Alloy**; an alloy of *aluminium* (Al) which contains 3.5–4.5% *copper* (Cu), 1.8–2.3% *nickel* (Ni), 1.2–1.7% *magnesium* (Mg), and remaining *aluminium* (Al); can maintain its strength at elevated temperatures, hence used in the casting of engine parts such as *pistons*; used largely in the form of *sheet* and *strip*.

Gangue A *Chemistry* term; also called **Matrix**; the earthy and siliceous impurities (in addition to the impurities of other minerals) present in an ore.

Gantt Chart An *Operations Management* term; an important *project management* tool used for scheduling work and monitoring its progress; developed by a famous *American* industrial engineer *Henry Laurence Gantt* (1861–1919) in 1914; a type of *bar chart* which plots activities (jobs) both planned and actual on *y*-axis against time on *x*-axis; shows a job's progress graphically or compares actual against planned performance over a period of time; used as a scheduling tool for planning production activities.

Gas Cutting A *Manufacturing* term; a process of splitting a workpiece by the heat of a gas flame.

Gas Cyaniding A *Manufacturing* term related to *heat treatment*; a *case hardening process* used to produce *hardened surface* in *low carbon steel* parts by heating the parts in an *ammonia atmosphere* at a temperature of 1700 °F and then *quenched* in *oil*; see also **Liquid Cyaniding**.

Gas Dynamics A *Fluid Mechanics* term; the study of a compressible fluid in motion.

Gas Metal Arc Welding A *Welding* term; see **Metal Inert Gas Welding**.

Gas Tungsten Arc Welding A *Welding* term; see **Tungsten Inert Gas Welding**.

Gas Turbine A *Power Plant* term; a work-producing device (turbomachine) which converts the heat energy of a compressible gas (the working fluid) into work, which is further used in generating electricity in power plants; also used in modern commercial jet engines; see also **Steam Turbine** and **Hydraulic Turbine**.

Gas Welding A *Welding* term; a *fusion welding process* which uses mainly heat of an oxyacetylene gas flame for heating the parts being joined; produces a temperature of around 3300 °C; see also **Arc Welding**.

Gas A *Physics* term; a state of matter which has *neither definite shape nor definite volume*; see also **Solid** and **Liquid**.

Gasket A *Design* term; an elastic or plastic device used to *prevent the flow of fluids* across mating surfaces of a mechanical assembly having no relative motions; for example, a gasket is placed between the cylinder head and the cylinder block of an engine to make the joint tight.

Gate A *Manufacturing* term related to *casting*; an *inlet* which runs into the mould cavity; a part of the *gating system*.

Gated Pattern A *Casting* term; a *type of pattern* consisting of many small patterns connected to each other through suitable channels, called gate formers; used to produce *multicavity moulds*, which are fed by the gate formers using a *single runner*; suitable for *mass production of small castings*.

Gating System A *Manufacturing* term related to *casting*; a network of channels used to deliver molten metal in the mould cavity; consists of *pouring basin*, sprue, *runner*, and *gate*.

Gauge Blocks A *Metrology* term; the rectangular-shaped steel blocks with highly polished and parallel surfaces; used to provide dimensional standards between a given range by stacking them one over another; available as a set of 86 or 45 pieces.

Gauge Length A *Strength of Materials* term; the specified length of a test specimen for which test is conducted and analysis carried out (usually 60 mm for conducting a tension test).

Gauge Pressure A *Fluid Mechanics* term; a pressure relative to the atmospheric pressure; the SI unit is *pascal* (Pa) which is equivalent to N/m^2; see also **Absolute Pressure** and **Vacuum**.

Gauss's Theorem A *Mathematics* term; see **Divergence Theorem**.

Gaussian Distribution Curve A *Statistics* term; see **Normal Distribution Curve**.

GDP An *Economics* term; see **Gross Domestic Product**.

Gear Drive A *Design* term related to a *gear*; a *positive drive* which uses a gear mechanism for power transmission; offers a *constant speed ratio* without any slippage; a highly *compact*, *efficient*, and *reliable* power transmission system; important parameters used in the design of gear drive mechanism include the pressure angle, pitch circle diameter, and diametral pitch.

Gear Pump A *Fluid Mechanics* term; a hydraulic control system consisting of a pair of identical spur gears, which revolve in opposite directions in a closely fitted casing; the theoretical discharge through a gear pump is given as:

$$Q = \frac{2ZVN}{60}\ (\mathrm{m}^3/\mathrm{s})$$

where

Z = number of teeth in one gear
V = volume of one tooth (m^3)
N = number of revolutions per minute (rpm).

Gear Ratio A *Theory of Machines* term related to *gear*; abbreviated as **GR**; in a gearset, a ratio of the number of revolutions (rotational speed) of the input (driving) gear to the number of revolutions (rotational speed) of the output (driven) gear; also a ratio of the number of teeth on the gear to the number of teeth on the pinion; gear ratio is always a positive number greater than 1 regardless of the direction in which the power flows through the gearset; see also **Velocity Ratio.**

Gear Train Value A *Theory of Machines* term related to *gear*; see **Train Value.**

Gear Train A *Theory of Machines* term related to *gear*; a mechanism consisting of two or more gears in mesh, which is used to increase or decrease rotational speed and torque and change or maintain rotational direction between the input and output shafts of a mechanical system; used to obtain a specific train value while transmitting motion or power between two shafts; used in *clocks*, *watches*, and *lathes*; classified into *simple*, *compound*, *reverted*, and *epicyclic* gear trains; see also **Train Value.**

Gear A *Theory of Machines* term; a rotating circular machine part (*wheel*) having teeth cut over its periphery; a device used to transmit *rotation*, *torque*, and *power* without any loss between closer shafts with a constant speed ratio; the most common *means of power transmission* in many machines such as *automobiles*, *tractors*, *metal cutting machines tools*, *rolling mills* and *marine engines*; the shape of a gear's tooth is mathematically defined and precisely machined according to well-defined codes and standards.

Gearset A *Theory of Machines* term related to a *gear*; a pair (set) of two gears having complementary teeth profiles that mesh with one another; forms the building block of a transmission system which transmits *rotation*, *torque*, and *power* between shafts; smaller of the two gears is called *pinion* (driving gear), and the larger one is called simply *gear* (driven).

General-Purpose Machines A *Manufacturing* term; machines which can perform basic manufacturing functions such as *turning*, *drilling*, and *facing*; for example, lathe is a general-purpose machine.

Generalized Hooke's Law A *Strength of Materials* term; a set of six equations which gives the relationships between a general state of stress and strain; applicable only to *homogeneous isotropic materials* under *multiaxial loading condition*; obtained using the principle of superposition from the simpler stress–strain equations; given as:

$$\begin{aligned}
\epsilon_x &= \frac{\sigma_x}{E} - \upsilon\frac{\sigma_y}{E} - \upsilon\frac{\sigma_z}{E} \\
\epsilon_y &= \frac{\sigma_y}{E} - \upsilon\frac{\sigma_z}{E} - \upsilon\frac{\sigma_x}{E} \\
\epsilon_z &= \frac{\sigma_z}{E} - \upsilon\frac{\sigma_x}{E} - \upsilon\frac{\sigma_y}{E} \\
\gamma_{xy} &= \frac{\tau_{xy}}{G} \\
\gamma_{yz} &= \frac{\tau_{yz}}{G} \\
\gamma_{zx} &= \frac{\tau_{zx}}{G}
\end{aligned}$$

where

ε_x = axial strain in *x*-direction
ε_y = axial strain in *y*-direction
ε_z = axial strain in *z*-direction
σ_x = normal stress in *x*-direction
σ_y = normal stress in *y*-direction
σ_z = normal stress in *z*-direction
γ_{xy} = shear strain in *xy*-plane
γ_{yz} = shear strain in *yz*-plane
γ_{zx} = shear strain in *zx*-plane
τ_{xy} = shear stress in *xy*-plane
τ_{yz} = shear stress in *yz*-plane
τ_{zx} = shear stress in *zx*-plane
E = modulus of elasticity
G = modulus of rigidity
υ = Poisson's ratio.

Generating A *Theory of Machines* term related to a *gear*; a *faster method* of gear manufacturing; useful for *mass production* to produce gear teeth on a gear blank using a hob or a pinion-type cutter; see also **Forming**.

Generator Efficiency An *Electrical* term; a ratio of the *electrical power output* to the *mechanical power input*; see also **Motor Efficiency**.

Generator An *Electrical* term; a device which converts mechanical energy into electrical energy; an important means of *power generation.*

Genetic Algorithm An *Operations Research* term; the computerized search and optimization algorithm based on the mechanics of natural genetics and natural selection; extensively used as an optimization technique.

Genichi Taguchi (1924–2012) A well-known *Japanese* quality expert who is known for a concept called '*design of experiments*' to product design; stressed on improving quality efforts during the '*design stage*' of a product; coined the word '*robust quality*' of a product which makes the product suitable to perform over a wide range of conditions; a winner of the *Deming Award.*

Geometric Centre A *Strength of Materials* term; see **Centroid.**

Geometric Similarity A *Fluid Mechanics* term; the similarity of shapes between a model and its prototype; see also **Dynamic Similarity** and **Kinematic Similarity.**

Geometrical Moment of Inertia A *Strength of Materials* term; see **Second Moment of Area.**

Geometrical Similarity A *Fluid Mechanics* term; the similarity of shapes between a model and its prototype; see also **Kinematic Similarity** and **Dynamic Similarity.**

George Elton Mayo (1880–1949) A Harvard Business School professor who was one of the key figures involved in the famous *Hawthorne studies* conducted at the *Hawthorne Plant* of the *Western Electric Company* of the *USA* during 1927–1932; see also **Hawthorne Studies.**

George Gabriel Stokes (1819–1903) An *English* physicist and pioneering investigator of *viscosity*; widely known for the *Navier–Stokes equations*; the unit of *kinematic viscosity* in the metric system (called *stoke*, which is equivalent to cm^2/s) is named after him; proposed *Stokes law* ($F = 6\pi\mu Rv$, where F = *drag force*, μ = *dynamic viscosity*, R = *radius of the spherical object*, and v = terminal flow velocity relative to the object) in 1851 related to *drag force* exerted on *spherical objects* moving through a *viscous fluid.*

Geostationary Satellite A *Physics* term; called so as the satellite appears to be at a fixed position at a definite height from the earth and appears stationary to an observer; also called **Communication Satellite** as it is used for the purpose of communication or **Geosynchronous Satellite** as it revolves around the earth in an orbit (concentric and coplanar with the equatorial plane) with the same angular speed in the same direction as is done by the earth around its own axis (from west to east), that is, both speeds are synchronized; the time period of revolution is 24 h, the orbital speed is 3.1 km/s, and it is placed at a height of 36,000 km above the equator of the earth; *Telstar* was the first communication satellite launched by the USA in 1962; see also **Polar Satellite.**

Geosynchronous Satellite A *Physics* term; see **Geostationary Satellite**.

German Silver A *Materials Engineering* term; an alloy of *copper* (Cu), *nickel* (Ni), and *zinc* (Zn) consisting of 60% Cu, 20% Ni, and 20% Zn; used in making *domestic utensils*.

Gib A *Design* term; a *mild steel piece* which finds use in a cotter joint to prevent the bending of the lower part of the socket end; also helps to use parallel holes in the spigot and socket; has the same thickness as that of cotter; increases the bearing area of contact between the mating surfaces; one or two gibs can be used to take care of both sides of the joint (one gib is used when the cotter has taper on one side, and two gibs are used when the cotter has taper on its both sides).

Gibbs Function A *Thermodynamics* term; a *thermodynamic property* used in the analysis of a *chemical system*; named after an *American* scientist *Josiah Willard Gibbs* (1839–1903), who received from Yale University in 1863 the first Ph.D. in engineering in *America*; denoted by *G*; expressed as:

$$G = H - TS$$

where

H = enthalpy of the working fluid
T = absolute temperature of the working fluid
S = entropy of the working fluid; see also **Helmholtz Function**.

Gigapascal A *Physics* term; the larger SI unit of *pressure* and *stress*, abbreviated as **GPa**; 1 GPa = 10^9 Pa; see also **Kilopascal** and **Megapascal**.

Glass A *Chemistry* term; an *amorphous material* obtained from *silica* (SiO_2); usually processed by melting and casting; when heated, it does not melt at a fixed temperature (melts into a liquid at a very high temperature), rather gradually softens to be moulded into any desired shape; a transparent material which can transmit visible light (as well as ultraviolet and infrared radiation); brittle and fragile and chemically inert; various colouring materials are used to prepare glasses of different colours; for example, ferric oxide (Fe_2O_3) is added to get yellow colour, chromic oxide (Cr_2O_3) to get green colour, manganese oxide (MnO) to get purple colour, and cobalt oxide (CoO) to get blue colour in glass; the ordinary glass (also called *soda glass* or *window glass*) has the approximate chemical formula of $Na_2O \cdot CaO \cdot 6SiO_2$; classified into *soda glass, potash glass, flint glass, crown glass, Pyrex glass, Jena glass, Crookes glass, bottle glass, ground glass*, and *safety glass*; extensively used in houses, cars, computer and television screens, and hundreds of other applications.

Global Positioning System An *information technology* tool, abbreviated as **GPS**; a *wireless technology* which uses satellite transmission to communicate exact locations; extensively used in logistics by transport companies to identify the exact locations of their vehicles carrying goods for distribution.

Globalization A *Management* term; the internationalization of markets and corporations in which activities are expanded and operated worldwide; characterized by an increased cross-border flow of trade in goods, services, and financial assets along with an increased international mobility of technology, information, and individuals; a few examples of globalization include *global markets*, *global operations*, *global financing*, and *global supply chains*.

Globular Transfer A *Welding* term; a *mode of metal transfer* in *gas metal arc welding* (GMAW) in which the consumable electrode is transferred across the arc in *large droplets*; see also **Spray Transfer**.

GMAW A *Welding* term; the short form of **Gas Metal Arc Welding**; see **Metal Inert Gas Welding**.

Gold A *Chemistry* term; a *yellow*, *heavy*, *highly ductile*, and *malleable noble metal* which has the specific gravity of 19.3 and the melting point of 1063 °C; usually alloyed with *silver* and *copper*; extensively used in making *jewellery* and *coinage*; *white gold* is an alloy of *gold* with *nickel*.

GOLF Analysis An *Operations Management* term related to *inventory*; a *selective inventory control technique* which classifies the inventories based on their source of availability into four categories: *G* (*government*) items, *O* (*ordinary*) items, *L* (*local*), and *F* (*foreign*) items; *G items* are imported by government agencies like State Trading Corporation, Metals Trading Corporation, etc. using special procedures, and normal rules of procurement are not applicable for them; *O items* are the most general items like groceries, and their procurement does not require special procedures; *L items* are locally available items confined to a particular place (e.g. Maner ladoo, Silav khaaja, and Mathura petha); *F items* are specially manufactured items procured from a distant source (foreign), and their procurement procedures are different than for local goods (e.g. excise duty-paid items).

Gotthilf Heinrich Ludwig Hagen (1797–1884) A *German* hydraulic engineer who conducted original studies of resistance in and transition between *laminar* and *turbulent* flow.

Governor A *Theory of Machines* term; a device used to maintain a constant mean speed of rotation of the crankshaft of an engine under variable load on the engine; does not control the cyclical fluctuations of speed; generally classified into two broad groups, namely *centrifugal* and *inertia* governors; see also **Flywheel**.

GPa A *Physics* term; see **Gigapascal**.

GPS An *information technology* tool; see **Global Positioning System**.

Gradient Operator A *Mathematics* term; also called **Del Operator**; a *vector operator*; denoted by ∇; defined in Cartesian coordinates (*i*, *j*, and *k* are the *unit vectors* along *x*-, *y*-, and *z*-directions, respectively) as:

$$\nabla = \left(\frac{\partial}{\partial x}, \frac{\partial}{\partial y}, \frac{\partial}{\partial z}\right) = \left(i\frac{\partial}{\partial x} + j\frac{\partial}{\partial y} + k\frac{\partial}{\partial z}\right)$$

Gradual Usage and Non-instantaneous Receipt Model An *Operations Management* term; see **Production Quantity Model.**

Gradually Applied Load A *Physics* term; see **Static Load.**

Grain Boundary A *Materials Engineering* term; the boundary which separates each grain from other grains in a polycrystalline microstructure.

Grain In *measurement* terms, a unit of weight equivalent to 64.8 mg or 1/7000 of a pound; in *materials* terms, any small, hard particle; a minutely small crystalline body that has definable limits; the individual crystallite in a polycrystaline microstructure; in *cereal* terms, a small, hard seed of a food plant.

Graph A pictorial representation which shows the relationship between two variable quantities in the form of a *straight line* or *curve*; represented by some mathematical relation.

Graphical Method A method based on the use of a graph for the solution of a problem; a method for solving linear programming problems involving only two variables; the method is based on drawing a graph to find a feasible region containing all possible feasible solutions to the problems that satisfy all the constraints of the problem; points lying within the feasible region satisfy all the constraints; optimal value of the objective function occurs at the corner points; in case of a *maximization* problem, corner points at which the objective function has a maximum value represent the optimal solution; in case of a *minimization* problem, corner points at which the objective function has a minimum value represent the optimal solution.

Graphite A *Chemistry* term; the greyish-coloured, the *purest* (like diamond), soft form of *carbon* (*C*) which contains 95–97% of *carbon*; has high density (2.25 g/cm^3); occurs in nature in free state but is also prepared artificially in large amounts using *Acheson process*; a *good lubricating agent* used as a *solid lubricant* (dry) in machines running at high speeds, where oil cannot be used as a lubricant; a *good conductor* of *heat* and *electricity* which makes it suitable for making *electrodes* and *carbon rods* for *dry cells*; used as a *moderator* for *fast-moving neutrons* in *atomic reactors*; also used in the manufacture of *crucibles* which can withstand high temperatures; the so-called *lead pencils* which mark paper black (called black lead), actually contain *graphite* mixed with a little *clay*.

Graphitic Carbon A *Materials Engineering* term; the *free, uncombined carbon* found in a metallic material in the form of *flakes*.

Gravimetric Analysis A *Chemistry* term; the analysis of a mixture of gases based on the mass (or weight) of each component; see also **Volumetric Analysis**.

Gravitation A *Physics* term; also called **Gravitational Force**; the force of attraction acting between any two bodies in the universe; discovered by *Sir Issac Newton* (1643–1727) in 1665 when he saw an apple falling down from a tree; plays an important role in controlling the entire structure of the universe; a conservative as well as a central force which acts through the line joining the centres of two bodies and can be zero, if separation between two bodies becomes infinity; see also **Gravity**.

Gravitational Field A *Physics* term; the space around a material body (*earth* or any *other planet* or *sun*) in which its gravitational pull can be experienced.

Gravitational Force A *Physics* term; see **Gravitation.**

Gravitational Mass A *Physics* term; the mass *related to the gravitational pull* of a body; defined by the *Newton's law of gravitation* as the magnitude of gravitational pull which is experienced by a body in a gravitational field of unit intensity; measured by a *spring balance*; equivalent to inertial mass; expressed in the same unit as inertial mass; see also **Inertial Mass.**

Gravitational Potential Energy A *Physics* term; the energy possessed by a body by virtue of its *position* above the surface of the earth; a function of height; given as:

$$\mathrm{P.E} = mgh$$

where

m = mass of the body
g = acceleration due to gravity and
h = height (altitude) of the body; see also **Potential Energy.**

Gravitational Potential A *Physics* term; the amount of work done in bringing a body of unit mass from infinity to a point in the gravitational field of the body without acceleration; a *scalar quantity* with a *negative value*; has the maximum value of zero at infinity; the SI unit is J/kg; given as:

$$\begin{aligned}\text{Gravitational potential} &= -\frac{GM}{r}\\ &= -\frac{GM}{R}\text{(for a body at the surface of the earth)}\end{aligned}$$

where

G = universal gravitational constant = $6.67 \times 10^{-11}\mathrm{N\,m^2\,kg^{-2}}$
M = mass of the gravitational body
r = distance between gravitational body and unit mass body
R = radius of the earth.

Gravitational Pull A *Physics* term; see **Gravity.**

Gravity Casting A *Manufacturing* term related to *casting*; see **Permanent Mould Casting.**

Gravity Die Casting A *Manufacturing* term related to *casting*; see **Permanent Mould Casting.**

Gravity Force A *Physics* term; see **Gravity.**

Gravity Pull A *Physics* term; see **Gravity**.

Gravity Segregation A *Manufacturing* term related to *casting*; a process of separation of the heavier component of a mixture material from its lighter component; usually found in *centrifugal casting* due to centrifugal force acting on the rotating mould.

Gravity A *Physics* term; also called **Gravitational Pull** or **Force of Gravity** or **Gravity Pull** or **Gravity Force**; the force that pulls an object towards the centre of earth; it is zero at the centre of the earth and is directed towards the centre; a measure of weight of a body.

Grey Iron A *Materials Engineering* term; see **Grey Cast Iron**.

Grease A *Fluid Mechanics* term; a highly *viscous*, *solid* or *semisolid* substance; generally consists of *soaps*, *mineral oil*, and *various additives*; extensively used as a *lubricant* to reduce friction and wear between contact surfaces.

Greenhouse Gas A gas which contributes to greenhouse effect by absorbing infrared radiation; includes gases such as carbon dioxide and chlorofluorocarbons.

Green Sand Mould A *Casting* term; a mould made of *green sand*.

Green Sand A *Casting* term; *the most commonly used moulding sand* which contains *silica*, *clay*, and *water*; named so because of *moisture* present in it.

Grey Cast Iron A *Materials Engineering* term; also called **Grey Iron**; one of the most widely used materials in casting as it can be easily cast; has a very low cost; *easily machinable* because of *graphite* present in the form of *thin flakes*, but is *brittle*; *cannot be readily welded* as it leads to the formation of crack which can be prevented by preheating the part before welding; the tensile strength varies from 100 to 400 MPa, and the compressive strength is 3–4 times its tensile strength.

Grinding A *Manufacturing* term; a *micromachining process* which uses tiny hard abrasive particles to remove very small amounts of material from a workpiece; produces better surface finish and close tolerances; used in cutting hard brittle materials.

Gross Domestic Product An *Economics* term, abbreviated as **GDP**; one of the two widely used measures of aggregate output of an economy other than the *gross national product* (GNP); a measure of all currently produced final goods and services evaluated at market prices; the most comprehensive measure of a nation's economic activity; policy makers use GDP figures to monitor short-run fluctuations in economic activity as well as long-run growth trends; usually measured on per year basis; changes with change in overall price level as well as when the actual volume of production changes; *stocks* and *bonds* are not part of GDP.

Gross Error A *Metrology* term; an error which occurs due to sheer carelessness of an observer; a few examples include *wrong recording of the observations, use of wrong values of observations in the calculations*, or *taking of reading without setting the instrument properly.*

Ground Glass A *Chemistry* term; a type of glass prepared by grinding ordinary soda glass by emery and turpentine oil or by sand blasting.

Group Cohesiveness A *Management* term; the degree to which group members are attracted to one another and share the group's goals.

Group Technology A *Manufacturing* term, abbreviated as **GT**; a manufacturing methodology in which parts are grouped and produced in families based on similarities in their design and manufacturing requirements.

Group A *Management* term; an association of two or more interacting and interdependent individuals who come together to achieve specific goals.

GT A *Manufacturing* term; see **Group Technology**.

GTAW A *Welding* term; the short form of **Gas Tungsten Arc Welding**; see **Tungsten Inert Gas Welding**.

Guided-Bend Test A *Strength of Materials* term; a *bending test* in which a specimen is bent to a desired shape by means of a *jig*.

Guidelines A *Management* term; a set of directions to follow.

H

Hacksaw A *Manufacturing* term; the simplest metal cutting tool which has a straight blade containing a number of cutting teeth; an important *fitting tool* used in *fitting shop*; can be operated manually or by using power (called *power hacksaw*); generally used for cutting off *bars*, *rods*, and *structural shapes*.

Haematite A *Chemistry* term; also called **Red Iron**; the *chief ore of iron* having the chemical formula of Fe_2O_3; contains about 70% iron; see also **Magnetite**.

Hagen-Poiseuille Flow A *Fluid Mechanics* term; often called simply **Poiseuille Flow**, named after the *German* hydraulic engineer *Gotthilf Heinrich Ludwig Hagen* (1797–1884) and the French physician *Jean Louis Poiseuille* (1799–1869); a *steady, incompressible, laminar flow* through a straight circular tube of constant cross section; has a *parabolic profile*.

Haigh's Criterion A *Strength of Materials* term related to *theory of material's failure*; see **Maximum Total Strain Energy Theory**.

Half Section An *Engineering Drawing* and a *Machine Drawing* term; the one-half section of an object or a machine or its component part either symmetrical about a horizontal line or about a vertical line.

Half-Life Period A *Chemistry* term related to *radioactivity*; a characteristic of every radioactive element; the time span during which the radioactivity of the element reduces to one-half of its original value; its value remains constant for each radioactive element; for example, the half-life period of radium (Ra) is 1590 years, whereas for other elements are a few seconds only; see also **Disintegration Constant**.

Hall Effect A *Physics* term; the appearance of a transverse voltage difference on a current carrying conductor when placed perpendicular to a magnetic field; the effect is directly proportional to the magnetic field strength; finds use in making *sensors*.

Hammer Forging A *Manufacturing* term; see also **Smith Forging**.

D. K. Singh, *Dictionary of Mechanical Engineering*,
https://doi.org/10.1007/978-981-99-1722-8_8

Hand File A *Manufacturing* term; one of the types of a file used in the fitting shop which has a *rectangular cross section* and is parallel for its full length; consists of double cut on its both faces and single cut on one edge, and there is no cut on other edge (safe edge); used in the filing of *square corners* of the work, without affecting its already filed adjacent surface.

Hand Rammer A *Manufacturing* term; a *foundry tool* and a *type of rammer* used to ram the sand mass in *bench moulding* which uses a bench; see also **Peen Rammer** and **Floor Rammer**.

Hard Automation An *Automation* term; a process of using *computer-driven automated machines* which is capable of producing a particular type of product only; cannot produce customized products; see also **Soft Automation**.

Hard Glass A *Chemistry* term; see **Potash Glass**.

Hard Mould Casting A *Manufacturing* term related to *casting*; see **Permanent Mould Casting**.

Hardenability A *Manufacturing* term; the ability (property) of a material which enables it to be easily *hardened*; classified into *core hardenability* and *surface hardenability*; *mild steel* has good hardenability.

Hardener A *Chemistry* term; a catalyst which causes polymer macromolecules in a resin to cross-link to each other.

Hardness Test A *Strength of Materials* term; a *destructive test* used to measure the resistance to penetration, indentation, or scratching of the surface of a material by a hard object; a qualitative measure of strength of a material which is based on finding a number, called a hardness number whose higher value is indicative of increased hardness of the material; three important hardness measuring tests include *Brinell*, *Rockwell*, and *Vickers* tests.

Hardness A *Strength of Materials* term; the resistance of a material against scratching or penetration by a pointed tool, an important mechanical property of materials.

Harmonic Mean A *Statistics* term; the inverse of the mean of the reciprocals of each individual item; for example, for *n* items of values from x_1 to x_n, the harmonic mean (a_h) is given as:

$$\frac{1}{a_h} = \frac{1}{n}\left(\frac{1}{x_1} + \frac{1}{x_2} + \cdots \frac{1}{x_n}\right)$$

or

$$a_h = \frac{n}{\sum_{r=1}^{n} \frac{1}{x_r}}$$

Harmonic Motion A *Physics* term; see **Harmonic Oscillation**.

Harmonic Oscillation A *Physics* term; also called **Harmonic Motion**; the oscillation (motion) which can be expressed in terms of *single harmonic function* like *sine function* or *cosine function*; see also **Non-harmonic Oscillation**.

Hart Mechanism A *Theory of Machines* term; a mechanism used to *generate exact straight-line motion*; consists of six links as compared to eight links contained in the *Peaucellier mechanism*, but requires more space; see also **Peaucellier Mechanism**.

Hastelloys A *Materials Engineering* term; a *nickel alloy* group; an alloy of *nickel*, *molybdenum*, and *iron*.

Hatching Lines An *Engineering Drawing* and a *Machine Drawing* term; the *continuous thin parallel lines normally drawn at 45° to the axis or to the main outline* of the section lines; used to show the areas of sections of a component; the separate areas of a section of the same component are shown by identical hatching lines, but the hatching lines with different directions or spacing can be used for showing the sections of adjacent components (usually drawn at 45°, but in opposite direction to those on the part sectioned earlier); the spacing between the hatching lines should be uniform and should be chosen in proportion to the size of the section.

Hawthorne Studies An *Operations Management* term; a series of studies conducted at the *Hawthorne plant* of the *Western Electric Company* of the *USA* during 1927–1932 aimed at improving the production efficiency by considering work environment (the impact of lighting), social group pressure and employee–management relationship which had an important effect on productivity; the study provided new insights into group behaviour of the workers.

HAZ A *Welding* term; see **Heat-Affected Zone**.

HCP Structure A *Chemistry* term; see **Hexagonal Close-Packed Structure**.

Head Loss A *Fluid Mechanics* term; the loss of energy associated with a *fluid flow*; the additional height which a fluid needs to be raised by a pump in order to overcome the frictional losses in a pipe; caused by viscosity; directly related to the wall shear stress of the pipe; expressed in terms of *height* of a liquid column.

Head on Collision A *Physics* term; see **One-Dimensional Collision**.

Head A *Fluid Mechanics* term; a unit of pressure in terms of a column of a liquid; usually expressed as column of *mercury* or *water*; for example, atmospheric pressure is equal to 760 mm of mercury (Hg).

Heading A *Manufacturing* term related to *forming*; a *cold working* process commonly used to produce heads of *screws* and *rivets*; also indicates the title of a topic in general terms.

Headstock A *Lathe* term; a principal part of a lathe which lies on the left side; contains a gear box for operating the spindle at various speeds while machining a workpiece; see also **Tailstock**.

Heart and Square A *Manufacturing* term; a *foundry tool* which resembles a heart-like structure at its one end and a square structure at the other end; used to finish the *mould cavity*.

Heat-Affected Zone A *Welding* term, abbreviated as **HAZ**; the area of a workpiece which lies in close proximity to the fusion zone, where metallurgical changes take place by the heat of welding or cutting.

Heat Capacity A *Physics* term; the product of mass of a body and its specific heat; classified into heat capacity at constant pressure and heat capacity at constant volume; the SI unit is J/K.

Heat Conductor A *Physics* term; also called **Thermal Conductor**; a material which conducts heat; for example, *all metals*; *aluminium* conducts heat about three times faster than *iron*.

Heat Engine A *Thermodynamics* term; a *work-producing device* which delivers net work output using heat input while operating in a *mechanical cycle* (in place of a thermodynamic cycle), as the working fluid does not undergo a complete cycle; examples include *gas turbines* and *car engines*; see also **Heat Pump**.

Heat Exchanger A *Thermodynamics* term; a device used to transfer heat between a heat source (or sink) and a fluid passing through the device or to transfer heat from one fluid to another; a few examples include *evaporators*, *boilers*, and *condensers*; classified into *parallel flow*, *counter flow*, and *crossflow heat exchangers*.

Heat Flux A *Heat Transfer* term; the amount of heat transferred per unit area per unit time from or to a surface of a body; the SI unit is W/m^2; an important parameter used in design systems like gas turbine blades, combustor walls in rocket motors, and locations of supersonic aircraft to avoid the development of excessive temperatures; denoted by q.

Heat Pump A *Thermodynamics* term; a *thermodynamic system* which while operating in a cycle removes heat from a low-temperature body by taking *work input*; see also **Heat Engine**.

Heat Reservoir A *Thermodynamics* term; a body of infinite heat capacity which is capable of absorbing or rejecting an unlimited quantity of heat without undergoing any change in its temperature, because of its extremely large mass; examples include *heat source* and *heat sink*; see also **Heat Source** and **Heat Sink**.

Heat Sink A *Thermodynamics* term; a low-temperature energy reservoir whose temperature remains unchanged, if some heat is added to or rejected from it; used for absorbing heat rejected by a system; a few examples include *atmospheric air*, *river*, and *sea*; see also **Heat Source**.

Heat Source A *Thermodynamics* term; a high-temperature energy reservoir whose temperature remains unchanged, if some heat is added to or rejected from it; used for supplying heat to a system; a few examples include *boiler furnace, combustion chamber, nuclear reactor*, and *sun*; see also **Heat Sink**.

Heat Transfer Coefficient A *Heat Transfer* term; also called **Film Coefficient** or **Surface Conductance**; a parameter which appears as a proportionality constant in the *Newton's law of cooling*; the heat flux from or to a surface which results from the unit temperature difference between the surface and a moving fluid in contact with it; used to find the heat transfer by convection; it is not a property of the fluid, but depends on heat transport properties of the fluid such as thermal conductivity, dynamic viscosity, density, and specific heat as well as on the velocity of flow and characteristic dimensions; denoted by h or f; the SI unit is W/m^2-K.

Heat Transfer Rate The time rate of heat transfer; the SI unit is J/s.

Heat Transfer An interdisciplinary subject of science and an important subject of mechanical engineering related to the transfer of heat energy from one system to another as a result of a temperature difference; most of the engineering design involves the consideration of heat transfer and its prediction for reliable operation of the designed system or product; the mathematical foundation for the study of heat transfer was laid by the work of *Newton, Fourier, Laplace*, and later by the work of *Prandtl, Nusselt*, and others; three basic modes of heat transfer include *conduction, convection*, and radiation.

Heat Treatment A *Manufacturing* term; a process of heating and cooling a metal to improve its structural or physical properties; usually performed to remove stresses caused by *welding, casting,* or *heavy machining*; four important variables which affect any heat treatment process include *carbon content, temperature of heating, time allowed for cooling*, and *cooling medium* (*water, oil,* or *air*); some important heat treatment processes include *annealing, normalizing*, and *tempering*.

Heat A *Thermodynamics* term; a form of energy which can be transferred between two systems (or a system and its surroundings) by virtue of a temperature difference; see also **Temperature**.

Heating Value A *Chemistry* term; the amount of heat released when a unit amount of a fuel is burnt completely at room temperature, and the combustion products are cooled to the room temperature.

Heating A *Thermodynamics* term; a process which raises the temperature of a body.

Heinrich G. Magnus (1802–1870) A *German* scientist who first observed the *Magnus effect* in 1852.

Heisenberg's Uncertainty Principle A *Chemistry* term; see **Uncertainty Principle**.

Helical Gear A *Theory of Machines* term related to a *gear*; a toothed gear used to transmit motion or power between *two parallel shafts*; an improvement over the spur gear, but is more complex and expensive to manufacture; *produces less noise and vibration at higher speeds*, because of smooth engagement and disengagement of teeth; *used in automobile's automatic transmission system*; can be subjected to greater torque and power compared to similarly sized spur gears, as the tooth-to-tooth forces are spread over more surface area that reduce contact stresses.

Helical Spring A *Strength of Materials* term related to a *spring*; also called **Torsion Spring** because of being subjected to torsion; a spring in the form of a helix made of metallic wires of usually circular cross section and shaped in cylindrical or conical form; helical springs having non-circular section find use in heavy-duty applications; conical helical spring finds use, where there is a space problem or variable stiffness is required; cylindrical helical springs are classified into close-coiled helical spring and open-coiled helical spring.

Helmholtz Function A *Thermodynamics* term; a thermodynamic property (other than Gibbs function) used in the analysis of a *chemical system*; named after a *German* physicist *Hermann Ludwig Ferdinand von Helmholtz* (1821–1894), who proposed it in 1882; denoted by A; expressed as:

$$A = U - TS$$

where

U = internal energy of the working fluid
T = absolute temperature of the working fluid
S = entropy of the working fluid; see also **Gibbs Function**.

Henri Fayol (1841–1925) A *French* engineer and the managing director of a large French coal-mining firm who proposed the *functional approach of management*; suggested '*14 principles of management*'; a contemporary of *F. W. Taylor* (1856–1915).

Henri Pitot (1695–1771) A *French* hydraulic engineer who constructed a double-tube device, called *Pitot tube*, used to indicate water velocity through a differential head.

Henry Ford (1863–1947) The founder of the US's *Ford Motor Company*; used the concept of *assembly lines* to produce *automobiles* which involved the chief elements of scientific management such as *standardized product designs, mass production, low manufacturing costs, mechanized assembly lines, specialization of labour*, and *interchangeable parts*.

Henry Gantt (1861–1919) One of the famous disciples of *F. W. Taylor* (1856–1915); an *American* mechanical engineer and a pioneer of *scientific management* who developed the famous *Gantt Chart*, extensively used for the scheduling and control of production operations; also introduced an *incentive system* (*bonus*) for workers to motivate them to complete jobs in less time than the allowed standard.

Heptagon A *Mathematics* term; a *polygon* having seven (7) sides; see also **Polygon**.

Heron's Formula A *Mathematics* term, named after *Hero of Alexandria*; used to find the area of a triangle in terms of its three sides, namely *a*, *b*, and *c*, and is given as:

$$\text{area of } \Delta \text{ ABC} = \sqrt{[s(s-a)(s-b)(s-c)]}$$

where $s = \frac{a+b+c}{2}$ = *semi-perimeter* of Δ ABC; this formula is applicable for all types of triangles, whether it is right-angled, equilateral, or isosceles.

Herringbone Gear A *Gear* term; also called **Double-Helical Gear**; a helical gear used to eliminate thrust loading on support bearings when high horsepower is required to be transmitted; consists of two sets of helical teeth of opposite hand cut side by side on the same gear blank and integrally manufactured on a single pinion.

Hertz A *Physics* term; the number of repetitions of a regular occurrence in 1 s; the SI unit of frequency is *hertz*, which is equivalent to cycles/sec.

Heterogeneous Joining A *Joining* term; see **Heterogeneous Welding**.

Heterogeneous System A *Chemistry* term; a system of *more than one phase*; a few examples include *a mixture of water and steam*, *a mixture of water and ice*, and *a mixture of water and oil*; see also **Homogeneous System**.

Heterogeneous Welding A *Joining* term; also called **Heterogeneous Joining**; a welding process which uses a filler metal whose composition is different than the materials being joined; for example, *soldering* and *brazing*; see also **Homogeneous Welding** and **Autogenous Welding**.

Heuristic Approach Also called **Hungarian Method** or simply **Heuristic**; an *algorithm* or a *probabilistic approach* in which a set of rules is systematically applied to find the solution of many business problems; useful when it is difficult to construct a mathematical model or when the constructed model has no exact solution.

Heuristic See **Heuristic Approach**.

HEV An *Electric Vehicle* term; see **Hybrid Electric Vehicle**.

Hexagon A *Mathematics* term; a *polygon* having six (6) sides.

Hexagonal Close-Packed Structure A *Chemistry* term, abbreviated as **HCP Structure**; one of the three basic types of atomic arrangements found in metals in which the unit cell consists of an average of *six atoms*; the atomic packing factor (APF) equals to 0.74; materials having hcp structure have poor formability (ability to be plastically deformed without fracture); a few examples include *beryllium* (Be), *cadmium* (Cd), *magnesium* (Mg), and *zinc* (Zn); see also **Body-Centred Cubic Structure** and **Face-Centred Cubic Structure**.

HGL A *Fluid Mechanics* term; see **Hydraulic Gradient Line**.

Hierarchy A *Management* term; a classification of people according to the authority and rank in an organization; useful in distributing the tasks and responsibility among the people for effective carrying of the works.

High Carbon Steel A *Materials Engineering* term; a plain carbon steel of high carbon content (0.6–1.75%, although the highest practical limit is 1.3%); has increased hardness and good wear resistance; used in making *cutting tools, cable, springs*, and *cutlery*; see also **Low Carbon Steel** and **Medium Carbon Steel**.

High Grade Energy A *Thermodynamics* term; the energy which can be completely transformed into shaft work without any loss; a few examples include *mechanical work, electrical work, wind energy, tidal energy*, and *kinetic energy of a jet*; see also **Low Grade Energy**.

High Head Turbine A *Fluid Mechanics* term; a hydraulic turbine operating at a high head (100 m and above); for example, *Pelton turbine*; see also **Low Head Turbine** and **Medium Head Turbine**.

High Level Manager A *Management* term; also called **Top Manager**; the individual at the top of an organization; formulates goals and objectives for the organization; responsible for making decisions about the direction of the organization; the different designations used for a high level manager include *chief executive officer* (CEO), *chief operating office*r (COO), *president, vice president, managing director, chancellor, vice chancellor* of a university, or *chairperson* of the board; see also **Middle Level Manager** and **Low Level Manager**.

High-Speed Steel A *Materials Engineering* term, abbreviated as **HSS**; a high-speed cutting tool material invented by *F. W. Taylor* (1856–1915); can maintain its hardness at elevated temperatures (around 650 °C) and has higher wear and abrasion resistance; contains *tungsten, molybdenum, chromium, cobalt*, and *vanadium* as alloying elements; classified into *molybdenum-based HSS* (*M-HSS*) or *tungsten-based HSS* (*T-HSS*); one of its popular varieties is called *18-4-1 HSS* with composition of 18% *tungsten*, 4% *chromium*, and 1% *vanadium*.

High-Temperature Materials A *Materials Engineering* term; see **Superalloys**.

Higher Level Needs A *Management* term related to *motivation*; the human needs which find mention in *Abraham Maslow's Need-Hierarchy Theory of motivation* and are lying at the upper (higher) level of hierarchy which cannot be completely satisfied; the higher level needs include *esteem* and *self-actualization* needs; see also **Lower Level Needs**.

Higher Pair A *Theory of Machines* term; a mechanism in which two links have *point* or *line contact* while in motion; found in a *motion combining sliding and turning*; a few examples include *ball and roller bearings, cams, gears*, and *belt, rope*, and *chain* drives; see also **Lower Pair**.

Hindalium A *Chemistry* term; an alloy of *aluminium* and *magnesium* with a small percentage of *chromium*; used mainly for the manufacture of *anodised utensils*.

Hinge Support A *Physics* and a *Strength of Materials* term; also called **Pin Support**; a type of simple support which prevents movement in horizontal and vertical directions but does not prevent rotation; has two reaction force components in two directions but has no reaction moment; see also **Roller Support**.

Hinged Beam A *Strength of Materials* term related to a *beam*; see **Simple Beam**.

Historical Analogy Method An *Operations Management* term related to *forecasting*; a *qualitative* method of demand forecasting based on analysis and comparison of the growth and development of similar products, called *analogous products*, which already exist; for example, the assumption can be made that four wheeler cars may follow the general sales pattern similar to two wheeler motorcycles; similarly the consumption of *CNG fuel* in automobiles may follow the similar consumption pattern as that of other petroleum fuels like *petrol* and *diesel*.

HML Analysis An *Operations Management* term related to *inventory*; a selective inventory control technique which classifies the inventories based on the unit value of components into three categories: *H* (*high unit-valued*) items, *M* (*medium unit-valued*) items, and *L* (*low unit-valued*) items; this technique lists inventory items in descending order of their unit values and exercises control accordingly; high unit-valued items require more control than low unit-valued item.

Hogging A *Strength of Materials* term related to *bending*; a *negative bending moment* which produces the downward concavity in a beam, as in case of a loaded cantilever beam; see also **Sagging**.

Holding Cost An *Operations Management* term; the cost of holding the *inventory*; holding cost includes costs associated with insurance, security, warehouse rentals, taxes, and losses due to breakage, spoilage, pilferage, and obsolescence; a part of total annual cost of inventory; increases with inventory size; one of the costs that finds use in determining economic order quantity (EOQ); see also **Ordering Cost** and **Setup Cost**.

Hollow Saddle Key A *Design* term related to a *key*; a saddle key which fits into the keyway provided in the hub of the mounting, while its underside which is hollow fits on to the curved surface of the shaft; see also **Flat Saddle Key**.

Homogeneous Body A *Chemistry* term; a body whose mass is distributed uniformly across its volume.

Homogeneous Joining A *Welding* term; see **Homogeneous Welding**.

Homogeneous System A *Chemistry* term; a system of *a single phase*; a few examples include *a mixture of air and water vapour*, *solution of ammonia in water* (*aqua ammonia*), and *dilute acids*; see also **Heterogeneous System**.

Homogeneous Welding A *Joining* term; also called **Homogeneous Joining**; a welding process which uses a filler metal whose composition is similar to that of parent materials being joined; for example, *gas*, *arc*, and *thermit welding processes*; see also **Heterogeneous Welding** and **Autogenous Welding**.

Homogeneous A *Chemistry* term; a state of a material which ensures that all the atoms within the material are uniformly positioned at their designated locations; removes the possibility of defects in a material.

Homopolymer A *Chemistry* term; a polymer made from only one type of monomer; for example, *polyethylene*; see also **Copolymer**.

Honing A *Manufacturing* term; an *abrasive micromachining process* primarily used to size and finish bored holes or to remove the tool marks left by the cutting tools using a rotating honing tool containing abrasive particles; see also **Lapping**.

Hooke's Joint A *Design* term; more frequently called **Universal Joint** or **U-Joint**; a joint used to connect two *non-parallel* shafts which intersect at angles larger than can be accommodated by flexible couplings; classified into *single universal joint* and *double universal joint*; typical applications of U-joints include automotive drive shafts, articulating farm tractors, industrial rolling mill drives, and mechanical control mechanisms.

Hooke's Law A *Strength of Materials* term; also called **Law of Elasticity** or **Theory of Elasticity**; a law proposed by *Robert Hooke* (1635–1703) in 1660 which defines the proportionality of stress and strain within elastic limit; states that within elastic limit (when strain is small), the stress is proportional to the strain; applies to all types of deformations such as tension, compression, bending, and twisting; the proportionality constant is called the modulus of elasticity of the body that depends upon the material of the body and can be different for different types of strain produced in the same material; evolved the definitions of *Young's modulus of elasticity* (E or Y), modulus of rigidity (G), and bulk modulus of elasticity (K).

Hoop Strain A *Strength of Materials* term; see **Circumferential Strain**.

Hoop Stress A *Strength of Materials* term; see **Circumferential Stress**.

Horizontal Centroidal Axis A *Strength of Materials* term; the axis which is horizontal to the cross section of a body; see also **Vertical Centroidal Axis**.

Horizontal Range A *Physics* term related to a *projectile*; the horizontal distance covered by a projectile between its point of projection and the point of hitting the ground; *independent of mass of the projectile*; denoted by R; given as:

$$R = \frac{u^2 \sin 2\theta}{g}$$

$$R_{\max} = \frac{u^2}{g} \quad (\text{for } \theta = 45^\circ)$$

where

u = velocity of projection
θ = angle of projection
g = acceleration due to gravity.

Horse Power A *Physics* term, abbreviated as ***hp***; a unit of power first adopted by a *Scottish* scientist *James Watt* (1736–1819); indicates the power exerted by a horse in pulling a load a non-SI unit of power which is equivalent to moving a 550 lb load a distance of 1 foot in 1 sec.

Hot Extrusion A *Manufacturing* term; an extrusion process carried out at elevated temperatures; used when metals and alloys being extruded lack ductility or a large extrusion force is required; commonly employed for low melting materials such as *aluminium, copper, magnesium, lead, tin,* and *zinc*; see also **Cold Extrusion**.

Hot Forging A *Manufacturing* term; see **Forging**.

Hot Hardness A *Manufacturing* term; see **Red Hardness**.

Hot Shortness A *Manufacturing* term; the ability of a metal to become brittle when it is heated at high temperatures; see also **Cold Shortness**.

Hot Tears A *Casting* term; a type of casting defect which occurs in the form of *ruptures*; produced by the tensile stresses on a casting; most commonly found in *malleable, steel,* and *nonferrous castings.*

Hot-Wire Anemometer A *Fluid Mechanics* term; a thermal device consisting of a very small diameter (a few μm) *tungsten or platinum wire* used to measure the *velocity of a hot fluid (air/gas);* particularly useful for measuring *turbulence* in fluid motion; also used to measure *velocity profile* in the boundary layer.

Hot Working A *Manufacturing* term; a *metal working process* in which the metal before shaping is heated above its recrystallization temperature; a few examples include *forging, hot rolling, hot extrusion,* and *hot pressing*; see also **Cold Working**.

Hot-Chamber Die Casting A *Manufacturing* term related to *casting*; a die casting process in which the melting pot is integral part of the machine setup; used to produce castings from *low melting point alloys* of *zinc, tin,* and *lead*; see also **Cold-Chamber Die Casting**.

Hot-Die Forging A *Manufacturing* term; also called **Isothermal Forging**; a hot metal working process used to shape a heated workpiece by pressing it in a heated die, both maintained at the same temperature.

House of Quality A *Quality Management* term; a part of *quality function deployment* (*QFD*) process which utilizes a planning matrix to relate customer 'wants' (*voice of the customer* (*VOC*)) to 'how' the firm is going to meet those 'wants'.

Household An *Economics* term; a consuming unit in an economy which may consist of a single person or any number of people.

Housekeeping An *Operations Management* term; a process of maintaining an orderly environment at a workstation in order to avoid confusion and errors; a management tool which is a part of 5S, which separates essential items from non-essential items, and thrusts upon the placement of only required items on the shop floor.

hp A *Physics* term; see **Horse Power**.

H-S Diagram A *Thermodynamics* term; see **Mollier Diagram**.

HSS A *Materials Engineering* term; see **High-Speed Steel**.

Hubbing A *Manufacturing* term; a *hot metal working process* used to produce multiple identical cavities economically in a flat metal blank by pressing it against a shaped-hub; useful for plastic moulding and die casting.

Human Error An incorrect act or decision by a person through either ignorance or accident.

Human Resource Planning A *Management* term; a process of determining human resource requirements of an organization.

Human Skill A *Management* term; see **Interpersonal Skill**.

Humid Specific Heat A *Thermodynamics* term related to *psychrometry*; the specific heat of moist air per kg of dry air.

Humidification A *Thermodynamics* term related to *psychrometry*; a process used to increase the moisture content of the air by using either liquid water spray or steam; see also **Dehumidification**.

Humidity Ratio A *Thermodynamics* term related to *psychrometry*; see **Specific Humidity**.

Hungarian Method See **Heuristic Approach**.

Hybrid Electric Vehicle An *Electric Vehicle* term, abbreviated as HEV; a vehicle which uses a combination of an electric motor driven by on-board batteries and a gasoline engine to drive the car; the battery of the vehicle is charged only through regenerative braking and can be used to drive the vehicle for shorter distances (1–2 miles), before the gasoline engine takes over; see also **Plug-In Hybrid Electric Vehicle**.

Hybrid Layout An *Operations Management* term related to *facilities planning*; also called **Combination Layout**; a combination of product and process layouts; has the merits of both layouts but eliminates their demerits; useful for a manufacturing activity involving fabrication and assembly, where fabrication uses process layout and assembly uses product layout; a *retail store* is another example of a hybrid layout; see also **Product Layout** and **Process Layout**.

Hybridization A *Chemistry* term; a process of redistribution of energy in different orbitals of an atom to give new orbitals of equivalent energy (new orbitals are *hybrid orbitals,* and the bond formed by such orbitals is called *hybrid bond*).

Hydel Energy A *Power Plant* term; the energy obtained from water stored in reservoirs at certain heights by constructing a dam on a river or any other sources of water; the falling water is allowed to strike and run the blades of a hydraulic turbine to start a generator for producing electrical power.

Hydraulic Efficiency A *Fluid Mechanics* term; a ratio of the *actual head* to the *theoretical head*; considered because of the internal losses of the machine (pump) occurring in the impeller and in the casing due to friction and separation.

Hydraulic Energy A *Fluid Mechanics* term; the energy associated with flowing water; includes *pressure energy*, *kinetic energy*, and *potential energy* (*datum energy*) of water.

Hydraulic Grade Line A *Fluid Mechanics* term; see **Hydraulic Gradient Line**.

Hydraulic Gradient Line A *Fluid Mechanics* term, abbreviated as **HGL**; also called **Hydraulic Grade Line**; a line which represents the *piezometric head* (the sum of *pressure head* ($P/\rho g$) and *elevation* (*datum*) *head* (z)) of a fluid; see also **Energy Grade Line**.

Hydraulic Jump A *Fluid Mechanics* term; a phenomenon of transition from a rapid flow to a slow flow of a liquid in a relatively short distance; involves an abrupt decrease in velocity, increase in depth, and a substantial head loss; comparable to some extent to that of sudden enlargement in a pipeline; may occur on an inclined or horizontal bed and in a channel of any shape; a hydraulic jump can occur only when the flow is *supercritical*, that is, when the *Froude number* is greater than one.

Hydraulic Press A *Fluid Mechanics* term; a device used to press a part or component of an assembly using *pressure of a fluid* to give it the desired shape; used in diverse applications such as *brakes of automobiles*, *hydraulic, and pneumatic controls of machinery* in industry, *hydraulic lifts*, and *automatic transmission in automobiles*; see also **Mechanical Press**.

Hydraulic Turbine A *Fluid Mechanics* term; also called **Hydro Turbine**; a work-producing device (*turbomachine*) which converts the potential energy of water (the working fluid) into work, which is further used in generating electricity in hydropower plants by rotating the shaft of a generator; the French engineer *Benoit Fourneyron* (1802–1867) developed the first commercially successful hydraulic turbine; a few examples include *Francis*, *Pelton*, and *Kaplan* turbines; see also **Gas Turbine** and **Steam Turbine**.

Hydraulics A *Fluid Mechanics* term; the physical science and technologies associated with liquids that are at rest or flowing under pressure; a subcategory of hydrodynamics which deals with the liquid flows in pipes and open channels.

Hydro Turbine A *Fluid Mechanics* term; see **Hydraulic Turbine**.

Hydrocarbon A *Chemistry* term; an organic compound or a molecule which contains only carbon and hydrogen atoms; produces carbon dioxide and water on burning; classified into *saturated* and *unsaturated* hydrocarbons; saturated hydrocarbons have single bonds between carbon atoms, whereas unsaturated hydrocarbons have double or triple bonds between carbon atoms; a few examples of saturated hydrocarbons include methane (CH_4), ethane (C_2H_6), propane (C_3H_8), etc.; a few examples of unsaturated hydrocarbons include ethylene (C_2H_4) and acetylene (C_2H_2).

Hydrodynamic Force A *Fluid Mechanics* term; a force which acts on a fluid in motion; see also **Hydrostatic Force**.

Hydrodynamic Lubrication A *Design* term; a type of lubrication which uses *oil pressure* to take up the load and does not allow metal-to-metal contact; for example, the *journal bearing* works on the principle of hydrodynamic lubrication.

Hydrodynamic Machining A *Nonconventional Machining* term; see **Water Jet Machining**.

Hydrodynamics A *Fluid Mechanics* term; see **Fluid Dynamics**.

Hydromagnetics A *Fluid Mechanics* term; see **Magneto-Fluid Mechanics**.

Hydrometer A *Fluid Mechanics* term; an instrument used to measure the *specific gravity of liquids*; operates on the principle of *buoyancy*.

Hydrostatic Extrusion A *Manufacturing* term; a cold metal working process used to shape a metal blank (*billet*) confined in a fluid chamber which exerts pressure on it through a ram; involves less extrusion force than conventional extrusion process due to nonphysical contact between billet and its chamber.

Hydrostatic Force A *Fluid Mechanics* term; a force which acts on a fluid at rest; see also **Hydrodynamic Force**.

Hydrostatic Pressure A *Fluid Mechanics* term; a pressure which depends on the reference level selected; equals to $\rho g h$ (where $\rho =$ density of the fluid, $g =$ acceleration due to gravity $= 9.81\ \text{m/s}^2$, and $h =$ elevation/depth from the reference level); increases with the fluid depth; see also **Static Pressure**, **Dynamic Pressure**, and **Stagnation Pressure**.

Hydrostatics A *Fluid Mechanics* term; see **Fluid Statics**.

Hygiene Factors A *Management* term related to *motivation*; see **Extrinsic Factors**.

Hygiene Theory A *Management* term related to *motivation*; a motivation theory proposed by *Frederick Herzberg* (1923–2000); also called **Motivator-Hygiene Theory** or **Two-Factor Theory** as this theory divides five levels of human needs as proposed by *Abraham Maslow* in his '*Need-Hierarchy Theory*' into two levels: the lower level needs and the higher level needs; the *lower level needs* (also called *extrinsic factors* or *dissatisfiers* or *hygiene factors*) are related to pay and working conditions such as *company policies, supervision, interpersonal relations, salary, status,* and *security*, whereas the *higher level needs* (also called *intrinsic factors* or *motivation factors* or *motivators* or *satisfiers*) include *achievement*, *recognition* (for work well done), *job interest*, *responsibility*, *growth*, *advancement*, and *belonging to a group.*

Hygrometer A *Thermodynamics* term related to *psychrometry*; an instrument used to measure the water vapour content in the air.

Hyper-Eutectoid Steel A *Materials Engineering* term; a state of phase of steel in *iron-carbon phase diagram* that contains more than 0.77% *carbon* (highest being 2.11%); see also **Hypo-Eutectoid Steel**.

Hypersonic Flow A *Fluid Mechanics* term; a fluid flow for which the *Mach number* (M) is greater than 7; see also **Supersonic Flow**.

Hypocycloid A *Theory of Machines* term related to a *gear*; the locus of a point on the circumference of a circle, which rolls without slipping inside another circle of a definite radius; *used in making tooth profile of a gear*; see also **Epicycloid**.

Hypo-Eutectoid Steel A *Materials Engineering* term; a state of phase of steel in *iron-carbon phase diagram* that contains less than 0.77% *carbon*; see also **Hyper-Eutectoid Steel**.

Hysteresis A *Physics* term; the difference in the output for a given input when this value is approached from the opposite direction.

I

IC Engine A *Thermodynamics* term; see Internal Combustion Engine.

Ice Point A *Thermodynamics* term; the temperature at which water freezes (or ice melts); a state of equilibrium between ice and water at a pressure of one atmosphere (101 kPa); equals to 0 °C or 32 °F; see also **Steam Point**.

Ideal Fluid A *Fluid Mechanics* term; a fluid which does not involve friction and is inviscid (its viscosity is zero); does not exist in reality; many fluids approximate frictionless flow at sufficient distances from solid boundaries and are considered ideal fluids; see also **Real Fluid**.

Ideal Gas Equation A *Chemistry* term; see **Equation of State**.

Ideal Gas A *Chemistry* term; see **Perfect Gas**.

Idler Gear A *Theory of Machines* term related to a *gear*; an intermediate wheel gear which has no effect on the gear train value or the velocity ratio of the gear train system; used in *simple gear trains* when distance between two gear shafts is large.

IHP An *IC Engine* term; the short form of **Indicated Horse Power**; see **Indicated Power**.

Immediate Predecessor A *Project Management* term; an activity which must immediately precede a given activity in a project.

Impact Extrusion A *Manufacturing* term; a cold metal working process used to produce *collapsible tubes* from a metal blank kept in a die using a strong single blow (through a punch), which causes the metal to flow plastically around the punch; suitable for metals having good ductility like *zinc*, *lead*, *tin*, and *aluminium*; typical impact extruded products include *medicine tubes*, *toothpaste tubes*, *shaving cream tubes*, and *food cans*.

D. K. Singh, *Dictionary of Mechanical Engineering*,
https://doi.org/10.1007/978-981-99-1722-8_9

Impact Load A *Physics* term; also called **Shock Load** as it produces shocks; a *dynamic load* which is applied and removed suddenly or a load falling freely through a certain height; produced when two objects collide or when a falling object strikes another object; the work done (W) by a falling load is given as:

$$W = P(h + \delta L)$$

where

P = free falling load
h = height through which the load falls
δL = linear deflection produced; see also **Fluctuating Load.**

Impact Test A *Strength of Materials* term; a *destructive* test used to measure the *fracture toughness* of solid materials; particularly useful in finding the ductile–brittle characteristic of solid materials; there can be two types of impact test: *Charpy test* and *Izod test*.

Impedance An *Electrical* term; the combination of *resistance* and *reactance* which opposes the flow of current in an *alternating current* (AC) circuit; denoted by Z; the SI unit is ohm (Ω).

Impeller Blade A *Fluid Mechanics* term related to a *pump*; also called simply **Impeller**; the rotating blade of a pump; see also **Bucket.**

Impeller A *Fluid Mechanics* term related to a *pump*; see **Impeller Blade.**

Impression-Die Forging A *Manufacturing* term; see **Closed-Die Forging.**

Improper Fraction A *Mathematics* term; a fraction whose numerator is either equal to or greater than the denominator; a few examples include $\frac{2}{2}$, $\frac{5}{4}$, $\frac{9}{4}$, etc.; see also **Proper Fraction.**

Improvement Curve A *Management* term; see **Learning Curve.**

Impulse Turbine A *Fluid Mechanics* term; a *hydraulic turbine* in which a high-velocity fluid jet coming out of a nozzle impinges on the blades of the rotor and exerts force on it by impulsive action under atmospheric pressure conditions; involves no pressure change across the rotor blades; fluid is not required to flow through the entire periphery of the runner as is required in a reaction turbine; *Pelton wheel* is the example of an impulse turbine; see also **Reaction Turbine.**

Impulse Wheel A *Fluid Mechanics* term; see **Pelton Turbine.**

Impulse A *Physics* term; the product of a force acting on a body and its time duration; equals to the change in momentum of the body; a *vector quantity*; the SI unit is newton-second (N-s); a cricketer stretches his hand behind to reduce the impact of a cricket ball so as to get less hurt, while catching it, by increasing its time duration, as impulse is conserved.

Impulsive Force A *Physics* term; also called **Blow**; a force which acts over a body for a very short interval of time and has a *very large magnitude.*

Incentive A *Management* term; an inducement or reward given to workers for their efficiency and hard works; used as a tool to motivate and encourage workers to produce more and better.

Inch Another FPS unit of length; the one-twelfth part of a foot, that is, 1 in. = $\frac{1}{12}$ foot = 2.54 cm; see also **Foot, Centimetre,** and **Millimetre**.

Inclined Plane A *Mathematics* and a *Physics* term; a plane inclined at a certain angle to the horizontal; a stationary flat surface set at an angle and used to bridge two planes which are offset by some vertical distance.

Incomplete Combustion An *IC Engine* term; a combustion process in which some unburnt fuel remains in the products even after the process is over, because of deficiency of oxygen in the reactants which leads to the formation of C, CO, H_2, or OH; see also **Complete Combustion**.

Incompressible Fluid A *Fluid Mechanics* term; a fluid which cannot be compressed and whose density is independent of its pressure; water or any other liquid is usually incompressible in nature; see also **Compressible Fluid**.

Inconels A *Materials Engineering* term; a *nickel alloy group* which is higher in *nickel* and *iron* content than the monels; see also **Monels**.

Incremental Cost An *Operations Management* term related to *cost analysis*; see **Marginal Cost**.

Independent Demand An *Operations Management* term; the demand for an item which is not related to the demand for any other item; for example, the demand of *finished products* and *spare parts*; see also **Dependent Demand**.

Independent Variable A *Mathematics* term; a variable quantity which can have any value within the specified limits, and that does not depend on other variables; see also **Dependent Variable**.

Indicated Horse Power An *IC Engine* term, abbreviated as **IHP**; see **Indicated Power**.

Indicated Power An *IC Engine* term, abbreviated as **IP**; also called **Indicated Horse Power**; the power developed inside the cylinder of an engine; greater than the brake power (brake horsepower); given as:

$$\mathrm{IP} = \frac{PLA\left(N \text{ or } \frac{N}{2}\right)n}{60{,}000}\,\mathrm{kW}$$

where

P = mean effective pressure (N/m^2)
L = stroke length of the piston or length of the cylinder (m)
A = cross-sectional area of the cylinder (m^2) = $\pi D^2/4$
D = diameter of the cylinder (m)
N = number of revolutions per minute (rpm) of the crankshaft (N in case of a two-stroke engine and $N/2$ in case of a four-stroke engine)
n = number of cylinders in the engine.

also, $IHP = \frac{PLA(N \text{ or } \frac{N}{2})n}{4500}$ (P in Kgf.m); see also **Brake Power**.

Indicator Diagram An *IC Engine* term; the plot between *pressure* and *volume* of a working fluid used in an engine; the area of the indicator diagram represents the magnitude of the net work done by the system in one engine cycle and is measured by an instrument called *planimeter*.

Indirect Cost An *Operations Management* term related to *cost analysis*; the indirect costs are not attributable to a specific output or work activity; for example, the *costs of common tools*, *general supplies* like *grease*, *oils*, etc., and *equipment maintenance* in a plant are treated as *indirect costs*; see also **Direct Cost**.

Indirect Extrusion A *Manufacturing* term; also called **Backward Extrusion** or **Reverse Extrusion** or **Inverted Extrusion**; an extrusion process in which a shaped-die is pushed towards the billet, and metal comes out in the opposite direction to the force applied in the form of required product; see also **Direct Extrusion**.

Induced Radioactivity A *Chemistry* term related to *radioactivity*; a process in which a heavy nucleus formed during a nuclear reaction starts disintegrating like a natural radioactive element as being unstable.

Induction Furnace A type of *electric furnace* which works on *induction principle* to produce heat; produces the highest temperature of 1750 °C; particularly suitable for *small foundries*; classified into *coreless induction furnace* and *core* (or *channel*) *induction furnace*.

Induction Motor An *Electrical* term; see also **Synchronous Motor** and **BLDC Motor**.

Inductor A *Physics* term; an electrical device which makes use of the magnetic effect of electric current.

Industrial Engineering A branch of engineering related to the *design*, *improvement*, and *installation* of integrated systems of *men*, *materials*, and *equipment* usually found in an industry.

Inelastic Collision A *Physics* term; the collision involving some loss of kinetic energy, that is, the kinetic energy is not conserved, but the linear momentum as well as the total energy of the system both are conserved; some or all the forces involved in an inelastic collision may be non-conservative in nature; the collisions we come across in our daily life are generally inelastic; see also **Elastic Collision.**

Inelastic Demand An *Economics* term; the demand for a product or service which does not proportionately increase or decrease with the change in its price; that is, the demand is relatively static even if the price changes; for example, the demand of the essential items such as *food grains*, *fruits* and *vegetables*, *clothing*, *drugs*, etc.; see also **Elastic Demand.**

Inelastic Strain Energy A *Strength of Materials* term; the strain energy lost in the process of permanently deforming a member; see also **Elastic Strain Energy**.

Inert Gas A *Chemistry* term; a gas which does not chemically combine with other elements; for example, *helium* (He), *argon* (Ar), *krypton* (Kr), and *xenon* (Xe); *helium* and *argon* are used in welding operations to provide a shielding atmosphere around the welding arc to protect the molten metal pool from atmospheric contamination.

Inert A *Chemistry* term; the condition of having limited ability to react chemically with other elements.

Inertia Governor A *Theory of Machines* term; a *speed-control device* which uses *suspended masses* (*balls*) that respond to speed variations due to their inertia.

Inertia A *Physics* term; a property of a body which enables it to resist any change in its present state (rest or uniform motion); evolved from Newton's first law of motion.

Inertial Mass A *Physics* term; the mass of a body *related to its inertia* in linear motion as defined by *Newton's second law of motion*; equals to the magnitude of external force required to produce unit acceleration in the body; opposes motion along a straight line caused by an external force; does not depend on *shape*, *size*, and *state* of the body, as well as, its temperature, but is proportional to the quantity of matter contained in the body; not affected by gravity or by the presence/absence of other nearby bodies, but increases with speed of the body, irrespective of its material; can be added by simple laws of algebra; given as:

$$m = \frac{m_0}{\sqrt{1 - \frac{v^2}{c^2}}}$$

where

m = inertial mass of the body
m_o = rest mass of the body
v = velocity of the body
c = velocity of light in vacuum; see also **Gravitational Mass.**

Inexact Differential A *Thermodynamics* term; the differential of a *path function* whose integral depends on the path between the end points; denoted by the symbol ∂; for example, work done in a process is represented by ∂W; see also **Exact Differential.**

Infinitesimal Element A *Design* term; an element of a body which has a very small area; used in finding the total force or moment on the body.

Inflexion Point A *Strength of Materials* term; see **Point of Contraflexure.**

Informal Organization A *Management* term; a network of interpersonal relationships which arise when people associate with each other; a few examples include the *machine shop group*, the fifth floor crowd, and the *morning coffee regulars*; see also **Formal Organization.**

Information System A system of information needed to be shared between different units of an organization for its successful operation; its elements include *hardware*, *software*, and *communication networks* such as *interne*t, *intranets* and *extranets*, and *data resources*; see also **Information Technology.**

Information Technology A technology which enables *communication, storage*, and *processing* of *information* within and between firms; see also **Information System.**

Infrared Thermometer A *Physics* term; a thermometer commonly used in *high-temperature applications* such as *furnaces* and *kilns*, where other types of thermometers would not be possible.

Infrasonic Wave A *Physics* term; the *longitudinal wave* having frequency less than 20 Hz; the human ear is not sensitive to this wave; usually produced by large vibrating bodies such as during an *earthquake*; see also **Ultrasonic Wave.**

Initiative A *Management* term; a process which allows to originate and carry out plans to ensure their success.

Innovation A *Management* term; a process of using a creative idea and putting it into practice to produce useful products, services, or work methods; science and technology play key role in innovation; also affected by the organizational design, for example, the way an organization's structure links its people in different specializations such as research and marketing, determines how fast the organization can introduce new products in the market; closely related to invention; see also **Invention.**

Inorganic Compound A *Chemistry* term; a chemical compound which lacks *carbon-hydrogen* bonds; see also **Organic Compound.**

Inorganic Materials A *Chemistry* term; the materials that *do not contain animal or vegetable cells* or *carbon compounds*; dissolve in water, and have better heat-resisting characteristics than organic materials; see also **Organic Materials.**

Inside Stakeholders A *Management* term; people who are closest to an organization and who have the strongest or the most direct claim on the organizational resources; include *shareholders, managers*, and the *workforce*; see also **Outside Stakeholders**.

Inspection A *Metrology* term; a process of observations and measurements of input and output parameters of a system.

Instantaneous Acceleration A *Physics* term; the acceleration calculated at a particular point or at a particular instant; when the acceleration is constant, the instantaneous acceleration is equal to the average acceleration.

Instantaneous Speed A *Physics* term; the speed of an object at a given instant of time; equals to its uniform speed in case of a uniform motion.

Insulator An *Electrical* and a *Heat Transfer* term; in *electrical* terms, a material which offers extremely high resistance to the current flow or has low electrical conductivity; for example, ceramic materials; in *heat transfer* terms, a material which does not allow the transfer of heat energy through it; for example, *glass wool, rock wool*, and *loose asbestos packing* are the effective heat insulators as air is entrapped in the small cells of these materials.

Intake Valve An *IC Engine* term; a valve through which air or air–fuel mixture is drawn into the cylinder of an engine; see also **Exhaust Valve**.

Intangible Factors A *Statistics* term; see **Qualitative Factors**.

Integer Linear Programming Problem An *Operations Research* term; see **Transportation Problem**.

Integer A *Mathematics* term; any positive or negative number including zero (0) such as − 3, − 2, − 1, 0, 1, 2, and 3; numbers such as 1, 2, and 3 are called *positive integers*, and numbers such as − 1, − 2, and − 3 are called *negative integers*.

Integrating A *Management* term; also called **Coordinating**; one of the important functions of a manager; a process of coordinating various tasks, functions, and divisions to make them directed towards the organizational goals; in the absence of it, the organization cannot achieve the desired results or outcome, even if a well-planned work has been nicely organized.

Intensive Property A *Thermodynamics* term; a property which is independent of mass of a system; a few examples include *temperature, pressure, density, velocity*, and *height*; see also **Extensive Property**.

Interference A *Design* term; used in the context of mating of cylindrical parts in which the internal member is larger than the external member; the opposite of *clearance*; see also **Clearance**.

Intermediate-Length Column A *Strength of Materials* term related to a *column*; see **Medium Column**.

Intermetallic Compound A *Chemistry* term; a compound formed by the combination of two or more metallic elements; is very hard and brittle; has mechanical properties similar to that of ceramic materials.

Intermittent Something which occurs at irregular intervals.

Intermolecular Force A *Chemistry* term; a *short-range force* acting over short distances between molecules of similar or dissimilar substances; classified into *adhesive* and *cohesive* forces.

Internal Chill A *Manufacturing* term related to *casting*; a metallic object placed *within the mould* cavity to absorb heat and promote more rapid solidification; must be made of the same material as that of casting, because it ultimately *becomes part of the final casting*; see also **External Chill.**

Internal Combustion Engine Abbreviated as **IC Engine**; a non-cyclic heat engine where combustion of the fuel–air mixture (the working fluid) takes place inside the cylinder of the engine; petrol and diesel engines are the two commonly used IC engines; used in *automobiles, scooters, motorbikes, power boats, aircraft, agricultural machinery,* and *earth moving equipment.*

Internal Energy A *Thermodynamics* term; the energy stored in a substance which results from the internal motion of its atoms and molecules; exists in many forms including *thermal*, *nuclear*, *chemical*, and *electrostatic*; the internal energy of an ideal gas is a function of *temperature* alone; denoted by U.

Internal Expanding Shoe Brake A *Theory of Machines* term; a type of brake which consists of two *semi-circular shoes* lined externally with a frictional material (ferodo) which are pressed against the inner flange of the drum at the time of braking; commonly used in *motor cars* and *light trucks*; see also **Band Brake, Block Brake,** and **Band and Block Brake**.

Internal Failure Cost A *Quality Management* term; a part of the cost of quality related to the defects *within* the system; includes *costs for scrap*, *repair*, and *rework*; see also **External Failure Cost.**

Internal Flow A *Fluid Mechanics* term; see **Duct Flow**.

Internal Force A *Physics* term; a force which results from the interaction between the particles of the constituent material; a force which acts in the connections between members; see also **External Force**.

Internal Friction A *Fluid Mechanics* term; see **Viscosity**.

Internet A *Computer* term; the short form of *International Network*; the computer network gateway to websites; *an extensively used tool of information technology.*

Interpersonal Skill A *Management* term; also called **Human Skill**; a *management skill* proposed by *Robert L. Katz* (1933-2010) which relates to a manager's ability to *interact* (work) with and understand other people (individually or in groups) in order to *communicate, motivate*, and *delegate.*

Interpolation A *Mathematics* term; the calculation of a parameter within the range of observations; see also **Extrapolation.**

Interstitial Atom A *Materials Engineering* term; also called **Interstitial** or **Interstitialcy**; an atom which occupies an interstitial site (position normally not occupied by an atom in a perfect crystal).

Interstitial Defect A *Materials Engineering* term; a defect in a material which occurs when an extra atom or ion is inserted into a crystal structure at a normally unoccupied position (*interstitial site*).

Interstitial A *Materials Engineering* term; see **Interstitial Atom.**

Interstitialcy A *Materials Engineering* term; see **Interstitial Atom.**

Intrinsic Factors A *Management* term related to *motivation*; also called **Motivation Factors** or **Motivators** or **Satisfiers**; one of the two factors of *Frederick Herzberg* (1923–2000)'s '*Two-Factor Theory*' of motivation; include *higher level needs* which form *internal factors* relating to human being's characteristics such as *achievement, recognition, job interest, responsibility, growth*, and *advancement*; their absence does not create satisfaction or dissatisfaction; see also **Extrinsic Factors.**

Intrinsic Semiconductor An *Electronics* term; an *ideal semiconductor* which is chemically pure with no lattice defects and no impurity; belongs to group VI in the periodic table; exhibits diamond cubic crystal structure; its conductivity strongly depends on temperature; examples include *germanium* (Ge) and *silicon* (Si); see also **Extrinsic Semiconductor.**

Intrusive Method A *Metrology* term; also called **Probe Method**; the measurement of a physical quantity which uses a *probe*, which is placed inside the system; see also **Non-intrusive Method.**

Invar A *Materials Engineering* term; an alloy of *nickel* (Ni) and *iron* (Fe) containing about 36% Ni; the coefficient of linear thermal expansion (α) of *invar* is approximately equal to zero which makes it suitable for being used in the best grades of *surveyor's tapes, watch springs, pendulums* for clocks, and other precision instruments.

Invention A *Management* term; a process of finding something new; pertains to new ideas and processes; see also **Innovation.**

Inventory File An *Operations Management* term; also called **Inventory Status File**; one of the three components of *material requirements planning* (MRP); a computerized file with a complete record of each material held in inventory.

Inventory Management An *Operations Management* term related to *inventory*; a process of managing the inventory in a firm in order to minimize the investment in inventory and also to ensure its optimal availability for the smooth functioning of the firm; more relevant for manufacturing organizations.

Inventory on Hand An *Operations Management* term related to *inventory*; the unused inventory carried over from the previous period.

Inventory Status File An *Operations Management* term; see **Inventory File**.

Inventory An *Operations Management* term: quantities of goods in stock; available in three forms, namely *raw materials*, *semi-finished goods* (*work-in-process*), and *finished goods*; an essential part of a production system.

Inversion Line A *Thermodynamics* term; a line which passes through the maximum temperature points of the constant enthalpy lines for a substance; the Joule–Thomson coefficient is positive to the left of the inversion line and is negative to the right of the this line; expansions that occur to the left of the inversion line between states of equal enthalpy result in a decrease in temperature, whereas expansions occurring to the right of the inversion line result in a temperature rise.

Inverted Extrusion A *Manufacturing* term; see **Indirect Extrusion**.

Investment Casting A *Manufacturing* term related to *casting*; also called **Lost-Wax Process** or **Lost-Pattern Casting** as the wax pattern is lost during the process or **Precision Casting** because of the high dimensional accuracy of the cast parts; a casting process which uses an *expandable wax pattern* which gets melted by the heat of the molten metal, thus creating a space for it to act as mould cavity; an expensive process used to produce *small precision parts of intricate design* from both *ferrous* and *nonferrous* alloys such as *jewellery items* and *dental caps*; the name '*investment*' is used to indicate the application of a *refractory material slurry* over the pattern consisting of very *fine-grained silica*, *ethyl silicate*, *water*, and *acids*.

Inviscid Fluid A *Fluid Mechanics* term; a fluid having *zero viscosity*; an ideal fluid is an inviscid fluid.

Involute A *Theory of Machines* term related to a *gear*; a parameter used to define the geometrical shape of a gear tooth; the locus of a point on a straight line which rolls on the circumference of a circle without slipping; see also **Cycloid**.

Ion A *Chemistry* term; a *charged particle* which results when an electron is added to or removed from a neutral atom or molecule; classified into *Anion* and *cation*; a few examples include *sodium ion* (Na^{+1}), *calcium ion* (Ca^{+2}), *chlorine ion* (Cl^{-1}), and *ammonium ion* ($NH_4{}^{+1}$).

Ionic Bond A *Chemistry* term; also called **Electrovalent Bond**; a chemical bond formed by the complete transfer of one or more electrons from the outermost energy shell (valence shell) of one atom to the outermost energy shell of the other atom;

a bond between a *cation* (a positive ion) and an *anion* (a negative ion); losing of electrons by an atom and gaining of these electrons by another atom help the two atoms to acquire the stable electronic configuration of the nearest noble gases; see also **Covalent Bond**.

Ionic Compound A *Chemistry* term; also called **Electrovalent Compound**; a compound formed as a result of an *ionic bond*; *does not conduct electricity in solid state, but conducts electricity in the molten state*; has a *high melting* and *boiling point*, but has a *low volatility*; freely soluble in water, but is only slightly soluble in organic solvents; see also **Covalent Compound**.

Ionization Energy A *Chemistry* term; the amount of *energy required to remove an electron* from an isolated gaseous atom of an element, resulting in the formation of a *positively charged ion* (*cation*); the smaller the value of ionization energy, the easier it is for the neutral atom to change into a cation; also there are second, third, or higher ionization energies of the element corresponding to elimination of one, two, or more electrons from the cations produced earlier; see also **Electron Affinity**.

IP An *IC Engine* term; see **Indicated Power**.

Iron Ore A *Chemistry* term; an ore of iron that is combined with oxygen and other elements and is mixed with rocks, clay, and sand; a few examples include *magnetite* (Fe_3O_4), *haematite* (Fe_2O_3), *limonite*, and *siderite*.

Iron A *Chemistry* term; a *metallic element* having symbol Fe with atomic number 26; the most abundant and *the most useful of all metals*; does not exist in free state except in *basalts* and *meteorites* in limited quantities; its important ores include *magnetite* (Fe_3O_4), *haematite* (Fe_2O_3), and *iron pyrites* (FeS_2); a greyish white lustrous ferrous metal which is *malleable* and *ductile* and has a very *high tensile strength*; melts at 1536 °C and boils at about 3000 °C; one of the heaviest metals with specific gravity 7.86; possesses strong magnetic properties (called ferromagnetism) which is lost when heated above Curie point (766 °C); its three allotropic forms include α-iron, δ-iron, and γ-iron which differ from one another in crystalline shape only; its three commercial varieties include *cast iron*, *wrought iron*, and *steel*; finds extensive industrial applications, perhaps most of all other metals; an *essential component of blood* (*haemoglobin*).

Ironing A *Manufacturing* term; a cold metal working process used to produce uniform thickness by straightening the walls of the drawn cup obtained in *deep drawing*.

Irreversible Cycle A *Thermodynamics* term; a cycle consisting of irreversible processes; see also **Reversible Cycle**.

Irreversible Engine A *Thermodynamics* term; an engine which operates on irreversible thermodynamic processes and has losses; see also **Reversible Engine**.

Irreversible Process A *Thermodynamics* term; a process in which a system passes through a sequence of non-equilibrium states; cannot be exactly restored to its initial state after the process has taken place as its initial and final states differ; occurs only in one direction; all the spontaneous processes are irreversible; friction, heat transfer across a finite temperature difference or spontaneous chemical reaction makes a process irreversible; a few examples include *throttling, free expansion, adiabatic mixing*, and *plastic deformation*; see also **Reversible Process.**

Irrotational Flow A *Fluid Mechanics* term; also called **Potential Flow**; a fluid flow in which the fluid particles do not rotate about their mass axes; the flow of an ideal fluid is an irrotational flow; for an irrotational flow, the vorticity is zero; the condition of irrotationality for a plane flow in *x–y* plane is given as:

$$\frac{\partial u}{\partial y} - \frac{\partial v}{\partial x} = 0$$

where u and v are the respective velocities of flow in *x*- and *y*-directions; see also **Rotational Flow.**

I-Section A *Strength of Materials* term; one of the most commonly used cross sections of structural members; resembles the capital English alphabet *I*.

Isentropic Compression A *Thermodynamics* term; a *reversible adiabatic compression* process which does not allow any heat transfer (neither heat addition nor heat rejection) and for which the entropy change, $\Delta S = 0$; follows the equation PV^{Υ} = constant (where P = pressure, V = volume, and Υ = adiabatic exponent = C_p/C_v = 1.4); see **Isentropic Expansion.**

Isentropic Efficiency A *Fluid Mechanics* and a *Thermodynamics* term; a ratio of the ideal work associated with an adiabatic and frictionless compression process to the actual work required to achieve the same total pressure rise.

Isentropic Expansion A *Thermodynamics* term; a *reversible adiabatic expansion* process which does not allow any heat transfer (neither heat addition nor heat rejection), and for which the entropy change, $\Delta S = 0$, and follows the equation PV^{Υ} = constant (where P = pressure, V = volume, and Υ = adiabatic exponent = C_p/C_v = 1.4); see **Isentropic Compression.**

Isentropic Process A *Thermodynamics* term; a *reversible adiabatic process* in which entropy remains constant, that is, the change in entropy for an isentropic process, $\Delta S = 0$; see also **Adiabatic Process.**

Ishikawa Diagram A *Quality Management* term; see **Cause-and-Effect Diagram.**

ISO 14000 A *Quality Management* term; an international quality standard for an *environment management system*; consists of five series of standards which cover a number of areas such as keeping track of the use of raw materials, their generation, treatment, and disposal of hazardous wastes.

ISO 9000 A *Quality Management* term; it is the *international standard for quality certification* developed in 1987 by the *International Organization for Standardization* (ISO), based in Geneva, Switzerland; it defines quality system standards and consists of a series of international quality management standards (from 9001 to 9004), which set uniform guidelines for processes to ensure that products conform to customer requirements; compliance with ISO 9000 standards says nothing about the actual quality of a product; the ISO 9000 standards were first revised in 1994 and again (significantly) in 2000, and the later was referred to as *ISO 9000:2000*; a major revision to ISO 9000 was completed in December 2008; the new standard is commonly referred to as *ISO 9000:2008*; plays a major role in setting quality standards for global manufacturers.

ISO 9001 A *Quality Management* term; the *most comprehensive* ISO 9000 *quality standards* difficult to attain; focuses on 20 aspects of a quality programme relating to *design, production, installation,* and *servicing* of products and includes *management responsibility, quality system documentation, purchasing, product design, inspection, training,* and *corrective action*; the latest modifications of the standard *ISO 9001:2015* follow a structure which makes it more compatible with other management systems; the modified version *ISO 9001:2015* gives greater emphasis to risk-based thinking, attempting to prevent undesirable outcomes.

ISO 9002 A *Quality Management* term; this standard covers the same areas as ISO 9001 for companies that produce the customer's designs or have their design and service activities at another location.

ISO 9003 A *Quality Management* term; this standard is used only for *inspecting* and *testing procedures.*

ISO 9004 A *Quality Management* term; the section of ISO 9000 quality standards which contains guidelines for *interpreting* the other quality standards and to provide *detailed guidance* to a company for the continual improvement of its quality management system in order to achieve and sustain customer satisfaction.

Isobaric Process A *Thermodynamics* term; also called **Isopiestic Process**; a process which occurs at constant pressure; the boiling of water to steam or the freezing of water to ice occur at constant pressure, which are examples of isobaric processes; the work done during an isobaric process undergoing between two states 1 (initial state condition) and 2 (final state condition) is given as:

$$W_{1-2} = \int_1^2 p\mathrm{d}V = p(V_2 - V_1).$$

Isochoric Process A *Thermodynamics* term; also called **Isometric Process**; a process which occurs at constant volume; the heat added or rejected during this process is only used in increasing or decreasing the internal energy of a system; the work done during an isochoric process is zero; the work done during an isochoric process undergoing between two states 1 (initial state condition) and 2 (final state condition) is given as:

$$W_{1-2} = \int_1^2 p\mathrm{d}V = p(V_2 - V_1) = 0.$$

Isolated System A *Thermodynamics* term; a thermodynamic system of fixed mass and fixed energy; does not allow mass or energy transfer across its boundary and also does not interact with its surrounding; see also **Open System** and **Closed System**.

Isometric Process A *Thermodynamics* term; see **Isochoric Process**.

Isopiestic Process A *Thermodynamics* term; see **Isobaric Process**.

Isosceles Trapezium A *Mathematics* term; a trapezium whose non-parallel sides are equal.

Isosceles Triangle A *Mathematics* term; a triangle whose any two sides are equal; see also **Scalene Triangle** and **Equilateral Triangle**.

Isothermal Forging A *Manufacturing* term; see **Hot-Die Forging**.

Isothermal Process A *Thermodynamics* term; a thermodynamic process which occurs at a constant temperature; follows the equation, PV = constant (where P = pressure and V = volume); the Boyle's law is valid for an isothermal process.

Isotropic Material A *Materials Engineering* term; exhibits uniform properties throughout in all directions; see also **Anisotropic Material** and **Orthotropic Material**.

Isotropic A *Materials Engineering* term; the characteristics of having properties (of a material) not varying with direction or having similar properties in every direction; for example, most crystalline materials exhibit isotropic properties; one of the important assumptions applied in the validity of *flexure formula* and *torsion formula*; see also **Anisotropic**.

Izod Test A *Strength of Materials* term; a least-used *impact test* which is conducted using a swinging pendulum to apply an impact load on a notched test specimen of size 75 mm × 10 mm × 10 mm supported as a cantilever beam, in which the load breaks the test specimen at its notched portion and the energy stored in the specimen is measured; see also **Charpy Test**.

J

J. B. Johnston Formula A *Strength of Materials* term related to *column*; see **Parabolic Formula**.

Jasper A *Chemistry* term; a *naturally occurring amorphous form of silica*; a coloured (red) variety of quartz used as a gem; see also **Amethyst** and **Opal**.

Jean le Rond d'Alembert (1717–1783) A *French* mathematician and philosopher who is known for *d'Alembert's paradox*; he was the first to show that the drag on bodies immersed in inviscid fluids is zero; originated concepts such as notion of *velocity* and *acceleration* and *differential expression of continuity*.

Jean Louis Poiseuille (1799–1869) A *French* physician who is known for *Poiseuille flow*; deduced experimentally the resistance laws for a laminar flow through capillary tubes.

Jena Glass A *Chemistry* term; a type of glass stronger than ordinary soda glass; obtained by decreasing the alkali content and increasing the alumina content in ordinary soda glass; contains oxides of barium and zinc (BaO and ZnO) along with boron trioxide (B_2O_3) in place of some silica; more resistant to the action of acid and alkali.

Jet Draught A *Power Plant* term related to a *boiler*; a *boiler accessory* used to produce a jet of *high-speed air* to be supplied to the grate to increase the heat supply to the boiler; fitted near the grate of a boiler; see also **Fan Draught**.

Jidoka An *Operations Management* term; a *Japanese* term used to mean the authority given to the workers to stop the production (assembly) line, when quality problems are encountered; prevents production of defective parts/products, and focuses on understanding the problem; *Taiichi Ohno* (1912–1990) used it as a quality improving tool.

Jig A *Manufacturing* term; a device used to hold and support a workpiece, and also to guide the path of a cutting tool during a process; used in conjunction with a fixture; need not be secured to the machine; see also **Fixture**.

D. K. Singh, *Dictionary of Mechanical Engineering*,
https://doi.org/10.1007/978-981-99-1722-8_10

JIT II An *Operations Management* term; a registered trademark of the *Bose Corporation*; an example of a close relationship between supplier and manufacturer; an outsourcing activity in which a supplier's representative functions as a member of the customer's purchasing department while being paid by the supplier and issues order to its own company on behalf of the customer; the representative is also involved in activities such as design, production planning, and value analysis.

JIT Inventory System An *Operations Management* term; a system based on JIT philosophy to provide the right materials, in right quantity, at right place, and at right time; a system of eliminating inventory which does not add any value to a product; conceptualizes on zero inventories.

JIT Manufacturing An *Operations Management* term; also called **JIT Production** or **Stockless Production** or **Zero Inventory System**; producing goods using JIT principle; a pull system of inventory management which emphasizes on the production of right products, in right quantity, and at right time; a zero inventory-based production system; see also **Lean Production**.

JIT Production An *Operations Management* term; see **JIT Manufacturing**.

JIT An *Operations Management* term; the short form of *just-in-time*; a Japanese management philosophy to control wastes and a technique to manage inventory effectively; evolved during the 1980s; connected to production of goods to meet a customer's demand on time; emphasizes on zero inventory and producing only those goods which are needed.

Job Analysis An *Operations Management* term; a process of assessing the work methods used in a job to see how it should be done, and the *knowledge*, *skills*, and *abilities* which individuals need to perform the jobs adequately; see also **Job Evaluation**.

Job Description An *Operations Management* term; a statement indicating what a job entails; see also **Job Analysis**.

Job Design An *Operations Management* term; an approach which specifies the content of each job including the *employee skills* and *training* needed to perform the job and the degree of specialization appropriate for the job; determines how work is distributed within an organization; helps to incorporate *flexibility* needed in the workforce; *job specialization* and *job expansion* are two important components of job design.

Job Enlargement An *Operations Management* term; the process of *expanding a job horizontally* to give the worker more variety of the same job (although not necessarily more responsibility); thrusts upon more skills of the workers; an example includes giving a production-line worker the task of building an entire product rather than a small subassembly; see also **Job Enrichment**.

Job Enrichment An *Operations Management* term; the process of *expanding a job vertically* to give the worker more responsibility which includes *planning, inspecting,* and *other management functions* necessary for job accomplishment; for example, an assembly worker may be given the added responsibility of testing a completed assembly (considering him as a quality inspector); see also **Job Enlargement**.

Job Evaluation An *Operations Management* term; a process which specifies the relative value of each job in an organization; see also **Job Analysis**.

Job Expansion An *Operations Management* term; another component of *job design* other than *job specialization*; classified into *job enlargement* and *job enrichment* (see them separately explained); see also **Job Specialization.**

Job Rotation An *Operations Management* term; a process of moving employees *horizontally* or *vertically* or the periodic exchange of a job among workers to expand their *skills, knowledge,* or *abilities*; a few examples include *rotating nurses among hospital wards* or *flight crews on different airline routes*; see also **Job Enrichment**.

Job Sequencing An *Operations Management* term; the process of determining the order in which jobs should be done at each workstation.

Job Specialization An *Operations Management* term; a component of *job design*; the specialized knowledge of a task which makes the range of a job narrow; requires a high degree of *repetition, higher efficiency,* and *high quality*; an example is a *heart specialist* who can diagnose and treat heart problems better than a general practitioner; see also **Job Expansion.**

Job A *Management* term; a group of related tasks or activities performed by an employee to meet an organization's objectives such as manufacturing of a product or delivery of a service.

Jobbing Production An *Operations Management* term; also called **One-Off Production** or **Job-Lot Production**; a type of a production system used to make one-off items according to a customer's specification where each product is unique and different in some aspect from previous ones; uses a wide range of general-purpose machines and equipment, where machines are arranged on function basis for maximum flexibility, and requires highly skilled personnel, but involves high level of work-in-progress (*WIP*); an example includes the *construction of a railway station* which will be different from other railway stations, even though there will be many similarities.

Job-Lot Production An *Operations Management* term; see **Jobbing Production**.

Johnson's Rule An *Operations Management* term; a *sequencing rule* or an approach used for sequencing any number of jobs on two machines/workstations in order to minimize the time required to complete all the jobs; gives an optimal sequence for jobs processed serially through two processes; also minimizes total idle time on the machines; from the list of unscheduled jobs, the job with the shortest processing time in either workstation is selected; if the shortest time is at the first workstation, the job is scheduled first otherwise scheduled in the last.

Joint Penetration A *Welding* term; the minimum depth a groove weld extends from its face into a joint.

Joint A *Design* term; the junction of members or the edges of members to be joined; two commonly used joints include *butt* and *lap* joints.

Joule Cycle A *Thermodynamics* term related to *air standard cycle*; see **Brayton Cycle**.

Joule A *Physics* term; the SI unit of work and energy, named after the *British* scientist *James P. Joule*, (1818–1889); the British scientist who established a relationship between work and heat and paved the way to the formulation of the first law of thermodynamics; denoted by J, which is equivalent to N-m; 1 J $= 10^7$ erg; see also **Erg**.

Joule–Thomson Coefficient A *hermodynamic coefficient* as well as a *property of a pure substance*; used to find the value of specific heat; denoted by μ; the symbols T, p and h stand for *temperature, pressure* and *enthalpy* respectively; the coefficient is given as:

$$\mu = \left(\frac{\partial T}{\partial p}\right)_h$$

Journal Bearing A *Design* term; a commonly employed bearing as well as the main crankshaft bearing used in the engine of an automobile which has no rolling elements, rather the shaft simply rotates (slides) within a polished sleeve lubricated by *oil* or any other fluid; see also **Rolling Contact Bearing**.

Judgement Methods An *Operations Management* term related to *forecasting*; see **Qualitative Methods**.

Junction Rule An *Electrical* term; see **Kirchhoff's Current Law**.

Junction Valve A *Power Plant* term related to a *boiler*; see **Steam Stop Valve**.

K

Kaizen An *Operations Management* term; a *Japanese* term for an organization's commitment to continuous improvement that helps to improve a product's quality and customer service; can be applied to *manufacturing* as well as *service* organizations.

Kanban An *Operations Management* term; a *Japanese* word coined by the *Toyota Motor Company* in the 1950s to indicate a *card* to be used as an *inventory management technique* in a pull production system to control the flow of materials on the assembly line; kanbans allow arrivals at a workstation to match the processing time; the absence of kanban indicates no production or no movement of materials in a production system; a simple system of planning and controlling production and a means of signalling to the upstream workstation that the downstream workstation is ready for the upstream station to produce another batch of parts; an integral part of a JIT manufacturing system; suitable for *repetitive manufacturing*; the most commonly used is dual kanban system which uses two types of kanban, namely *production kanban* and *withdrawal kanban.*

Kaoru Ishikawa (1915–1989) A well-known *Japanese* quality guru and one of the pioneers of *quality management* who is best known for his quality tool developed in 1943, called '*cause-and-effect diagram*' (used to diagnose quality problems), also called '*Ishikawa diagram*' or '*fishbone diagram*'; also evolved the concept of '*quality circle*'.

Kaplan Turbine A *Fluid Mechanics* term; a *hydraulic* turbine, named in honour of its inventor *Viktor Kaplan* (1876–1934); an *axial-flow*, *single-stage*, *reaction hydraulic turbine* with adjustable blades; more compact in design than Francis turbine; most suited for *low head* (15 m or less) and *high volume flow rate* conditions; see also **Propeller Turbine** and **Francis Turbine**.

KCL An *Electrical* term; see **Kirchhoff's Current Law**.

KDM The short form of **Kinematics and Dynamics of Machines**; also called **Theory of Machines** or **Mechanics of Machines**; a core subject of *mechanical*

D. K. Singh, *Dictionary of Mechanical Engineering*,
https://doi.org/10.1007/978-981-99-1722-8_11

engineering which deals with motions involved in a machine and forces acting on its components.

Kelvin Scale A *Thermodynamics* term; an absolute temperature scale as well as a thermodynamic temperature scale used in the SI system; named after its inventor *Lord Kelvin* (1824–1907); uses the unit *kelvin* (K) for temperature measurement; the lowest temperature on the Kelvin scale is absolute zero (0 K); see also **Celsius Scale**, **Fahrenheit Scale**, and **Rankine Scale**.

Kelvin A *Thermodynamics* term; the SI unit of temperature, named after *Lord Kelvin* (1824–1907) (originally called *William Thomson*); denoted by K; 1 K of temperature is equal to 1/273.16 of the thermodynamic temperature of the triple point of water.

Kelvin-Planck Statement A *Thermodynamics* term; the first statement of the *second law of thermodynamics*, named after *Lord Kelvin* (1824–1907) who proposed the Kelvin scale of temperature and *Max Planck* (1858–1947) whose work on radiation laid to the foundation of quantum theory; the statement relates to *heat engines*; states that no heat engine can produce a net amount of work while exchanging heat with a single reservoir; see also **Clausius Statement**.

Kerf A *Welding* term; the width of the cut made in a workpiece during a cutting operation.

Kernel A *Chemistry* term; the nucleus and electrons other than those present in the outermost shell.

Key A *Design* term; a *fastening element* usually made of *mild steel*; used to prevent relative rotation between a shaft and the shaft-supported elements such as *pulleys, gears, wheels,* or *couplings*; usually driven parallel to the axis of the shafts subjected to torsional or twisting stresses; can resist shear over a longitudinal section; classified into *saddle*, *sunk*, and *round* keys.

kgf A *Physics* term; see **Kilogram Force**.

Kilogram Force A *Physics* term, abbreviated as **kgf**; the MKS unit of force; exerted on a body of mass 1 kg and produces an acceleration of 9.81 $\frac{m}{s^2}$; kilogram force and newton are related as: 1 kgf = 9.81 N; see also **Newton**.

Kilojoule A *Physics* term; the bigger SI unit of energy; denoted by kJ; equivalent to kN m; 1 kJ = 10^3 J.

Kilopascal A *Physics* term; the larger SI unit of *pressure* and *stress*; abbreviated as **kPa**; 1 kPa = 10^3 Pa; see also **Megapascal** and **Gigapascal**.

Kilowatt-hour An *Electrical* term, abbreviated as **kWh**; the *commercial unit of power*; corresponds to a power consumption of 1 kW in 1 h; 1 kWh = 3600 kJ.

Kinematic Coefficient of Viscosity A *Fluid Mechanics* term; see **Kinematic Viscosity**.

Kinematic Similarity A *Fluid Mechanics* term; the similarity of motions between a model and its prototype; indicates that both model and prototype must undergo similar *velocities* and *accelerations*; see also **Geometric Similarity** and **Dynamic Similarity**.

Kinematic Viscosity A *Fluid Mechanics* term; also called **Kinematic Coefficient of Viscosity**; a ratio of the absolute viscosity (μ) to the density (ρ) of a fluid; a measure of the relative magnitudes of absolute viscosity and inertia of the fluid; its two common units include m^2/s and *stoke*, named after *Sir George Stokes* (1819–1903), an English physicist and pioneering investigator of viscosity; 1 *stoke* = 1 $cm^2/s = 0.0001\ m^2/s$; has dimension $[L^2T^{-1}]$; see also **Absolute Viscosity**.

Kinematics and Dynamics of Machines See **KDM**.

Kinematics A *Physics* term; a branch of *dynamics* which deals with the motion of a body without considering the forces causing the motion; see also **Kinetics**.

Kinetic Energy of Rotation A *Physics* term; the energy possessed by a body on account of its rotation about a given axis; given as:

$$\text{K. E. of rotation} = \frac{1}{2} I \omega^2$$

where

I = moment of inertia of the body about an axis
ω = angular velocity of the body.

Kinetic Energy A *Physics* term; the energy associated with the motion of a body.

Kinetic Friction A *Physics* term; also called **Dynamic Friction**; the friction occurring between two surfaces of the bodies in motion; has lesser value than the static friction; classified into sliding and rolling friction; see also **Static Friction** and **Limiting Friction**.

Kinetic Head A *Fluid Mechanics* term; see **Velocity Head**.

Kinetic Viscosity A *Fluid Mechanics* term; **Dynamic Viscosity**.

Kinetics A *Physics* term; a branch of dynamics which deals with both the motion of a body as well as the forces causing the motion; see also **Kinematics**.

Kirchhoff's Current Law An *Electrical* term; an electric circuit law, abbreviated as **KCL**; also called **Junction Rule**; the algebraic sum of all the currents meeting at a node is zero; see also **Kirchhoff's Voltage Law**.

Kirchhoff's Law of Radiation A *Heat Transfer* term; states that the emissivity and the absorptivity of a surface are equal at the same temperature and wavelength.

Kirchhoff's Voltage Law An *Electrical* term; an electric circuit law, abbreviated as **KVL**; also called **Loop Rule**; the algebraic sum of the voltage drops taken in a specified direction around a closed path is zero; see also **Kirchhoff's Current Law**.

Knocking An *IC Engine* term; an undesirable phenomenon in which a *diesel engine* vibrates badly with a loud noise due to *high compression ratio* and *incomplete combustion* of the fuel in the engine cylinder; adversely affects the performance of the engine; see also **Detonation**.

Knuckle Joint A *Design* term; a *temporary fastening device* used to connect two circular rods or bars whose axes are collinear or coplanar and intersecting and are subjected to tensile forces; permits a small amount of flexibility or angular movement and is very common in both machines and structures; a few examples involving knuckle joint include valve and eccentric rods, tension links in bridge structures, link of a cycle chain, and link of suspension chain.

kPa A *Physics* term; see **Kilopascal**.

KVL An *Electrical* term; see **Kirchhoff's Voltage Law**.

kWh An *Electrical* term; see **Kilowatt-hour**.

L

Labour Cost An *Operations Management* term related to *cost analysis*; the price paid in return for the work done by an employee; a few examples includes *wages, salaries, commissions, bonus*, etc. of an employee; classified into *direct labour cost* and *indirect labour cost*; *direct labour cost* is directly related to the manufacture of a product, and its examples include *wages of a welder fabricating a structure* or *wages of construction workers*; *indirect labour cost* is not directly related to the production of a product but is essential for the progressive movement and handling of the product to the point of dispatch, and its examples include the *cost incurred on maintenance men, helpers in a machine shop or foundry shop or machine setters,* etc.

Labour Index An *Operations Management* term; a ratio of the quality cost to the direct labour hours; see also **Cost Index** and **Sales Index**.

Labour Productivity A *Management* term; a ratio of the units of goods and services produced to the number of hours worked (labour input); a measure of *performance* of an organization; see also **Capital Productivity** and **Material Productivity**.

Labour Standards An *Operations Management* term related to *work study*; the amount of time required to perform a job (or part of a job) under ordinary operating conditions; helps a firm to determine the *labour cost*, the *staffing requirements* to meet production targets and estimate time and cost for making make-or-buy decisions; used to plan and control operations.

Labour An *Economics* term; the *physical* and *intellectual* services of the people including the *training*, *education*, and the *abilities* of the individuals in a society; the economic growth of a country depends on the size and quality of the labour force, which in turn, is a function of the size of the working-age population.

Labour-Intensive Technology An *Economics* term; a technology which relies heavily on *human labour* instead of capital; see also **Capital-Intensive Technology**.

D. K. Singh, *Dictionary of Mechanical Engineering*,
https://doi.org/10.1007/978-981-99-1722-8_12

Ladder Logic An *Electronics* term; a *computer programming language* which resembles a ladder with two vertical lines and horizontal rungs that is often used for a programmable logic controller (PLC).

Lagrangian Method A *Fluid Mechanics* term; named after *Joseph Louis Lagrange* (1736–1813), an *Italian* mathematician and astronomer who proposed it; a mathematical method used to describe the motion of individual fluid particles at various instants of time in a fluid flow; involves a lot of difficulties in solving the flow equations, hence is not preferred; see also **Eulerian Method**.

Lame's Theory A *Strength of Materials* term related to a *pressure vessel*; used in the analysis of thick-walled pressure vessels for finding the hoop and radial stresses across the section of the vessels.

Lami's Theorem A *Physics* term; states that if three coplanar forces acting at a point are in equilibrium, then each force is proportional to the sine of the angle between the other two forces; finds use in determining magnitude and direction of a force when other two forces are known, assuming that all forces are in equilibrium.

Laminar Flow A *Fluid Mechanics* term; an *orderly fluid flow* in which fluid particles move in a smooth path in *layers* or *laminas*, with one layer sliding smoothly over an adjacent layer; usually occurs when the velocity of flow is small or when the fluid viscosity is high; the Reynolds number (Re) for a laminar flow is less than 2000; the shear stress in a laminar flow can be calculated by using Newton's law of viscosity; the flow of oil through a pipe is laminar; see also **Turbulent Flow**.

Laminated Spring A *Strength of Materials* term related to a *spring*; also called **Leaf Spring** or **Carriage Spring**, because being used for load carrying applications or **Bending Spring**, because of being subjected to bending only; consists of a number of thin curved plates of different lengths (called leaves) and uniform thickness which are placed over one another and bent to the same curvature and clamped together at the centre; may be of *cantilever*, *semi-elliptical* or *elliptical type* of which semi-elliptical leaf spring is the most common one; finds application in suspension system of front or rear axle of cars, buses, trucks, trains, etc.

Lamp Black A *Chemistry* term; also called **Carbon Black**; the almost *pure form of amorphous carbon* (C) which contains 98–99% of *carbon*; obtained as soot when hydrocarbons such as natural gas and other substances rich in carbon (e.g. *kerosene oil*, *petroleum*, *turpentine oil* or *acetylene*) are burnt in a limited supply of air; used as a *black pigment* in making *printing inks*, *paints*, and *shoe polish*; also extensively used as a *filler material* in making *rubber tyres* in automobile industries.

Lancashire Boiler A *Power Plant* term related to a *boiler*; the most commonly used *fire tube boiler*; categorized as an *internally fired low-pressure* boiler which is horizontally placed, stationary, and involves natural circulation; has low evaporative capacity and produces steam at low pressures; occupies more space, but is economical and involves low maintenance costs; see also **Babcock and Wilcox Boiler**.

Lancing A *Manufacturing* term; a *sheet metal operation* used to produce a line cut (*slit*) in a blank.

Land An *Economics* term; all the *natural resources* such as *minerals, forests,* and *water* as well as the *land* itself which can be combined with *labour* and *capital* to produce *goods* and *services.*

Lap Joint A *Joining* term; a joint produced by placing the two parts one over another; a stronger joint than butt joint because of the large area in contact; see also **Butt Joint**.

Lapping A *Manufacturing* term; an abrasive *micromachining process* used to produce a high degree of surface finish with close dimensional tolerances ($\geq 0.5\ \mu$m) by means of a rotating tool (called *lap*) containing abrasive particles; see also **Honing**.

Laps A *Forging* term; the *forging defects* which occur *on the surface of the forged parts* in the form of *small sharp cavities.*

Laser Beam Cutting A *Manufacturing* term, abbreviated as **LBC**; a process which uses a laser beam to cut materials precisely; see also **Laser Beam Drilling** and **Laser Beam Welding**.

Laser Beam Drilling A *Manufacturing* term, abbreviated as **LBD**; a process which uses a laser beam to produce an accurate hole in materials; see also **Laser Beam Cutting** and **Laser Beam Welding**.

Laser Beam Machining A *Nonconventional Machining* term, abbreviated as **LBM**; a nonconventional machining process which uses a high-energy, coherent laser light beam to melt and evaporate the materials from *metallic* and *non-metallic* workpieces; capable of producing very accurate holes as small as 0.005 mm in *refractory metals, ceramics,* and *composite materials* without warping the workpieces; can be used for *cutting, welding,* and *marking*; used extensively in *electronic* and *automotive industries.*

Laser Beam Welding A *Manufacturing* term, abbreviated as **LBW**; a welding process which uses a *laser beam* to fuse a joint; see also **Laser Beam Cutting** and **Laser Beam Drilling**.

LASER A *Physics* term; the short form of *Light Amplification by Stimulated Emission of Radiation*; a highly concentrated and coherent beam of light used for very precise work in many fields including medical and engineering.

Latent Heat of Fusion A *Physics* term; the quantity of heat required to melt unit mass of a solid into liquid or to freeze unit mass of a liquid into solid; the latent heat of fusion of ice at atmospheric pressure is equal to 80 cal/gm; see also **Latent Heat of Sublimation** and **Latent Heat of Vaporization**.

Latent Heat of Sublimation A *Physics* term; the quantity of heat required to convert unit mass of a solid directly into vapour without converting it into liquid or to convert unit mass of a vapour directly into solid without converting it into liquid; the SI unit is J/kg; see also **Latent Heat of Vaporization** and **Latent Heat of Fusion**.

Latent Heat of Vaporization A *Physics* term; the quantity of heat required to vaporize unit mass of a liquid into vapour or to condense unit mass of a vapour into liquid at a given temperature or pressure; highly sensitive to pressure; decreases as pressure or temperature increases and becomes zero at the critical point; the latent heat of vaporization of water at atmospheric pressure is equal to 540 cal/gm; see also **Latent Heat of Sublimation** and **Latent Heat of Fusion**.

Latent Heat A *Physics* term; the quantity of heat required to produce a phase change in a substance in its unit mass at constant pressure and temperature; the SI unit is J/kg; see also **Sensible Heat**.

Lateral Deformation A *Strength of Materials* term; the deformation which occurs in a solid body in its lateral dimensions (*breadth, height* or *diameter*) on being subjected to an external force; see also **Linear Deformation** and **Angular Deformation**.

Lateral Strain A *Strength of Materials* term; the strain produced in the lateral dimensions of a body such as breadth, thickness, or diameter; quantified as a ratio of the change in lateral dimension of a body to its original lateral dimension; connected to the longitudinal strain through the Poisson's ratio; its value is smaller than the longitudinal strain; see also **Longitudinal Strain**.

Latest Finish Time (LFT) A *Project Management* term; the latest time an activity can be completed without delaying the critical path time of the project; finds use in *critical path method* (CPM); see also **Earliest Finish Time (EFT)**.

Latest Start Time (LST) A *Project Management* term; the latest time an activity can start without delaying critical path time of the project; finds use in *critical path method* (CPM); equals to the latest finish time (LFT) of the immediately succeeding activity minus activity time: $LST = LFT - t$; when more than one activity immediately follows the activity, the minimum of the LFT of all the following activities is considered as the LST of the activity; see also **Earliest Start Time (EST)**.

Lathe A *Manufacturing* term; the most basic and one of the oldest machine tools used to remove the material from a workpiece during its machining; the various operations performed by a lathe include *turning, facing, drilling, boring, reaming, threading, cutting, knurling*, etc.

Lattice Energy A *Chemistry* term; the amount of energy released when *cations* and *anions* are brought from infinity to their respective equilibrium sites in the crystal lattice to form *one mole of an ionic compound*; *the higher the magnitude of the lattice energy, the greater is the tendency to form an ionic bond*; for higher lattice energy, the force of electrostatic attraction between the constituent ions (given by Coulomb's Law) should be high, which in turn, depends on the small sizes of cations and anions.

Lattice Parameter A *Materials Engineering* term; a parameter which describes the *size and shape of a unit cell* of a crystal structure; includes the dimensions of the sides of the unit cell and the angles between the sides; for example, a *cubic crystal* requires only length of one of the sides of the cube to completely describe the unit cell.

Law of Action and Reaction A *Physics* term; see **Newton's Third Law of Motion.**

Law of Conservation of Linear Momentum A *Physics* term; see **Principle of Conservation of Linear Momentum.**

Law of Demand An *Economics* term; states that the quantity of a good or service which people can demand decreases or increases with rise or fall in the price of that good or service, everything held constant; concludes that people purchase more of something when the price of that item falls and vice versa.

Law of Dimensional Homogeneity A *Physics* term; see **Principle of Dimensional Homogeneity.**

Law of Elasticity A *Strength of Materials* term; see **Hooke's Law.**

Law of Gravitation A *Physics* term; see **Newton's Law of Gravitation.**

Law of Inertia A *Physics* term; see **Newton's First Law of Motion.**

Law of Momentum A *Physics* term; see **Newton's Second Law of Motion.**

Law of Parallelogram of Forces A *Physics* term; also called **Parallelogram Law of Forces**; states that if two coplanar forces acting at a point are represented in magnitude and direction by the two adjacent sides of a parallelogram, then their resultant is given in magnitude and direction by the diagonal of the parallelogram passing through the same point.

Law of Polygon of Forces A *Physics* term; also called **Polygon Law of Forces**; states that if three or more coplanar forces acting at a point are represented in magnitude and direction by the sides of a polygon taken in order, then their resultant is given in magnitude and direction by its closing side taken in opposite order; see also **Law of Triangle of Forces.**

Law of Triangle of Forces A *Physics* term; also called **Triangle Law of Forces**; states that if three coplanar forces acting at a point are represented in magnitude and direction by the three sides of a triangle taken in order, then they are said to be in equilibrium; gives the condition of equilibrium of a body which is acted upon by three coplanar forces.

Law of Viscosity A *Fluid Mechanics* term; see **Newton's Law of Viscosity.**

Laws of Coulomb Friction A *Physics* term; see **Laws of Limiting Friction.**

Laws of Dry Friction A *Physics* term; see **Laws of Limiting Friction.**

Laws of Limiting Friction A *Physics* term; also called **Laws of Coulomb Friction** or **Laws of Dry Friction** or **Laws of Solid Friction**; the five laws connected to dependency of the frictional force are stated below:

- The frictional force is directly proportional to the normal reaction between the two bodies in contact.
- The frictional force is independent of sliding velocity.
- The frictional force acts in a direction opposite to that in which motion is to impend.
- The frictional force depends upon the nature of material of the surfaces in contact, that is, force of adhesion and their roughness and smoothness.
- The frictional force does not depend on the shape and area between contact surfaces.

Laws of Motion A *Physics* term; also called **Newton's Laws of Motion** or **Fundamental Laws of Motion**; proposed by *Sir Isaac Newton* (1642–1727); relate to the linear motion of a solid body; there are three laws of motion, namely the first law of motion (also called the *law of inertia*), the second law of motion (also called *law of momentum*), and the third law of motion (also called the *law of action and reaction*).

Laws of Solid Friction A *Physics* term; see **Laws of Limiting Friction**.

LBC A *Manufacturing* term; see **Laser Beam Cutting**.

LBD A *Manufacturing* term; see **Laser Beam Drilling**.

LBM A *Nonconventional Machining* term; see **Laser Beam Machining**.

LBW A *Manufacturing* term; see **Laser Beam Welding**.

Lead Burning A *Welding* term; a misnomer for the welding of lead.

Lead Time An *Operations Management* term; the time required to complete a job from start to finish; manufacturing involves two types of lead time, namely *ordering* lead time and *manufacturing* lead time.

Lead A *Chemistry* and a *Gear* term; in *chemistry* terms, a greyish coloured *nonferrous metallic element* having symbol Pb with atomic number 82 and atomic weight 207.19; extracted from *galena* (PbS) which contains about 6–8% of lead along with small quantities of silver (up to 0.1%); shows a bright metallic lustre on being freshly cut, but becomes dull on exposure to air due to the formation of a thin layer of the oxide on its surface; a soft, malleable (but not very ductile) metal which can mark paper, and can be cut with a knife and scratched with the finger-nail; a heavier metal (specific gravity 11.34) which melts at 327.4 °C and boils at 1725 °C, and is not attacked by pure air-free water; used in making lead pipes (for carrying water), bullets, lead accumulators, and in making lead alloys such as *type metal*, *solder*, and *pewter*, and in making pigments like *chrome yellow*, *chrome red*, *red lead* and *white lead*; also used in making telegraph and telephone wires buried in the earth; a *gear screw thread* term which indicates the axial distance which a nut moves on the thread in one turn.

Leading A *Management* term; a management function related to directing the work of others.

Leaf Spring A *Strength of Materials* term related to a *spring*; see **Laminated Spring**.

Lean Manufacturing An *Operations Management* term; see **Lean Production**.

Lean Mixture An *IC Engine* term; a mixture of fuel and air where air is in excess.

Lean Production An *Operations Management* term; also called **Lean Manufacturing**; a management philosophy developed by *Toyota Motor Company* of Japan which focuses on elimination of all the wasteful activities from a production system; uses lean tools such as *kaizen, Kanban,* and *Poka-yoke.*

Learning Curve A *Management* term; also called **Wright's Law**, named after a US aeronautical engineer *Theodore Paul Wright* (1895–1970) who introduced it in an article in the *Journal of Aeronautical Sciences* in 1936; also called **Improvement Curve** or **Experience Curve**; a graphical plot between the number of units produced (x-axis) and the cycle time per unit (y-axis); a tool related to improvement in performance by way of learning over time (of workers); concludes that repetition of the same job results in decrease in cycle time; organizational costs are reduced due to experience and learning effects that result from processing a higher volume; usually expressed in terms of 100% minus the percentage rate of improvement; an 80% learning curve indicates an improvement rate of 20% each time production doubles, and a 90% learning curve indicates a 10% improvement rate, and so on.

Learning Organization A *Management* term; see **Boundaryless Organization**.

Least Count Error A *Metrology* term; a type of *systematic error* associated with the resolution of a measuring instrument; for example, in the measurement of length, the reliability is up to $\frac{1}{10}$ cm using a metre scale, up to $\frac{1}{100}$ cm using a vernier scale, up to $\frac{1}{1000}$ cm using a screw gauge; all the readings or values measured by the instrument are good only up to its least count, and deviation from these values produces least count errors.

Least Count A *Metrology* term; the smallest value which can be accurately measured by a measuring instrument; the least count of the *vernier calliper* is 0.01 cm and that of a *spherometer* or *screw gauge* is 0.001 cm.

Le Chatelier's Principle A *Chemistry* term; states that if the number of moles of reactants exceeds that of the products, then increasing the pressure at which the reaction occurs forces the reaction towards completion.

LED An *Electronics* term; see **Light Emitting Diode**.

Lee and Shaffer Theory A *Manufacturing* term related to *machining*; a metal cutting theory which explains the process of orthogonal cutting and is based on the stress–strain distribution in a perfectly plastic material; given as:

$$\phi = 45^\circ + \alpha - \beta$$

where

Φ = shear angle
β = angle of friction
α = rake angle of the cutting tool; see also **Merchant's Theory**.

Left Hand/Right Hand Chart A *Motion Study* term; see **Operation Chart**.

Length A *Mathematics* term; the size of a line segment.

Leonardo da Vinci (1452–1519) A famous *Italian* painter who is widely considered to be one of the greatest painters of all time and perhaps the most diversely talented person ever to have lived; widely known for his painting '*Mona Lisa*'; expressed elementary *principle of continuity*; observed and sketched many *basic flow phenomena*; suggested design for *hydraulic machinery*.

Leonhard Euler (1707–1783) A pioneering *Swiss* mathematician and physicist who is known for the *Euler number* (a non-dimensional group); pioneered work on the relationship between pressure and flow; introduced the concept of *cavitation* and *principle of centrifugal machinery*.

Level 1 Charging An *Electric Vehicle* term; a type of charging device which uses a 120 V electrical outlet for charging the battery; a common household charging outlet; uses a 8–20 amperes current; takes around 17–25 h to fully charge a 100-mile battery; see also **Level 2 Charging** and **Level 3 Charging**.

Level 2 Charging An *Electric Vehicle* term; a type of charging device which uses a 208–240 V installed electrical outlet for charging the battery; uses up to a 80 amperes current; offers faster charging than Level 1 charging, and takes around 4–5 h to fully charge a 100-mile battery; see also **Level 2 Charging** and **Level 3 Charging**.

Level 3 Charging An *Electric Vehicle* term; also called **DC Charging**; a commercial charging device which uses a 480 V direct current electrical outlet installed at the charging station for charging the battery; takes about 30 min to fully charge a 100-mile battery; see also **Level 1 Charging** and **Level 2 Charging**.

Lever A *Physics* term; a rigid bar that is allowed to rotate at some angle about a pivot point called a *fulcrum*; used to move a load when an effort force is applied to the bar; depending on the position of the fulcrum on the lever, the lever may be used to change the direction of an effort force, multiply an effort force or increase the distance that is traveled by a load; levers are divided into three categories: first-class, second-class, and third-class levers.

Levitation Casting A *Manufacturing* term related to *casting*; see **Electromagnetic Casting**.

Lewis Equation A *Theory of Machines* term related to *gear*; a gear design equation proposed by *Wilfred Lewis* (1854–1929); used to find the bending stress in a gear tooth; given as:

$$\sigma = \frac{6Wl}{bt^2}$$

where

σ = bending stress
W = force on the gear tooth
l = height (length) of the gear tooth
b = face width of the gear tooth
t = tooth thickness.

Lewis Ferry Moody (1880–1953) A famous *American* scientist who proposed a method of correlating pipe resistance data, better called *Moody chart*, widely used in *fluid mechanics.*

LFT An *Operations Management* term; see **Lot-for-Lot**.

Lift Force A *Fluid Mechanics* term; see **Lift**.

Lift A *Fluid Mechanics* and a *Cam* and *Follower* term; in *fluid mechanics* terms, also called **Lift Force**; a force that a fluid flow exerts on a body at right angle to the direction of flow; a vertical force for a vehicle or object in level flight; created by an unbalanced pressure acting on the top and bottom of the object; pressure on a wing, for example, is much higher on the bottom surface than on the top surface, and the total lift developed by the wing supports the weight of an aircraft; see also **Drag**; in *cam and follower* terms, the difference between the maximum distance of the lowest point of the follower from the axis of rotation of the cam and the least radius.

Light Emitting Diode An *Electronics* term, abbreviated as **LED**; a diode which gives visible light on being energized by a current; used as an indicator to display the functions of the electronic and electrical components.

Light Year A *Physics* term; an *astronomical unit* of measurement of distance; one light year is the distance travelled by the light in vacuum in one year and is equal to 9.46×10^{15} m.

Lignite A *Chemistry* term; a type of coal which contains 67% of *carbon.*

Like Fractions A *Mathematics* term; the fractions having same denominators; a few examples include $\frac{2}{11}, \frac{4}{11}, \frac{7}{11}$ etc.; see also **Unlike Fractions**.

Like Vectors A *Physics* term; two or more parallel vectors having the same direction; see also **Unlike Vectors**.

Limestone A *Chemistry* term; a *sedimentary rock* which consists largely of *calcium carbonate* ($CaCO_3$) in varying degrees of purity; used as a *flux* in the blast furnace to separate impurities from the iron ore to convert it into *pig iron.*

Limit of Proportionality A *Strength of Materials* term; a limit till stress and strain are linearly proportional to each other within elastic limit defined by the Hooke's law; see also **Elastic Limit.**

Limiting Friction A *Physics* term; the maximum value of the static friction; see also **Static Friction** and **Kinetic Friction.**

Limits A *Metrology* term; the stated maximum and minimum dimensions of an object.

Limonite A *Chemistry* term; an iron ore having the chemical formula $2Fe_2O_3.H_2O$; contains about 52–66% *iron.*

Line and Functional Organization A *Management* term; a variation of the line and staff organization which empowers the staff executives on the basis of their functional expertise in decision-making and gives them authority and responsibility for their tasks; the problems of work interference and jealousy among the line and staff executives are non-existent (which is commonly found in the line and staff organization); suitable for *large size organizations*; see also **Line and Staff Organization.**

Line and Staff Organization A *Management* term; the organizational structure which consists of both line executives and staff executives (supporting staffs helping line executives); combines the advantages of both line organization (speedy action) and functional organization (specialities); the staff executives relieve the line executives of their additional tasks to allow them to concentrate on their main functions; the staff executives do not have the decision-making authority; the work quality improves due to work speciality (of staff executives), but the production costs increase due to salary payments to the staff executives; negative qualities such as jealousy and work interference among line and staff executives exists; suitable for *large size organizations*; see also **Line Organization**, **Functional Organization**, and **Line and Functional Organization.**

Line Authority A *Management* term; the relationship in which a superior exercises direct supervision over a subordinate; the power associated with a person in his supervisory position.

Line Balancing An *Operations Management* term; a procedure of assigning the tasks among the workers at the assembly line stations in a manner so that their performance times are made as equal as possible and reduce their idleness.

Line Force A *Physics* term; see **Linear Force.**

Line Layout An *Operations Management* term related to *facilities planning*; see **Product Layout.**

Line Measurement A *Metrology* term; the measurement of distance between *two lines* or *edges* with the help of a measuring instrument such as a *steel rule*; see also **End Measurement**.

Line Organization A *Management* term; also called the **Military Organization**; the simplest organizational structure arranged in a linear way and has no functional categories; suitable for a *small, entrepreneur-dominated company* with 1 or 2 product lines; lacks works speciality; the employees in line organization tend to be generalists and jack-of-all trades; ensures speedy action, but encourages dictatorial approach; used in *automated* and *continuous process industries* such as *sugar, textile, pharmaceutical,* and *chemical.*

Line Role A *Management* term; a manager who has the direct responsibility for the production of goods and services; see also **Staff Role**.

Linear Defect A *Materials Engineering* term; see **Dislocation**.

Linear Deformation A *Strength of Materials* term; the deformation which occurs in a solid body in its linear direction (*length*) on being subjected to an external force; see also **Lateral Deformation** and **Angular Deformation**.

Linear Elastic Structures A *Strength of Materials* term; the structures for which the *principle of superposition* is valid, which in turn, is applicable when the material follows Hooke's law and the deformations produced are small.

Linear Force A *Physics* term; also called **Line Force**; a distributed force which *acts along a line* of a body; usually expressed as force per unit length; for example, a *surface tension force.*

Linear Momentum A *Physics* term; the product of mass of a body and its linear velocity; a *vector quantity* whose direction is same as that of velocity of the body; the SI unit is kg-m/s; has dimension $[MLT^{-1}]$; see also **Angular Momentum**.

Linear Motion A *Physics* term; see **Rectilinear Motion**.

Linear Programming An *Operations Research* term; an *optimization technique* in which the *objective function* and the *constraints* are expressed in the *linear form*; the complete mathematical statement of a linear programming problem includes a set of simultaneous linear equations that represent the *conditions of the problem* and a linear function that represents the *objective* of the problem; used in the allocation of *scarce resources* in an organization with an objective of making their *optimal use.*

Linear Regression A *Statistics* term; a process of analysis in which a set of data points best fits a straight line whose equation is given by

$$y = mx + c$$

where

m = slope of the line on x-axis
c = intercept of the line on y-axis; see also **Nonlinear Regression**.

Link A *Theory of Machines* term; also called **Element**; a part of a machine which has motion relative to some other part; may consist of several separately manufactured parts, but on being assembled can act as a single link; for example, in case of a steam engine, *piston rod* and *crosshead* are assembled together to act as a single link; may not necessarily be a rigid body, rather a resistant body, and examples in this category include liquids (which are resistant to compressive forces and are used as links in devices like hydraulic presses, hydraulic brakes, and hydraulic jacks) and chains, belts, and ropes (which are resistant to tensile forces), which are used for transmitting motion and force.

Liquefaction A *Thermodynamics* term; the process of converting a gas into its liquid form; usually carried out at very low temperatures.

Liquid Cyaniding A *Manufacturing* term related to *heat treatment*; a *case hardening process* used to produce *hardened surface* in *low carbon steel* parts by heating the parts in a bath of *cyanide salts* at a temperature of 1550–1600 °F, allowing them to stay at that temperature for about two hours and then *quenched* in *brine*, *water* or *oil*; see also **Gas Cyaniding**.

Liquid A *Physics* term; a substance which can flow, and cannot resist shear forces; a state of matter which has a *definite volume* but *no definite shape*; takes the shape of a container in which it is kept; see also **Solid** and **Gas**.

Liquid–Metal Forging A *Manufacturing* term related to *casting*; see **Squeeze Casting**.

Liquid-Penetrant Testing A *Materials Engineering* term; a *nondestructive testing* in which *surface defects* are observed by the presence of a high-visibility liquid which has previously penetrated into the defects and is subsequently drawn out by the capillary action of a fine powder.

Liquidus A *Materials Engineering* term; the line above which a single liquid phase is present in a *phase diagram*; see also **Solidus**.

Lithography An *Electronics* term; a *print-making technique* used for the processing of *integrated circuits*.

Lithopone A *Chemistry* term; a mixture of *Zinc sulphide* (ZnS) and *Barium sulphate* ($BaSO_4$); finds use in making *paints*; has superior painting property (spreads quickly to cover more area) over white zinc and white lead.

LMTD A *Thermodynamics* term; see **Log Mean Temperature Difference**.

Load Voltage A *Welding* term; the voltage at the output terminals of a welding machine when an arc is in progress; equals to the sum of the *arc voltage* and the *voltage drop* in the welding circuit; see also **Open Circuit Voltage**.

Load A *Strength of Materials* term; the amount of a force applied to a material or structure; the SI unit is newton (N); see also **Force**.

Loading An *Operations Management* term; also called **Shop Loading** or **Machine Loading**; the distribution of workload to be carried out at each workstation on a shop floor according to the planning period.

Loam Mould A *Casting* term; the mould made of *loam sand*; used for making *extremely large size castings.*

Loam Moulding A *Casting* term; the process of making *loam moulds* used for making *extremely large size castings.*

Loam Sand A *Casting* term; the moulding sand which contains *silica* and *clay*, where *clay* is in higher amounts as high as 50%.

Log Mean Temperature Difference A *Thermodynamics* term, abbreviated as **LMTD**; a temperature difference which involves a *natural logarithm*; used in the design of a *heat exchanger* to find the heat transfer; given as:

$$\Delta T_m = \frac{\Delta T_1 - \Delta T_2}{\ln \frac{\Delta T_1}{\Delta T_2}}$$

where

$\Delta T_1 =$ temperature difference between hot and cold fluids at one end of heat exchanger
$\Delta T_2 =$ temperature difference between hot and cold fluids at another end of heat exchanger.

Logarithmic Strain A *Strength of Materials* term; see **True Strain**.

Logistics An *Operations Management* term related to a *supply chain*; a business function responsible for *transporting* and *delivering* products to the right place at the right time throughout the supply chain; a part of supply chain management which is related to *movement* and *positioning* of *inventory* in a supply chain; see also **Reverse Logistics**.

Long Column A *Strength of Materials* term related to a *column*; a column of long length having its slenderness ratio greater than 120; long columns fail mainly by buckling and not by direct compressive stress; see also **Short Column** and **Medium Column**.

Longitudinal Strain A *Strength of Materials* term; also called **Axial Strain** or **Normal Strain** or **Direct Strain**; the strain produced on account of change in length of a body in the longitudinal direction, that is, along the axis of the body under uniaxial tension or compression; quantified as a ratio of the change in length of a body to its original length; a dimensionless quantity, but is expressed as mm/mm or m/m; classified into *tensile strain* and *compressive strain*; see also **Lateral Strain**.

Longitudinal Stress A *Strength of Materials* term; see **Normal Stress**.

Long-Term Forecasts An *Operations Management* term related to *forecasting*; have a time frame of over two years and generally about three years; these forecasts are used in production planning to support strategic decisions about planning *products, processes, technologies*, and *facilities*; causal model methods and management judgement methods are primarily used for *long-term forecasts*; see also **Short-Term Forecasts** and **Medium-Term Forecasts**.

Long-Term Liabilities A *Financial Management* term; the liabilities which do not need to be met over the coming year rather to be settled over long periods such as long-term *loans* and *mortgages*; see also **Short-Term Liabilities**.

Loop Rule An *Electrical* term; see **Kirchhoff's Voltage Law**.

Loop An *Electrical* term; see **Mesh**.

Loose Piece Pattern A *Casting* term; a type of pattern used to produce moulds using its different-sized loosely attached parts of which the main pattern (the bigger part) is first taken out of the mould and the loose pieces are taken out through the space created by the main pattern; used to produce moulds which create withdrawal problem for the pattern.

Lost Mass A *Chemistry* term; see **Mass Defect**.

Lost Sale An *Operations Management* term; see **Stockout Cost**.

Lost-Foam Casting A *Manufacturing* term related to *casting*; see **Full Mould Casting**.

Lost-Pattern Casting A *Manufacturing* term related to *casting*; see **Investment Casting**.

Lost-Wax Process A *Manufacturing* term related to *casting*; see **Investment Casting**.

Lot Size Inventory An *Operations Management* term; see **Cycle Inventory**.

Lot Sizing An *Operations Management* term; a process of determining the size of the order quantities of each of the components in a product.

Lot An *Operations Management* term; see **Production Lot**.

Lot-for-Lot An *Operations Management* term, abbreviated as **LFT**; the simplest lot sizing rule in which a lot size is equal to the net requirement during every period in the planning horizon; see also **Fixed Order Quantity** and **Periodic Order Quantity**.

Louis Marie Henri Navier (1785–1836) A *French* mathematician who is widely known for the *Navier–Stokes equations*; extended the equations of motion used in fluid mechanics to include the molecular forces.

Low Carbon Steel A *Materials Engineering* term; also called **Mild Steel** because of its mild (low) carbon content (less than 0.3%); a highly *ductile* material which can be easily drawn into wires; possesses good *formability* and *weldability*, but lacks hardness; the most widely used material in production activities; finds use in making *bolts*, *nuts*, *sheets*, *plates*, *tubes*, and *machine components* which do not require high strength; see also **Medium Carbon Steel** and **High Carbon Steel**.

Low Grade Energy A *Thermodynamics* term; the energy which cannot be completely transformed into mechanical work; a few examples include *heat energy*, *nuclear energy*, and *heat produced by the combustion of various fuels* (such as *coal*, *wood*, *oil*, etc.); see also **High Grade Energy**.

Low Head Turbine A *Fluid Mechanics* term; a hydraulic turbine operating at a low head (between 3 and 50 m); for example, *Kaplan turbine* and *propeller turbine*; see also **Medium Head Turbine** and **High Head Turbine**.

Low Level Manager A *Management* term; also called **First Line Manager** or **Supervisor** as he is the first to be contacted by the workers for any problem and he also supervises the works of the workers; a person who is actually translating the organizational goals into reality and is responsible for directing the day-to-day activities of operative employees in an organization; see also **High Level Manager** and **Middle Level Manager**.

Low-Pressure Casting A *Manufacturing* term related to *casting*; also called **Low-Pressure Permanent Mould Casting** or **Low-Pressure Die Casting** as the molten metal is fed into the mould cavity of a metallic mould (die) at a low pressure or **Pressure Pouring**; a casting process which does not use riser; the molten metal pressure lies between gravity die casting and die casting; the process is economical as more number of castings can be produced with the same die because of its increased life resulting from low-pressure application; produces dense castings made from *aluminium*, *steel*, and *cast iron* with improved mechanical properties; the typical products made by this method include *engine blocks* (aluminium) and *railroad wheels* (steel).

Low-Pressure Die Casting A *Manufacturing* term related to *casting*; see **Low-Pressure Casting**.

Low-Pressure Permanent Mould Casting A *Manufacturing* term related to *casting*; see **Low-Pressure Casting**.

Lower Critical Temperature A *Materials Engineering* term; the temperature at which the transformation of an alloy from one solid structure to another completes, as it is cooling; see also **Upper Critical Temperature**.

Lower Level Needs A *Management* term related to *motivation*; the human needs which find mention in *Abraham Maslow's Need-Hierarchy Theory of motivation* and are lying at the base (lower) level of hierarchy which can be satisfied; the lower level needs include *physiological*, *safety*, and *social* needs; see also **Higher Level Needs**.

Lower Pair A *Theory of Machines* term; a mechanism in which two links have *surface contact* while in motion; found in *pure turning* or *sliding*; a few examples include *a shaft revolving in a bearing*, *a universal joint*, *an automobile steering gear*, and *a straight line motion mechanism*; see also **Higher Pair**.

LPT An *Operations Management* term; one of the priority rules for sequencing jobs which stands for *longest processing time*; jobs with the longest processing time are assigned first.

Lubricant A *Fluid Mechanics* term; a viscous substance used for lubricating moving parts of a machine in order to reduce friction and wear between contact surfaces; a few examples include *oil* and *grease*.

Lubrication A *Fluid Mechanics* term; a process of using a lubricant to reduce friction and wear between contact surfaces.

Ludwig Prandtl (1875–1953) A *German* professor who is called the '*father of present-day fluid mechanics*'; introduced the concept of a '*fluid boundary layer*', which laid the foundation for the unification of the *theoretical* and *experimental* aspects of *fluid mechanics*.

Luminescence A *Physics* term; the re-emission of photons of visible light in association with photon absorption.

Lumpy Demand An *Operations Management* term; the demand for a material which varies irregularly from time to time.

M

Macaulay's Method A *Strength of Materials* term related to *deflection*; an analytical deflection method devised by *W. H. Macaulay* (1853–1936); a double integration method which uses the *differential equation of flexure* to find the deflection of a loaded beam under discontinuous loading conditions; the method uses a step function which formulates a single equation of moments for all the loads acting on a beam for which the constants of integration apply uniformly to all the sections of the beam.

Mach Number A *Fluid Mechanics* term; a *nondimensional group*, named after *Ernst Mach* (1838–1916), an *Austrian* physicist and philosopher and one of the pioneers in the field of supersonic aerodynamics; a ratio of the *inertia force* to the *compressibility force*; a ratio of the *fluid velocity* (or the velocity of a body through a stationary fluid) to the *velocity of sound* in that fluid; denoted M; a commonly used parameter in *compressible flow problems*, particularly in the fields of *gas dynamics* and *aerodynamics*; used to classify the types of flow on the basis of its value as given below:

Mach Number (M)	Types of Flow
$M < 0.3$	Incompressible Flow
$0.3 < M < 1$	Subsonic Flow
$0.9 < M < 1.1$	Transonic Flow
$M = 1$	Sonic Flow
$1 < M < 7$	Supersonic Flow
$M > 7$	Hypersonic Flow

Machinability A *Manufacturing* term; the ability (property) of a material which enables it to be easily *machined*; *mild steel* and *grey cast iron* have good machinability.

Machine Design An important subject of *mechanical engineering* and part of mechanical design which involves the selection of *proper materials*, *dimensions*,

D. K. Singh, *Dictionary of Mechanical Engineering*,
https://doi.org/10.1007/978-981-99-1722-8_13

and *shapes* of the components of a *machine* in order to devise new or improved machines to accomplish a specific purpose; see also **Mechanical Design**.

Machine Drawing A visual representation of the component parts of a machine for their assembly or manufacturing in *industries*; an indispensable communicating medium employed in industries.

Machine Loading An *Operations Management* term; see **Loading**.

Machine Moulding A *Casting* term; the process of making a sand mould using various machines, which make the *moulding process faster*, *better*, and *more efficient*.

Machine Reliability A *Statistics* term; the likelihood of a machine not breaking down, malfunctioning or needing repairs in a given time period or number of hours of use; see also **Product Reliability**.

Machine Tools A *Manufacturing* term; machines which are used to shape a workpiece to its proper dimensions by removing unwanted material from it through the cutting action of sharpened blades; a few examples include *drill presses*, *lathes*, and *milling machines*.

Machine A multi-force member structure which contains moving parts having motion relative to each other; a device designed to transmit and modify forces or to convert energy from one form to another or to produce useful work and power; classified into *power generating machines* (*IC engines* and *turbines*), *production machines* (*lathe*, *milling machine*, and *shaper*), *transportation machines* (*aeroplanes*, *automobiles*, and *trains*), *computing machines* (*computers*), and others.

Machining Allowance A *Casting* term; a type of pattern allowance provided in a sand-mould pattern in the form of increased size over its actual size to take care of machining of rough surfaces of the castings; varies between 2 and 5 mm for small size castings and more than 25 mm for large size castings.

Machining A *Manufacturing* term; a process of gradually removing the undesired material from a workpiece (usually in the form of *small chips*), while giving it the proper shape and size.

Macroeconomics An *Economics* term; a branch of economics which deals with the study of behaviour of an economy as a whole, that is, the economic behaviour of aggregates; includes variables such as *total output* in the economy, *aggregate price level*, *inflation*, *employment* and *unemployment*, *interest rates*, *wage rates*, and *foreign exchange rates*; the macroscopic variables are affected by *government policies*; see also **Microeconomics**.

Macroscopic Properties A *Thermodynamics* term; the properties connected to the overall behaviour of a system; has nothing to do with the composition of the matter; changes in macroscopic properties can be perceived by human senses; a few examples include *pressure*, *volume*, and *temperature*; finds applications in *classical thermodynamics*; useful in the study of *heat engines* and *engineering thermodynamics*.

Macrostructure A *Materials Engineering* term; the structure of a material for which the length scale is greater than 1,00,000 nm (1 nm = 10^{-9} m); described by features such as *porosity*, *surface coatings*, as well as *internal* or *external micro-cracks*; see also **Microstructure**.

Magnetic Field A *Physics* term; the region of physical attraction produced by an electrical current.

Magnetism A *Physics* term; a physical phenomenon associated with the attraction of certain materials; exhibited by *iron* (Fe), *cobalt* (Co), and *nickel* (Ni).

Magnetite A *Chemistry* term; a brownish ore of iron having the chemical formula of Fe_3O_4; contains about 65–70% iron; has magnetic behaviour and is least commonly used; see also **Haematite**.

Magneto Fluid Mechanics A *Fluid Mechanics* term; also called **Magnetogasdynamics** or **Hydromagnetics**; the study of motion of an electrically charged conducting fluid in the presence of a magnetic field.

Magnetogasdynamics A *Fluid Mechanics* term; see **Magneto Fluid Mechanics**.

Magnus Effect A *Fluid Mechanics* term, named after the *German* scientist *Heinrich G. Magnus* (1802–1870) who first observed it in 1852; a phenomenon of producing lift by the rotation of a solid body placed in the free stream of a fluid.

Major Diameter A *Theory of Machines* term related to *gear*; the largest diameter of a screw thread; see also **Minor Diameter**.

Make-to-Order An *Operations Management* term; also called **Produce-to-Order**; a manufacturing approach or strategy in which products and services are designed and produced according to a customer's specifications; involves extensive interaction with a customer to provide a high degree of customization and has the ability to satisfy the unique needs of customers but has a long delivery time; a few examples include *wedding invitations*, *custom-built homes*, *custom-tailored clothes*, *charter airline flights*, and *most professional services* such as *medical*, *legal*, and *financial services*; see also **Make-to-Stock** and **Assemble-to-Order**.

Make-to-Stock An *Operations Management* term; also called **Produce-to-Stock**; the oldest manufacturing approach of producing products ahead of time and placing them in inventory until customers demand them; used in *mass production* where similar products are produced in large volume; acts as a safeguard against any fluctuation in demand or lead time but suffers from huge costs of holding the inventories; not useful for producing customized products; see also **Assemble-to-Order** and **Make-to-Order**.

Malachite A *Chemistry* term having the chemical formula ($Cu(OH)_2 . CuCO_3$); an important ore of *copper* (Cu).

Malcolm Baldrige National Quality Award A *Quality Management* term; see **Baldrige Award**.

Malleability A *Manufacturing* term; a mechanical property that defines the ability of a material to deform permanently under compression without breaking; enables the material to be easily rolled or hammered into thin sheets; *mild steel* has good malleability.

Malleable Cast Iron A *Materials Engineering* term; a type of cast iron having better *ductility* because of its high tensile strength (more than 350 MPa); has elongation as much as 18% as compared to less than 1 percent elongation of *grey cast iron*; is very *machinable*; obtained by the heat treatment (annealing) of *white cast iron* which produces round-shaped clumps of graphite.

Mallet A *Manufacturing* term; a wooden hammer used for *light load applications* in *sheet metal operations.*

Management by Objectives A *Management* term, abbreviated as **MBO**; an organization-wide approach which integrates many key management activities in a systematic manner toward desired objectives; consciously directed towards the effective and efficient achievement of organizational and individual objectives; links organizational objectives and the behaviour of individuals; a widely used organizational tool for *performance appraisal* and *employee motivation.*

Management Competencies A *Management* term; a cluster of related *knowledge, skills,* and *attitudes* related to effective managerial performance; affects the organizational atmosphere, activities and productivity; a good manager can turn *straw* to *gold*, whereas a poor manager can do the reverse.

Management Skill A *Management* term; the abilities or behaviours crucial to success in a managerial position; the three important management skills proposed by *Robert L. Katz* (1917-) include *conceptual, interpersonal* (human), and *technical.*

Management Task A *Management* term; also called **Task of a Manager**; the task of a manager which broadly includes *planning, organizing, integrating*, and *monitoring.*

Management The process of getting works done effectively and efficiently through and with other people; an important discipline to tackle the real-life problems; an art as well as science; no golden rule or definite answers exist in management, but the experience acquired over a period of time helps in good decision-making; important divisions of management include *financial, personnel* (*human resource*), *operations,* and *marketing.*

Manager A *Management* term; a person who directs the activities of other people in an organization and does not get directly involved in the execution of a task; classified into *top level* manager, *middle level* manager and *low level* manager (*supervisor*).

Manganese A *Chemistry* term; a chemical element having symbol Mn and atomic number 25; an alloying element used in steel (in large amounts up to 14%) to increase its *hardness* and *toughness*.

Man–Machine Chart A *Motion Study* term; see **Activity Chart**.

Manning Equation A *Fluid Mechanics* term; used to find the *fluid velocity* in an *open channel*; given as:

$$V = \frac{1}{n} R^{2/3} S^{1/2}$$

where

V = velocity of the fluid
R = hydraulic radius
n = roughness coefficient
S = slope of the bed of the channel with horizontal.

Manometer A *Fluid Mechanics* term; a device used to measure *small* and *moderate* pressures; a commonly used manometer is *U-Tube manometer*.

Manometry A *Fluid Mechanics* term; a technique used for pressure measurement by employing the hydrostatic law $\frac{dp}{dz} = \rho \boldsymbol{g}$, where p stands for pressure, z for datum head, ρ for density of fluid, and $\boldsymbol{g}$ for acceleration due to gravity.

Manual Transmission An *Automobile* term; a mechanism used in *automobiles* which uses a clutch pedal to shift gears for motion transmission; see also **Automatic Transmission**.

Manufacturing Approach An *Operations Management* term; an approach needed to start a manufacturing activity; some examples include *make-to-stock*, *assemble-to-order*, and *make-to-order*.

Manufacturing Cell A *Manufacturing* term; also called simply **Cell**; a workstation where several machines are kept together to work upon a workpiece according to operation needed.

Manufacturing Facilities The facilities needed to start a manufacturing activity; include human and material resources such as *workforce*, *space*, *raw materials*, *machines*, and *equipment*.

Manufacturing Flexibility A *Manufacturing* term; the capability of a manufacturing system to adapt successfully to the changing environmental conditions and process requirements in order to produce a broader range of products.

Manufacturing Lead Time An *Operations Management* term; the time required to process a part through the sequence of machines specified on the route sheet; includes both the operation time and the non-productive time (changeover time of machines); see also **Ordering Lead Time**.

Manufacturing Organization A *Management* term; the organization involved in the production of goods (finished products) from raw materials; employs a huge workforce and is a vital parameter of economic growth; a few examples include *automobile companies* and *customers' goods producing companies*; see also **Service Organization**.

Manufacturing Resource Planning An *Operations Management* term; see **MRP II**.

Manufacturing See **Production**.

Margin of Safety A *Strength of Materials* term; used in the aircraft industry in place of factor of safety; given as:

$$\frac{\text{Ultimate load}}{\text{Design load}} - 1$$

Marginal Cost An *Operations Management* term related to *cost analysis*; also called **Incremental Cost** or **Differential Cost**; the additional cost which results from increasing the output of a system by one (or more) units; the fixed cost remains unchanged by increasing output by one (or more) units.

Market Equilibrium An *Economics* term; a marketing condition where the quantity supplied and the quantity demanded are equal; at this stage, there is no tendency for price to change.

Market Research Method An *Operations Management* term related to *forecasting*; a *qualitative* forecasting method used to determine a customer's interest in a product or service by creating and testing hypothesis through data-gathering surveys; uses a questionnaire based on the interest and liking or disliking of the customer, and then analyses the information using judgement and statistical tools to interpret the responses; a very effective method for *short-term forecasting*.

Market An *Economics* term; the institution through which buyers and sellers interact and engage in exchange; it may be a local market where people go to trade food and clothing or it may be an international telecommunications network over which people buy and sell financial securities.

Marketing Management A *Management* term; a discipline of management which deals with sales forecasts (of products), customer orders, customer feedback and information on product promotion and their development; helps a firm to understand its customer's requirements and accordingly react to fulfill them.

Martensite A *Materials Engineering* term; a *supersaturated solid solution* of carbon in *ferrite*; the *hardest* and *strongest* form of steel.

Maslow's Theory A *Management* term related to *motivation*; see **Need-Hierarchy Theory**.

Mass Balance A *Fluid Mechanics* term; a condition in which mass of the components on input and output sides of a device is balanced; the concept is used in the *continuity equation*; see also **Energy Balance**.

Mass Concentration A *Chemistry* term; the concentration expressed in terms of *mass* of a substance per unit volume of a *gas* or *liquid*.

Mass Customization An *Operations Management* term; a marketing goal of providing tailored solutions of specific requirements for a large number of customers; *computer-aided design* and *computer-aided manufacturing* (CAD/CAM) and *flexible manufacturing systems* (FMS) are the means for mass customization.

Mass Defect A *Chemistry* term; also called **Lost Mass**; the difference in mass between the actual mass of an isotope of an element (invariably less) and the sum of the mass of protons, neutrons, and electrons present in it; the loss in mass is equivalent to the energy released in the formation of a given stable nucleus of an element from individual protons and neutrons; quantified by the famous *Einstein's mass-energy relationship*, $E = mc^2$, where m represents the mass defect and c the velocity of light in vacuum.

Mass Density A *Physics* term; see **Density**.

Mass Flow Rate A *Thermodynamics* and a *Fluid Mechanics* term; also called **Mass Flux**; the amount of mass of a fluid which flows through a cross section per unit time; the SI unit is kg/s.

Mass Flux A *Thermodynamics* and a *Fluid Mechanics* term; see **Mass Flow Rate**.

Mass Moment of Inertia A *Physics* term; also called **Rotational Inertia**; a property of a rotating body which enables it to resist any change in its rotational motion about an axis; plays the same role in rotational motion as the mass plays in linear motion, that is, mass moment of inertia is rotational analogous of mass in linear motion.

Mass Production An *Operations management* term; the high-volume production of a standardized product for a large market; developed in *North America* to accommodate three major goals of the US economy such as the need to export, the need to provide employment for a massive, largely unskilled workforce and the need to establish itself as a world player in industry; see also **Batch Production** and **Continuous Production**.

Mass A *Physics* term; the quantity of matter contained in a body; can never be zero; constitutes the weight of a body; the SI unit is kilogram (kg).

Mass-Energy Equivalence A *Physics* term; a relationship proposed by the German scientist *Albert Einstein* (1879–1955) in 1905; states that mass (m) and energy (E) can be converted to each other by the equation $E = mc^2$, where c = speed of light in vacuum; one kilogram of a matter is equivalent to 9×10^{16} J of energy.

Master Production Schedule An *Operations Management* term, abbreviated as **MPS**; a time-phased plan which specifies how many and when a firm plans to build each end item; an important tool of production planning and control (PPC) and an input of material requirements planning (MRP); derived from the aggregate production plan and is based on *demand forecasts*, *customer orders*, and *capacity limitations* of a plant; the manufacturing firms generally work towards monthly delivery schedules.

Match Plate Pattern A *Casting* term; a *type of pattern* used to produce moulds for large-scale production of small castings by using a *wooden or metal match plate* on which several small patterns (equally divided into cope and drag parts of the moulds) are mounted on its both sides; see also **Cope and Drag Pattern**.

Material Cost An *Operations Management* term related to *cost analysis*; the cost of commodities; classified into *direct material cost* and *indirect material cos*t; *direct material cost* is the cost incurred on the material used in the production of a product, and its example includes cost of *high speed steel* (HSS) used in making of a cutting tool; *indirect material cost* is not directly used in making of the product but is used in the production process, and its examples *grease*, *oils*, *cotton waste*, etc.

Material Handling An *Operations Management* term; a process involving the movements of materials (manually or mechanically) in batches or as a single item within a plant using horizontal, vertical or combined motion; the material handling cost is roughly 20–60% of the total production cost.

Material Productivity A *Management* term; a ratio of the output to the materials input; a measure of *efficiency* of a manufacturing organization; see also **Labour Productivity** and **Capital Productivity**.

Material Requirements Planning An *Operations Management* term, abbreviated as **MRP**; a *computerised technique* used to determine the optimum quantity and timing of materials requirement for production operations according to master production schedule (MPS); its three basic components include *bill of materials* (BOM) file, *inventory file*, and *master production schedule* (MPS); see also **Manufacturing Resource Planning** and **Capacity Requirement Planning**.

Materials Handling System An *Operations Management* term; the entire network of transportation to receive materials, store them in inventories, move them between various workstations, and finally load the finished products into vehicles for delivering them to customers.

Materials Management A *Management* term; an important area of *operations management* related to planning and management of flow of materials into and through an organization.

Materials Science and Engineering An *interdisciplinary branch* of engineering as well as a *core subject of mechanical engineering*; related to inventing new materials and improving the existing ones by developing a deeper understanding of the microstructure-composition-synthesis-processing relationships.

Mathematical Model A model consisting of entirely abstract symbols and numerical values; a few examples include *operations research models* and *financial system models*.

Matrix Organization A *Management* term; an organizational structure which combines functional and project or product patterns of departmentation in the same organization structure; commonly found in engineering and research & development; also widely used in product marketing organization.

Matrix A *Chemistry* term; see **Gangue**.

Max Weber (1864–1920) A *German* sociologist and a political economist who proposed the '*bureaucracy*' (bureaucratic structure) in management that clearly defined authority and responsibility of the people in an organization, and was based on division of labours, clearly defined hierarchy, detailed rules and regulations and interpersonal relationships, and was aimed to provide stability and certainty in an organization.

Maximum Distortion Energy Theory A *Strength of Materials* term related to *theory of material's failure*; also called **von Mises Yield Criterion**, named after the German-American mathematician *Richard von Mises* (1883–1953) who proposed it or also called **Octahedral Shear Theory**; a failure theory which states that a material begins yielding, when the maximum shear strain energy (called *distortion energy*) per unit volume equals to the shear strain energy per unit volume at the yield point in the *uniaxial tension test*; for a material being subjected to three mutually perpendicular principal stresses σ_1, σ_2, and σ_3, this theory can be mathematically expressed as:

$$\frac{1}{12G}\left[(\sigma_1 - \sigma_2)^2 + (\sigma_2 - \sigma_3)^2 + (\sigma_3 - \sigma_1)^2\right] = \frac{\sigma_{yp}^2}{6G} \tag{1}$$

$$\left[(\sigma_1 - \sigma_2)^2 + (\sigma_2 - \sigma_3)^2 + (\sigma_3 - \sigma_1)^2\right] = 2\sigma_{yp}^2 \tag{2}$$

where

G = modulus of rigidity of material
σ_{yp} = yield stress.

Maximum Height A *Projectile* term; the maximum height attained by a projectile above the point of projection during its flight; *independent of the mass of the projectile*; *equals to one-fourth of the maximum range of the projectile*; denoted by *h*; given as:

$$h = \frac{u^2 \sin^2 \theta}{2g}$$

where

u = velocity of projection
θ = angle of projection
g = acceleration due to gravity.

Maximum Normal Strain Theory A *Strength of Materials* term; also called **Maximum Principal Strain Theory** or **Saint Venant's Criterion** named after a French elasticity theorist *Saint Venant* (1797–1886); this theory gives better result for *brittle materials* under biaxial tension, but is not used these days; states that a material fails, when the maximum principal strain reaches the strain at the yield point in the simple tension test; for a body subjected to three mutually perpendicular stresses, σ_x, σ_y and σ_z acting along *x, y,* and *z* directions respectively, and σ_{yp} being the yield point stress, this theory can be mathematically expressed as:

$$\frac{\sigma_x}{E} - \frac{\nu}{E}(\sigma_y + \sigma_z) = \frac{\sigma_{yp}}{E}$$

$$\text{or} \quad \sigma_x - \upsilon(\sigma_y + \sigma_z) = \sigma_{yp}$$

where

ν = Poisson's ratio
E = Young's modulus of the material.

Maximum Normal Stress Theory A *Strength of Materials* term; also called **Maximum Principal Stress Theory** or **Rankine's Theory** named after *William John Macquoorn Rankine* (1820–1872); the oldest and the simplest of all the theories of failure developed in 1857; used in predicting the failure of *brittle materials*; states that a material fails when the maximum principal stress developed due to external load reaches the ultimate strength of the material under uniaxial tension test; for a plane stress condition, this theory can be expressed as:

$$\sigma_1 = \sigma_u$$

where

σ_u = ultimate strength in uniaxial tension test
σ_1 = maximum principal stress

$$= \frac{\sigma_x + \sigma_y}{2} + \frac{\sqrt{(\sigma_x + \sigma_y)^2 + 4\tau^2}}{2}$$

σ_x = normal stress in *x*-direction
σ_y = normal stress in *y*-direction
τ = shear stress in *x*-*y* plane

the normalized form of the theory can be written as:

$$\frac{\sigma_1}{\sigma_u} = \pm 1 \; or \; \frac{\sigma_2}{\sigma_u} = \pm 1 (\sigma_2 = \text{minimum principal stress}).$$

Maximum Principal Strain Theory A *Strength of Materials* term; see **Maximum Normal Strain Theory**.

Maximum Principal Stress Theory A *Strength of Materials* term; see **Maximum Normal Stress Theory**.

Maximum Shear Stress Theory A *Strength of Materials* term related to *theory of material's failure*; also called **Tresca's Yield Criterion**, named after the French engineer *Henri Edouard Tresca* (1814–1885) who proposed it; *one of the widely used laws of plasticity* and a failure theory which states that a material fails, when the maximum shear stress developed in the material equals to the maximum shear stress at the yield point in the *uniaxial tension test*; used to predict the failure of *ductile materials*; for a body being subjected to two mutually perpendicular normal stresses σ_x and σ_y and a shear stress τ, this theory can be expressed as:

$$\pm \frac{1}{2}\sqrt{\left[\left(\sigma_x - \sigma_y\right)^2 + 4\tau^2\right]} = \frac{\sigma_{yp}}{2} \quad (1)$$

$$\pm \sqrt{\left[\left(\sigma_x - \sigma_y\right)^2 + 4\tau^2\right]} = \sigma_{yp} \quad (1a)$$

$$\text{alternatively,} \quad \pm \frac{\sigma_1 - \sigma_2}{2} = \frac{\sigma_{yp}}{2} \quad (2)$$

$$\pm \sigma_1 - \sigma_2 = \sigma_{yp} \quad (2a)$$

where

σ_1 = maximum principal stress
σ_2 = minimum principal stress
σ_{yp} = yield stress.

Maximum Total Strain Energy Theory A *Strength of Materials* term related to *theory of material's failure*; also called **Haigh's Criterion**; a failure theory which states that a material fails, when its total strain energy per unit volume reaches its total strain energy at the yield point in the simple *tension test*; for a body being subjected to a three-dimensional stress system consisting of three mutually perpendicular normal stresses σ_x, σ_y, and σ_z, this theory can be expressed as:

$$\frac{1}{2E}[\sigma_x 2 + \sigma_y 2 + \sigma_z 2 - 2\upsilon(\sigma_x\sigma_y + \sigma_y\sigma_z + \sigma_z\sigma_x)] = \frac{\sigma_{yp}^2}{2E}$$

where

$\sigma_x, \sigma_y, \sigma_z$	= three mutually perpendicular normal stresses in x, y, and z directions respectively
σ_{yp}	= yield stress
E	= modulus of elasticity of material
ν	= Poisson's ratio.

Maxwell Relations A *Thermodynamics* term; relate the partial derivatives of state properties such as *pressure* (*p*), *volume* (*v*), and *temperature* (*T*) and thermodynamic property *entropy* (*s*) of a simple compressible substance to each other; useful in determining the *change in entropy* (*ds*), which cannot be measured directly, by simply measuring the changes in properties *P*, *v*, and *T*; expressed as:

$$\left(\frac{\partial T}{\partial v}\right)_s = -\left(\frac{\partial P}{\partial s}\right)_v$$

$$\left(\frac{\partial T}{\partial P}\right)_S = \left(\frac{\partial v}{\partial s}\right)_P$$

$$\left(\frac{\partial s}{\partial v}\right)_T = \left(\frac{\partial P}{\partial T}\right)_v$$

$$\left(\frac{\partial s}{\partial P}\right)_T = -\left(\frac{\partial v}{\partial T}\right)_P$$

MBO A *Management* term; see **Management by Objectives.**

Mean Absolute Error A *Metrology* term; the arithmetic mean of the magnitudes of absolute errors in the measurement of a physical quantity; denoted by Δa_{mean}; given as:

$$\Delta a_{\text{mean}} = \frac{|\Delta a_1| + |\Delta a_2| + \cdots + |\Delta a_n|}{n} = \frac{1}{n}\sum_{i=1}^{i=n} |\Delta a_i|$$

In the above expression $\Delta a_1, \Delta a_2, \ldots \Delta a_n$ are the individual absolute errors in the measured $a_1, a_2, \ldots a_n$ of the physical quantity defined as:

$$\Delta a_1 = a_m - a_1$$
$$\Delta a_2 = a_m - a_2$$
$$\cdots\cdots\cdots\cdots$$
$$\Delta a_n = a_m - a_n$$

where a_m is the arithmetic mean of the measured values, given as:

$$a_m = \frac{a_1 + a_2 + \cdots + a_n}{n} = \frac{1}{n} x \sum_{i=1}^{i=n} a_i$$

Mean Deviation A *Statistics* term; the average of all the deviations; used as one of the measures of dispersion; calculated by taking the mean of the difference between each figure and the arithmetic mean or the median of the series (ignoring sign); given as:

$$\text{Mean deviation} = \frac{\sum_{r=1}^{n} |(a_a - x_r)|}{n}$$

where

a_a = arithmetic mean (average)
n = number of items in the series
x_r = individual item.

Mean Free Path A *Chemistry* term; the average distance that an electron wave can travel without deflection or a molecule travels before it collides with another molecule.

Mean A *Statistics* term; see **Arithmetic Mean**.

Measurand A *Metrology* term; the physical parameter being measured; for example, *force*, *pressure*, or *length*.

Measurement A *Metrology* term; a process of obtaining a quantitative comparison between a predetermined standard and an unknown magnitude of a parameter; widely used in the fields of science and engineering.

Mechanical Advantage A *Theory of Machines* term; a ratio of the output force (load lifted) to the input force (effort required); a measure of the effectiveness of a machine; the higher the mechanical advantage, the higher is the *mechanical efficiency* of a machine; has a value always higher than unity; see also **Velocity Ratio**.

Mechanical Design An important subject of *mechanical engineering* which involves the selection of *proper materials*, *dimensions*, and *shapes* of the components of a structure or machine so that they will support given loads and function without failure; a systematic process for devising a product or system to meet society's technical needs; see also **Machine Design**.

Mechanical Efficiency A *Physics* term; a ratio of the *power output* to the *power input* of a machine; see also **Thermal Efficiency**.

Mechanical Energy A form of energy which can be converted into mechanical work completely and directly by an ideal mechanical device such as an *ideal turbine*; the *kinetic* and *potential* energies are the familiar forms of mechanical energy; the pressure of a flowing fluid is also associated with its mechanical energy.

Mechanical Engineer A technical person who is involved in the research, design, development, manufacture and testing of *tools*, *engines*, *machines*, *structures*, and other *mechanical devices*; deals with a wide range of machines including power-producing machines such as *generators*, *internal combustion engines*, *steam* and *gas turbines*, and *jet* and *rocket engines*; also deals with power-using machines such as *refrigerators* and *air conditioners*, *robots* used in manufacturing, *machine tools*, *materials handling systems*, and *industrial production equipment*.

Mechanical Engineering A discipline of engineering which encompasses the elements like *properties of forces*, *materials*, *energy*, *fluids*, *and motion*, and their application to devise useful products that advance society and improve people's lives; also deals with research, design and manufacturing of power-producing and power-consuming machines like *turbines*, *refrigerator*, and *air conditioner*.

Mechanical Equilibrium A *Physics* term; a state of equilibrium which occurs when there are no unbalanced forces or there is a uniformity of pressure in the system; see also **Thermal Equilibrium** and **Chemical Equilibrium**.

Mechanical Press A *Design* term; a device used to press a part or component of an assembly *using a mechanical means* to give it the desired shape; see also **Hydraulic Press**.

Mechanical System A *Physics* term; a system consisting of forces and their actions on a body.

Mechanical Torsion Meter A *Metrology* term; a device used to measure *torque* by measuring the angle of twist over a fixed length of a shaft using *mechanical means* (pointer and a scale); see also **Optical Torsion Meter**.

Mechanical Work A *Thermodynamics* term; see **Shaft Work**.

Mechanics of Deformable Bodies See **Strength of Materials**.

Mechanics of Machines See **KDM**.

Mechanics of Materials See **Strength of Materials**.

Mechanics of Solids See **Strength of Materials**.

Mechanics A *Physics* term; a branch of physics which deals with force and its application on different bodies (*particles*, *rigid bodies*, *liquids*, etc.); the principles of mechanics are used in *propulsion of rockets*, *propagation of water waves* and *sound waves*, etc.

Mechanism A *Design* term; an assemblage of moving mechanical components which are supported by a rigid structure.

Mechanization A *Management* term; the replacement of human effort in the execution of an activity by a mechanism; a tool to improve industrial performance and increase efficiency.

Mechatronics A recently explored subject which combines and uses the principles of *computer engineering*, *electronics engineering*, and *mechanical engineering*.

Median A *Statistics* term; the middle value of a data obtained after placing the numbers in an ascending or descending order; if there are an even number of items, then the arithmetic mean of the two central numbers is taken as the median; also a triangle term which represents a line segment that connects the vertex of a triangle to the middle point of the opposite side; every triangle has three medians and their point of intersection is called centroid of the triangle; see also **Arithmetic Mean** and **Mode**.

Medium Carbon Steel A *Materials Engineering* term; a plain carbon steel of medium carbon content (0.3% to 0.6%); has higher strength than low-carbon steel (*mild steel*), because of its higher carbon content; used in making machine parts such as *gears*, *axles*, *connecting rods*, *crankshafts*, *railroad equipment*, and parts for metal working machinery; see also **Low Carbon Steel** and **High Carbon Steel**.

Medium Column A *Strength of Materials* term related to a *column*; also called **Intermediate-Length Column**; a column whose length lies between short and long columns; its slenderness ratio varies between 30 and 120; medium columns fail due to the combined effect of buckling and direct compressive stress; see also **Short Column** and **Long Column**.

Medium Head Turbine A *Fluid Mechanics* term; a hydraulic turbine operating at a medium head (between 30 and 500 m); for example, *Francis turbine*; see also **Low Head Turbine** and **High Head Turbine**.

Medium-Term Forecasts An *Operations Management* term related to *forecasting*; have a time frame of three months to two years; these forecasts are commonly used for *aggregate production planning*, including decisions which affect short-term capacity such as *subcontracting* and *overtime*; causal model methods are commonly used for *medium-term forecasts*; see also **Short-Term Forecasts** and **Long-Term Forecasts**.

Megapascal A *Physics* term; the larger SI unit of *pressure* and *stress*; abbreviated as **MPa**; 1 MPa = 10^6 Pa; see also **Kilopascal** and **Gigapascal**.

Melting Furnace A furnace used to melt metals for various applications, especially for casting purposes; a few examples include *electric furnace*, *induction furnace*, *crucible furnace*, and *cupola furnace*.

Melting Point Temperature A *Physics* term; see **Melting Point**.

Melting Point A *Physics* term; also called **Melting Point Temperature**; a *physical property* of a substance; a temperature at which the solid phase of a substance changes into its liquid phase and the two phases co-exists; the higher melting point indicates the greater amount of heat needed to melt the given volume of a metal; a parameter which finds consideration in *heat treatment, hot working,* and *casting* operations; also a key factor during the selection of tool and die materials in manufacturing operations (the higher the melting point of the material, the more difficult the operation becomes); *tungsten* has the highest melting point (3410 °C) in the metal group, making it suitable for *non-consumable electrodes.*

Melting Range A *Materials Engineering* term; the difference between the *solidus* and *liquidus* temperatures.

Melting A *Thermodynamics* term; a process associated with the phase change of a substance from *solid* to *liquid* as a result of absorbing heat.

Meniscus A *Fluid Mechanics* term; a curved liquid surface formed near the place of contact of a liquid with a solid; the nature of curvature depends on the relative magnitude of the cohesive forces existing between the molecules of the liquid, and the adhesive forces existing between the molecules of the liquid and those of the solid; for example, *a concave surface is formed when water touches a glass surface, and a convex surface is formed when mercury touches a glass surface.*

Mer A *Chemistry* term; the repeating unit or the building block of a long chain in the formation of a polymeric molecule.

Merchant's Theory A *Manufacturing* term related to *machining*; a metal cutting theory proposed by *M. E. Merchant*; explains the process of orthogonal cutting and is based on the relationship of shear stress and normal stress acting on the shear plane of a workpiece; expressed as:

$$2\phi + \beta - \alpha = 90^{\circ} \tag{1}$$

The theory was found suitable for certain non-metallic materials but did not work well with steel, so a new modified equation was suggested as:

$$2\phi + \beta - \alpha = C_{\mathrm{m}} \tag{2}$$

where

Φ = shear angle
β = angle of friction
α = rake angle of the cutting tool
C_m = machining constant; depends on the workpiece material; varies between 70° and 80°; for steel, $C_m = 80°$; see also **Lee and Shaffer Theory**.

Mercury Thermometer A device which uses mercury encapsulated in glass to measure temperature; commonly employed to measure human body temperature; usually calibrated in degree Fahrenheit.

Mesh An *Electrical* term; also called **Loop**; a closed path in an electrical circuit.

Metacentre A *Fluid Mechanics* term; a fixed point in a fluid about which a floating body when disturbed from its initial position starts oscillating before it comes to rest again; the point of intersection of the line of action of the force of buoyancy and the normal axis of the body, when the body is given a small angular displacement.

Metacentric Height A *Fluid Mechanics* term; the vertical distance between the *centre of gravity* of a floating body and the *metacentre*; a measure of stability for a floating body; the larger the length of the metacentric height, the more stable is the floating body.

Metal Inert Gas Welding A *Welding* term, abbreviated as **MIG Welding**; also called **Gas Metal Arc Welding**, abbreviated as **GMAW**; an arc welding process which uses a shielding atmosphere of inert gases such as *argon, helium, nitrogen,* or *carbon dioxide* (either alone or in combination) to protect the arc produced between a consumable electrode (bare electrode) acting as a filler metal and the workpiece; uses *reverse polarity* that ensures stable arc and smooth metal transfer resulting in good quality weld; used previously for the joining of aluminium and stainless steels, but currently also used for other *ferrous* and *non-ferrous* metals; see also **Tungsten Inert Gas Welding**.

Metallic Bond A *Chemistry* term; *a variation of the covalent bond*; the force of attraction which binds a metal atom (or Kernel) to a number of electrons within its sphere of influence; *non-directional as well as weaker than a covalent bond.*

Metallic Materials A *Chemistry* term; the materials having metallic bonds.

Metallic Property A *Chemistry* term; the property which characterizes a metal; a few examples include *electrical conductivity, thermal conductivity, bright lustre, ductility, malleability,* and *elasticity.*

Metallography A *Materials Engineering* term; a process of preparing a metallic sample to a mirror-like finish, exposing it to etching (a chemical attack) and then subjecting it to a light radiation from an optical microscope for the purpose of observing or recording sample's microstructure.

Metalloid A *Chemistry* term; also called **Semi-Metal**; a *chemical element* which shows the properties of both metal and non-metal; includes *silicon* (Si), *germanium* (Ge), *arsenic* (As), *antimony* (Sb), and *tellurium* (Te).

Metallurgy A *Materials Engineering* term; the science and technology of *extracting metals* from their ores present in the earth, refining them and preparing them for use; some important metallurgical operations include *crushing* and *grinding* of ore, *concentration* or *benefaction* of ore, *extraction* of crude metal from the concentrated ore, and *refining* of the crude metal.

Metals A *Materials Engineering* term; the materials having a regular arrangement of constituent atoms; have relatively *high strength, high stiffness, ductility* or *formability,* and *shock resistance*; lack the desired properties in pure form, hence is used in the combined form (*alloy*); particularly useful for structural or load-bearing applications; have wide engineering and domestic applications; find extensive use in making *alloys*; a few examples include *iron, copper, aluminium,* and *zinc*; see also **Non-Metals.**

Method of Joints A *Strength of Materials* term related to a *truss*; a method of finding forces in different members of a truss by isolating the joint (under consideration) from the entire truss and applying the equations of static equilibrium, $\Sigma F_x = 0$ and $\Sigma F_y = 0$; involves only two equations for every joint; useful for finding forces in every member of the truss; see also **Method of Sections.**

Method of Sections A *Strength of Materials* term related to a *truss*; a method of finding forces in different members of a truss using a section which divides the entire truss into two parts and considering equilibrium of either of the parts treating it as a free body; the selected section should cut minimum number of members, must cut the desired member and the members must have maximum three unknowns; a horizontal section is usually not selected as it will cut many members of unknown forces; a faster and efficient method of finding forces in few members of a truss; see also **Method of Joints.**

Method Study An *Operations Management* term related to *work study*; the systematic and scientific evaluation of existing and proposed methods of doing a work as a means of developing the proposed method through analysis and critical examination; helps in improving the methods of production and results in more effective use of *material, machine,* and *manpower*; see also **Motion Study**.

Metre A *Physics* term; the SI unit of length; denoted by m; and 1 m = 100 cm = 1000 mm; see also **Centimetre** and **Millimetre.**

Metric System A *Physics* term; a measuring system in which the fundamental units are metre, litre, and kilogram.

Microchronometer A *Work Study* term; also called **Wink Counter**; a precision time measuring instrument developed by *Frank Gilbreth* (1868–1934); can measure 1/2000 of a minute (called *wink*).

Microcontroller An *Electronics* term; an electronic device which consists of a microprocessor, memory unit, input and output ports, a crystal oscillator for timing, analog-to-digital converters (ADC) and digital-to-analog converters (DAC) all on one circuit board.

Microeconomics An *Economics* term; the study of economics at the level of the individual economic entity like the individual firm, the individual consumer, or the individual worker; see also **Macroeconomics.**

Micromanometer A *Fluid Mechanics* term; a device used to measure small pressure difference between two points or to measure the large pressure difference precisely.

Micrometer A *Metrology* term; a precision measuring instrument used to measure the thickness of flat stocks or the diameter of round objects.

Micromotion Analysis An *Operations Management* term related to *work study*; see **Micromotion Study**.

Micromotion Study An *Operations Management* term related to *work study*; also called **Micromotion Analysis**; the micro level study of the fundamental motions involved in an operation performed by an operator; originated by *Frank Gilbreth* (1868–1934) and *Liiian Gilbreth* (1878-1972); the study uses a motion picture camera and a time measuring device to accurately indicate the time intervals on the motion picture film in order to find one best pattern of motions involving *less efforts*, *time*, and *fatigue* while performing a task.

Microprocessor An *Electronics* term; a single-chip computer element which contains a control unit, central processing circuitry and all the necessary logic functions to serve as a central processing unit (CPU) of a microcomputer or as a dedicated automatic control system.

Microscopy The use of a microscope for investigation.

Microstructure A *Materials Engineering* term; the detailed structure of a material which is revealed during its microscopic examination, for which the length-scale varies between 10 and 1000 nm (1 nm = 10^{-9} m); described by features such as *average grain size*, *grain size distribution*, *grain shape*, *grain orientation*, and other features related to defects in materials; see also **Macrostructure**.

Middle Level Manager A *Management* term; also called **Middle Manager**; the individual at levels of management between the low level manager (the *supervisor*) and the top level manager; responsible for translating the goals set by the top level manager into specific details which lower level managers can perform; the different designations used for a middle level manager include *project leader*, *unit chief*, *district manager*, *dean*, *head of department*, *bishop*, or *divisional manager*; see also **High Level Manager** and **Low Level Manager**.

Middle Manager A *Management* term; see **Middle Level Manager**.

MIG Welding A *Welding* term; see **Metal Inert Gas Welding**.

Mild Steel A *Materials Engineering* term; see **Low Carbon Steel**.

Military Organization A *Management* term; see **Line Organization**.

Mill A *Manufacturing* term; see **Milling Machine**.

Miller Indices A *Materials Engineering* term; a *notation system* consisting of a three-digit set of integers used in crystallography for lattice planes in crystal lattices; named after a British mineralogist *William Hallowes Miller* (1801–1880) who introduced it in 1839 and laid the foundations of *modern crystallography*; define the coefficients of imaginary planes in a crystal, thus used to characterize a crystalline plane; see also **Miller–Bravais Indices**.

Millimetre A *Physics* term; the most widely used unit of length for *engineering applications*; used to measure very small length in the integer form; the one-tenth part of a centimetre; the one-thousandth part of a metre (m); denoted by mm; 1 mm $= 10^{-1}$ cm $= 10^{-3}$ m, that is, 1 cm = 10 mm and 1 m = 1000 mm; see also **Metre** and **Centimetre**.

Milling Machine A *Manufacturing* term; also called **Mill**; a versatile machine used for the machining of rough surfaces of a workpiece to make it flat and smooth; also used for *cutting slots*, *grooves*, and *holes*.

Milling A *Manufacturing* term; a *versatile machining process* which can be used to produce a variety of shapes; the typical products include *moulds* and *dies*; the most common types of milling operations include *slab milling*, *face milling*, and *end milling*.

Mineral A *Chemistry* term; a naturally occurring material in the form of which a metal occurs in the earth along with impurities; see also **Ore**.

Minor Diameter A *Theory of Machines* term related to a *gear*; the smallest diameter of a screw thread; see also **Major Diameter**.

Minor Losses A *Fluid Mechanics* term; the smaller head losses associated with a fluid during its flow in a pipeline which result from changes in flow velocity, either due to change in its magnitude or its direction or a change in both; include losses of heads due to *sudden contraction*, *sudden enlargement* or *bending of the pipeline* or due to *various pipe fittings* (*valves*, *elbows*, etc.) along with losses of heads at the *entrance* or *exit* of the pipeline.

Misalignment A *Design* term; a condition in which axes of the two shafts do not lie in the same line; may lead to failure of bearings or fatigue failure of shafts.

Misrun A *Casting* term; a type of *casting defect* which occurs *due to lack of molten metal fluidity* (because of low metal temperature), which prevents the molten metal to reach the desired part of the mould cavity and thus solidifying the metal in the midway; see also **Cold Shut**.

Mistake-Proofing An *Operations Management* term; see **Poka-Yoke**.

Mixed Cycle A *Thermodynamics* term related to *air standard cycle*; see **Dual Cycle**.

Mixed Dislocation A *Materials Engineering* term; a type of *linear defect* in a material which has characteristics from both *edge* and *screw* dislocations; the Burgers vector is neither perpendicular nor parallel to the dislocation line, but remains at a fixed orientation, that is, remains the same for all portions of the mixed dislocation; the *most common linear defect found in actual materials*; see also **Edge Dislocation** and **Screw Dislocation**.

Mixed Flow Turbine A *Fluid Mechanics* term; a turbine in which a fluid flows partly along the radius and partly along the axis of the shaft of the turbine; for example, *Kaplan turbine* is a mixed flow turbine; see also **Axial Flow Turbine** and **Radial Flow Turbine**.

Mixed Flow A *Fluid Mechanics* term; a fluid flow which consists of partly radial flow and partly axial flow; occurs in *Kaplan turbine*; see also **Axial Flow** and **Radial Flow**.

Mixed Fraction A *Mathematics* term; a combination of a whole number and a proper fraction; a few examples include $1\,\frac{1}{4}$, $2\,\frac{1}{2}$ etc.; see also **Proper Fraction** and **Improper Fraction**.

Mixed-Flow Pump A *Fluid Mechanics* term; a *rotodynamic pump* which is intermediate between centrifugal pump and axial-flow pump; a fluid enters axially but leaves at some angle between radially and axially; see also **Axial-Flow Pump** and **Centrifugal Pump**.

MNC A *Management* term; see **Multinational Company**.

Mode A *Statistics* term; the most frequently occurring value in a group of measurements; in case of data consisting of many numbers, its mode will be that number which will repeat more times; a multimode situation is that, when different items appear the same number of times; there may be no mode, if no item is repeated; see also **Arithmetic Mean** and **Median**.

Model A *Design* term; a *representation* or *abstraction* of an actual object or situation; used to study and analyse the behaviour of a system for the purpose of improving its performance; quantitative techniques are used to investigate the relationship among the variables in a model; may be classified into *physical* and *mathematical* models; see also **Prototype**.

Modular Design A *Manufacturing* term; the creation of products from some combination of basic pre-existing subsystems; a product design concept which provides a range of product options to a consumer in a shorter lead time.

Modular Organization A *Management* term; see **Boundaryless Organization**.

Module A *Theory of Machines* term related to a *gear*; a ratio of the *pitch circle diameter* to the *number of teeth* of the gear; the *reciprocal of the diametral pitch*; denoted by *m*; expressed in inch per tooth or mm per tooth; used to measure the size of a gear tooth in the metric system.

Modulus of Elasticity A *Strength of Materials* term; see **Young's Modulus of Elasticity**.

Modulus of Resilience A *Strength of Materials* term related to a *spring*; the energy which a material can absorb per unit its volume without undergoing permanent deformation; equals to the area under the stress–strain curve up to the elastic limit; see also **Resilience**.

Modulus of Rigidity A *Strength of Materials* term; also called **Shear Modulus**; a ratio of the shear stress to the shear strain within elastic limit; an elastic constant denoted by G; connected to solid materials only; the shear modulus of solids is always less than their Young's modulus (E); for most of the solid materials, the value is G is approximately one-third of the value of E; the shear modulus of an ideal liquid is zero, since a liquid at rest cannot sustain a shear force; the SI unit is pascal (Pa), which is equivalent to N/m^2; the values of G for some materials are: steel (80 GPa), iron (50 GPa), copper (40 GPa), glass (30 GPa), aluminium (25 GPa), and wood (10 GPa); see also **Young's Modulus of Elasticity** and **Bulk Modulus of Elasticity**.

Modulus of Toughness A *Strength of Materials* term; the energy which a material can absorb per unit its volume without fracture; equals to the total area under the stress–strain curve up to the fracture point; see also **Toughness**.

Mohr's Circle Method A *Strength of Materials* term; a graphical method represented in the form of a circle and named in honour of a *German* civil engineer *Christian Otto Mohr* (1835–1918); used for finding the stresses on any inclined (oblique) plane of an element subjected to plane stresses; especially useful for finding the principal stresses, and the maximum and minimum shear stresses; the radius of the Mohr's circle gives the value of maximum shear stress.

Mohr's First Theorem A *Strength of Materials* term related to *deflection*; named after a famous *German* civil engineer Otto Mohr (1835-1918); also called **First Moment-Area Theorem**; a theorem used to find the slope of a loaded beam; states that the change in slope between any two points of a loaded beam is equal to the net area of the bending moment diagram between these two points divided by the flexural rigidity of the beam; see also **Mohr's Second Theorem**.

Mohr's Second Theorem A *Strength of Materials* term related to *deflection*; named after a famous *German* civil engineer Otto Mohr (1835-1918); also called **Second Moment-Area Theorem**; a theorem used to find the deflection of a point with respect to another point of fixed position (reference point) of a loaded beam; states that the vertical deflection of a point on a straight beam, measured from the tangent drawn at another point on the beam, is equal to the moment of area of bending moment diagram between these two points about the point where this deflection occurs, divided by the flexural rigidity of the beam; see also **Mohr's First Theorem**.

Mohr's Theorems A *Strength of Materials* term related to *deflection*; named after a famous *German* civil engineer *Otto Mohr* (1835–1918); used to find *deflection* of loaded beams; see **Mohr's First Theorem** and **Mohr's Second Theorem**.

Moissan's Method A *Chemistry* term; a method used to produce *artificial diamond.*

Moist Air A *Thermodynamics* term related to *psychrometry*; the air containing both *dry air* and *water vapour.*

Moisture A *Thermodynamics* term; see **Water Vapour**.

Molar Mass A *Chemistry* term; also called **Molecular Weight**; the mass of one mole of a substance (molecule or compound) expressed in grams; also the mass of one kmol of the substance expressed in kilograms; for example, the molecular weight of *water* (H_2O) is 18 g.

Molar Specific Entropy A *Thermodynamics* term; the entropy per unit number of moles of a working fluid; the SI unit is J/kg mol K.

Molar Specific Property A *Thermodynamics* term; a property based on the number of moles of a substance in place of its mass.

Molar Specific Volume A *Thermodynamics* term; the volume of a body per unit of its number of moles; the SI unit is m^3/kg mol; see also **Specific Volume**.

Molecular Range A *Chemistry* term; the maximum distance up to which a molecule can exert its influence to attract other molecules.

Molecular Weight A Chemistry term; see **Molar Mass**.

Mollier Chart A *Thermodynamics* term; see **Mollier Diagram**.

Mollier Diagram A *Thermodynamics* term, named after the *German* scientist *Richard Mollier* (1863–1935); also called **Mollier Chart** or **Enthalpy-Entropy Diagram** or **H–S Diagram**; a plot drawn between *enthalpy* (h) on the vertical axis and *entropy* (s) on the horizontal axis; commonly employed in the analysis of a *vapour power cycle* (the *Rankine cycle*) because of relative ease with which the pertinent property values can be determined for the various fluid states in the cycle.

Molybdenum A *Chemistry* term; a metallic element having symbol Mo with atomic number 42; a fairly hard silvery white metal which melts at 2610 °C; an important alloying element which is added to steel to increase its hot hardness; also used for the *filament supports* of electrical lamps and as *windings* for resistance furnaces.

Moment of Inertia A *Strength of Materials* term; see **Second Moment of Area**.

Moment of Momentum A *Physics* term; see **Angular Momentum**.

Moment A *Physics* term; also called **Torque**; the product of magnitude of a force and the perpendicular distance from the line of action of the force; produces *turning effect* in a body about an axis; classified into *clockwise* and *counterclockwise* moments; counterclockwise moment is considered *positive* and clockwise moment *negative*;

the SI unit is N-m; a *vector quantity*, which is defined as the *cross product* of force and perpendicular distance, and is expressed as:

$$\boldsymbol{M} = \boldsymbol{r} \times \boldsymbol{F} = rF \sin\theta\, \boldsymbol{n}$$

where

$\boldsymbol{F}$ = force vector corresponding to the force F acting on the body
$\boldsymbol{r}$ = perpendicular distance (vector) of the force from the axis of rotation
θ = angle between $\boldsymbol{r}$ and F
$\boldsymbol{n}$ = unit vector perpendicular to the plane containing $\boldsymbol{r}$ and $\boldsymbol{F}$.

Moment-Area Method A *Strength of Materials* term related to *deflection*; also called **Area-Moment Method**; a deflection method proposed by a famous *German* civil engineer *Otto Mohr* (1835–1918); uses Mohr's theorems based on *bending moment diagram* to find slope and deflection of loaded beams; useful for beams of *constant flexural rigidity*.

Monel Metal A *Materials Engineering* term; an alloy of *nickel* (Ni) and *copper* (Cu) consisting of about 60% Ni and 40% Cu; used in making *turbine blades* and *propellers* because of its acid-proof nature; also finds use in making decorative pieces because of its shining surface.

Monitoring A *Management* term; see **Controlling**.

Monomer A *Chemistry* term; an individual molecule which combines with similar molecules to form a *polymeric molecule*; for example, an *ethylene* is a monomer for *polythene*.

Monopoly An *Economics* term; a market structure in which there is just one firm which prevents the entry of other firms; discourages competition and dominates price.

Monorail An *Operations Management* term; an I-section beam attached to the ceiling having a trolley or carrier which moves along it; used to transfer material from one place to another in machine shops and other shops.

Monte Carlo Method An *Operations Research* term; a method of simulation that uses random numbers; used to solve problems which involve conditions of uncertainty and where mathematical evaluation is not possible; yields a solution which is very close to the optimal, but not necessarily the exact solution; restricted only to models involving uncertainty.

Moody Chart A *Fluid Mechanics* term; a chart named after an *American* scientist *Lewis Ferry Moody* (1880–1953); the chart relates the friction factor as a function of Reynolds number (a non-dimensional group) and relative roughness for round pipes; universally valid for all steady, fully developed incompressible pipe flows.

Moritz Weber (1871–1951) A *German* professor of naval mechanics who is known for the *Weber number* (a non-dimensional group); emphasized the use of the principles of *similitude* in fluid flow studies; formulated a capillarity similarity parameter.

MOS See **Strength of Materials**.

Most Likely Time A *Project Management* term; the most probable time to complete a project activity; used to calculate the *expected time of completion* of an activity in *program evaluation and review technique* (PERT); see also **Optimistic Time** and **Pessimistic Time**.

Motion Study An *Operations Management* term related to *work study*; a part of *work study* developed by *Frank Gilbreth* (1868–1934) and *Lillian Gilbreth* (1878–1973); the study of motions involved in a work to differentiate between useful and wasteful motions and accordingly eliminate wasteful motions; an important part of *job analysis*; see also **Method Study**.

Motion A *Physics* term; the state of a body with some velocity; the state of a body in which it changes its position with respect to time; a few examples include *a man walking on a road* and *a train moving on rails*; see also **Rest**.

Motivation Factors A *Management* term related to *motivation*; see **Intrinsic Factors.**

Motivation A *Management* term; an inner force which excites and drives a person to action and achieve something; subjective and qualitative in nature rather than objective and quantitative; a psychological entity which cannot be measured; varies with time such as age of the person, mood at the time, immediate history and with person's characteristics like attitude; a means used by the managers to induce their subordinates to act in a desired manner by fulfilling their desires, needs, wishes, and other requirements.

Motivator-Hygiene Theory A *Management* term related to *motivation*; see **Hygiene Theory**.

Motivators A *Management* term related to *motivation*; see **Intrinsic Factors**.

Motor Efficiency An *Electrical* term; a ratio of the *mechanical power output* (shaft work output) to the *electrical power input*; see also **Generator Efficiency**.

Mould Cavity A *Manufacturing* term related to *casting*; a cavity which lies within a mould which holds molten metal used for making the castings.

Mould A *Manufacturing* term related to *casting*; a *void* created in a *compact sand mass* which, when filled with molten metal, will produce a casting; an impression created by a pattern after its withdrawal from the mould; also an assembly of two or more metal blocks used to make castings of good surface finish and intricate details.

Moulding Sand A *Casting* term; the sand used in making moulds; principally contains *silica, clay,* and *water* in various proportions; the important properties of the moulding sand include *permeability, refractoriness, cohesiveness, flowability,* and *collapsibility*; some of its types include *parting sand, facing sand,* and *core sand.*

Movement Inventory An *Operations Management* term related to *inventory*; the inventory in transit or in the pipeline; a few examples include *oil* and *gasoline*; see also **Buffer Inventory**, **Cycle Inventory**, and **Anticipation Inventory**.

Moving Average Method An *Operations Management* term related to *forecasting*; a *short-range quantitative* forecasting method which averages the data from a few recent past periods to form the forecast for the next period; the *average moves with time* with the *omission of older data* used in calculating the average, hence called moving average method; the greater the number of periods used, the more accurate the forecast is; each data is assigned *equal weightage*; does not provide good forecasts in case of *biased demand trend* or *seasonal influences*; see also **Weighted Moving Average Method**.

MPa A *Physics* term; see **Megapascal**.

MPS An *Operations Management* term; see **Master Production Schedule**.

MRP II An *Operations Management* term which is the acronym of **Manufacturing Resource Planning**; an extension of the material requirements planning (MRP); a system of *planning* and *controlling* the *operational, engineering,* and *financial resources* of a *manufacturing unit*; see also **Material Requirements Planning**.

MRP An *Operations Management* term; see **Material Requirements Planning**.

M-Series A *Materials Engineering* term; one of the two series of *high-speed steel* (HSS) that contains higher content of *molybdenum* (up to about 10%), see also **T-Series**.

Muda An *Operations Management* term; a *Japanese* word used to indicate wasteful activity which does not add any value; seven wastes were identified by *Taiichi Ohno* (1912–1990), which include *overproduction, overprocessing, transportation, waiting, inventory, motion*, and *defects.*

Muff Coupling A *Design* term; also called **Sleeve Coupling**; a type of *rigid coupling* and a *temporary fastening device* used to connect *two coaxial shafts*; consists of a hollow cast iron cylinder (called *muff* or *sleeve*) fitted over the ends of the two shafts to be connected and a *sunk key* inserted.

Muffler A *Physics* term; a device used to reduce noise in an instrument.

Muller A *Manufacturing* term; a *foundry tool* used to mix the ingredients of the moulding sand such as *sand, binders*, and *water* thoroughly.

Multi-factor Productivity A *Management* term; see **Total Factor Productivity**.

Multimeter A Metrology term; a measuring instrument which combines three separate meters, namely an *ammeter* to measure current, a *voltmeter* to measure voltage, and an *ohmmeter* to measure resistance.

Multinational Company A *Management* term, abbreviated as **MNC**; a company which has expanded its business operations in more than one country.

Multiple Sampling A *Quality Control* term; the acceptance sampling of a lot based on many small samples; used when the results of a single sampling requires more scrutiny; see also **Single Sampling**.

Multi-skilled Worker An *Operations Management* term; a worker who is cross-trained on several jobs to become multi-skilled; has the advantage of being moved to other jobs as per need to eliminate any job imbalance in work flows which may occur because of quality problems or machine breakdowns.

N

N-2 Technique A *Quality Control* term; a *Japanese* method of inspection in which the first and the last piece in a lot are inspected for the acceptance or rejection of the lot; the entire lot is accepted, if two pieces are found meeting the prescribed standards.

Nanotechnology A set of technologies which are based on *physical*, *chemical*, and *biological* phenomena occurring at a nano scale, that is, at a length-scale of 10^{-9} m.

Nappe A *Fluid Mechanics* term; a sheet of water in the downstream of a weir springing freely over the weir.

National Income An *Economics* term; a measure of income earned by a nation from the current production of goods and services; includes incomes of factors of production such as *land*, *labour*, and *capital*.

Natural Convection A *Heat Transfer* term; see **Free Convection**.

Natural Rubber A *Chemistry* term; see **Rubber**.

Natural Satellite A *Physics* term; a satellite made by the nature or a *celestial body* which is revolving continuously around another celestial body; for example, *earth* is a natural satellite of the *sun*, and *moon*, in turn, is a natural satellite of the *earth*; in our planet system, each planet is a natural satellite of the sun, and some of the planets also have their own natural satellites similar to that of earth; Jupiter has 16 and Saturn has 18 natural satellites which are called their moons; see also **Artificial Satellite**.

Natural Strain A *Strength of Materials* term; see **True Strain**.

Natural Variation A *Statistics* term; also called **Common Cause Variation** or **Chance Variation**; occurs at random and individual sources or causes cannot be identified or explained; occurs as a result of complex interactions of *variations* in *materials*, *tools*, *machines*, *information*, *workers*, and the *environment*; example includes variations occurring in various processes; forms a pattern which can be described as the *normal distribution*; see also **Assignable Variation**.

D. K. Singh, *Dictionary of Mechanical Engineering*,
https://doi.org/10.1007/978-981-99-1722-8_14

Navier–Stokes Equations A *Fluid Mechanics* term; named after a *French* mathematician *Louis Marie Henri Navier* (1785–1836) and an *English* mechanician *Sir George Gabriel Stokes* (1819–1903); used to provide a complete mathematical description of the flow of *incompressible* Newtonian fluids when used with the *conservation of mass equation*; expressed as:

along *x*-direction

$$\rho\left(\frac{\partial u}{\partial t}+u\frac{\partial u}{\partial x}+v\frac{\partial u}{\partial y}+w\frac{\partial u}{\partial z}\right)=-\frac{\partial p}{\partial x}+\rho g_x+\mu\left(\frac{\partial^2 u}{\partial x^2}+\frac{\partial^2 u}{\partial y^2}+\frac{\partial^2 u}{\partial z^2}\right)$$

along *y*-direction

$$\rho\left(\frac{\partial v}{\partial t}+u\frac{\partial v}{\partial x}+v\frac{\partial v}{\partial y}+w\frac{\partial v}{\partial z}\right)=-\frac{\partial p}{\partial y}+\rho g_y+\mu\left(\frac{\partial^2 v}{\partial x^2}+\frac{\partial^2 v}{\partial y^2}+\frac{\partial^2 v}{\partial z^2}\right)$$

along *z*-direction

$$\rho\left(\frac{\partial w}{\partial t}+u\frac{\partial w}{\partial x}+v\frac{\partial w}{\partial y}+w\frac{\partial w}{\partial z}\right)=-\frac{\partial p}{\partial z}+\rho g_z+\mu\left(\frac{\partial^2 w}{\partial x^2}+\frac{\partial^2 w}{\partial y^2}+\frac{\partial^2 w}{\partial z^2}\right)$$

where

ρ = density of the fluid
p = pressure of the fluid
g_x, g_y, g_z = components of the acceleration of gravity along *x*, *y*, and *z* direction, respectively
u, v, w = components of the fluid velocity along *x*, *y*, and *z* direction, respectively
μ = dynamic viscosity of the fluid.

NC A *Manufacturing* term related to *automation*; see **Numerical Control**.

Near-Net Shape Casting A *Manufacturing* term related to *casting*; see **Permanent Mould Casting**.

Near-Net-Shape Forging A *Manufacturing* term; see **Precision Forging**.

Neck Formation A *Strength of Materials* term; also called **Necking**; a stage in ductile material's (mild steel) deformation process when the rate of increase of strain is very high, which produces a visible decrease in cross-sectional area where a material will eventually fracture.

Necking Region A *Strength of Materials* term; the region on an engineering stress–strain diagram between the ultimate stress and the point of fracture.

Necking A *Strength of Materials* term; see **Neck Formation**.

Need-Hierarchy Theory A *Management* term related to *motivation*; a motivation theory proposed by *Abraham Harold Maslow* (1908–1970), hence also called **Maslow's Theory**; based on the hierarchy of five levels (from lower to higher) of human needs, namely *physiological need*, *safety need*, *social need*, *esteem need*, and *self-actualization need*, taken in that order; suggests that a higher level need arises when the needs below it are satisfied and an unsatisfied need acts as a motivator; also suggests that lower level needs can be satisfied, but higher level needs are never completely satisfied.

Needle File A *Manufacturing* term; one of the types of a file used in the fitting shop which has a needle-like sharp point; used in the filing of *very fine works* such as *pierced design of sheet metal.*

Negative Gauge Pressure A *Fluid Mechanics* term; see **Vacuum**.

Negative Integers A *Mathematics* term; any negative number such as − 1, − 2, − 3, − 4, etc.; see also **Positive Integers**.

Net Positive Suction Head A *Fluid Mechanics* term related to a *pump*; abbreviated as **NPSH**; the difference between a pump's inlet *stagnation pressure head* and the *vapour pressure head*; used as a flow parameter to test the condition of *cavitation* in a pump.

Net Working Capital A *Financial Management* term; the amount by which a firm's current assets exceeds its current liabilities; a common measure of a *firm's overall liquidity.*

Net-Shaped Forging A *Manufacturing* term; see **Precision Forging**.

Network Organization A *Management* term; see **Boundaryless Organization**.

Neutral Axis A *Strength of Materials* term; an imaginary line along which the neutral plane of a beam (or any other member) intersects the plane of bending and where the longitudinal stress is zero; it passes through the axis of the beam containing its centroid, and is perpendicular to the plane of symmetry; see also **Plane of Symmetry**.

Neutral Equilibrium A *Physics* term; a type of *translational static equilibrium* which ensures a body to stay in equilibrium even after being slightly displaced and released; for example, a ball rolling on the ground; see also **Stable Equilibrium** and **Unstable Equilibrium**.

Neutral Flame A *Chemistry* term; a type of oxyacetylene gas flame which contains equal proportion of fuel gas (*acetylene*) and oxygen in the mixture; finds use in gas welding to join *mild steel*, *alloy steel*, *grey cast iron*, and others; the most extensively used gas flame among all flames; see also **Oxidizing Flame** and **Carburizing Flame**.

Neutral Plane A *Strength of Materials* term; also called **Neutral Surface**; an imaginary plane lying in a beam (or any other member) subjected to bending where the longitudinal stress is zero; this plane contains the neutral axes of all the cross sections of a beam (or any other member) under bending and passes through the centroids of the cross sections.

Neutral Refractories A *Materials Engineering* term; refractories which do not combine with either acidic or basic refractories, and are produced by the combination of the two; a few examples include *silicon carbide, chromite* (Cr_2O_3), and *carbon*; see also **Acidic Refractories** and **Basic Refractories.**

Neutral Surface A *Strength of Materials* term; see **Neutral Plane.**

Neutron A *Chemistry* term; one of the three fundamental particles of an atom of a chemical element discovered by *James Chadwick* (1891–1974), a *British* physicist in 1932, while bombarding a thin sheet of *beryllium* (Be) by *alpha particles*; a *neutral* particle having a mass equal to 1.6748×10^{-27} kg (slightly greater than that of proton); resides in the nucleus of the atom whose mass is almost entirely due to the presence of protons and neutrons; see also **Electron** and **Proton.**

newton A *Physics* term; the *SI unit of force* named after the famous *British* scientist *Sir Isaac Newton* (1642–1727); denoted by *N*; 1*N* is a force which acts on a body of mass 1 kg and produces an acceleration of 1 m/s^2; newton and dyne are related as 1 N = 10^5 dyne; see also **Dyne.**

Newton's First Law of Motion A *Physics* term; one of the *laws of motion* proposed by a famous *British* scientist *Sir Issac Newton* (1642-1727); also called **Law of Inertia**; states that everybody continues to remain in a state of rest or of uniform motion along a straight line unless an external force tries to change its state; provides the condition of static equilibrium of a particle and the definition of force, and lays the foundation for the study of statics.

Newton's Law of Cooling A *Heat Transfer* term; a law proposed by a famous *British* scientist *Sir Issac Newton* (1642-1727); states that the rate of heat transfer by *convection* is proportional to the surface area of a body through which the heat transfer takes place and to the temperature difference between the surface temperature and the fluid temperature; given as:

$$Q = hA(T_s - T_f)$$

where

Q = heat transfer rate
h = convection heat transfer coefficient
A = surface area of the body
T_s = surface temperature of the body
T_f = fluid temperature.

Newton's Law of Gravitation A *Physics* term; a law proposed by a famous *British* scientist *Sir Issac Newton* (1642-1727); also called **Law of Gravitation**; states that everybody in the universe attracts every other body with a force (F), which is directly proportional to the product of their masses, and inversely proportional to the square of the distance between them, and acts along the line joining the two bodies; the force is always *attractive*; for two bodies having their masses m_1 and m_2 and separated by a distance r, this law can be mathematically expressed as:

$$F = G\frac{m_1 m_2}{r^2},$$

where

G = universal gravitational constant = 6.67×10^{-11} N.m^2.kg^{-2}.

Newton's Law of Viscosity A *Fluid Mechanics* term; also called **Law of Viscosity**; a law proposed by the famous *British* scientist *Sir Isaac Newton* (1642–1727); states that the shear stress of a fluid varies directly proportional to its velocity gradient; expressed as:

$$\tau = \mu\frac{\mathrm{d}v}{\mathrm{d}y}$$

where

τ = shear stress
μ = absolute viscosity
$\frac{\mathrm{d}v}{\mathrm{d}y}$ = velocity gradient

Newton's Laws of Motion A *Physics* term; see **Laws of Motion**.

Newton's Second Law of Motion A *Physics* term; one of the *laws of motion* proposed by a famous *British* scientist *Sir Issac Newton* (1642-1727); also called **Law of Momentum**; the time-rate of change of linear momentum of a body is equal to the net force acting on the body and takes place in the direction of the force; provides the quantitative definition of force which equals to mass times acceleration and has the same direction as the acceleration.

Newton's Third Law of Motion A *Physics* term; one of the *laws of motion* proposed a famous *British* scientist *Sir Issac Newton* (1642-1727); also called **Law of action and Reaction**; to every action there is an equal and opposite reaction; used in both statics and dynamics.

Newtonian Fluid A *Fluid Mechanics* term; a fluid which obeys the *Newton's law of viscosity*; named after a famous *British* scientist *Sir Issac Newton* (1642-1727); for a newtonian fluid, the shear stress is directly proportional to the velocity gradient; the most common examples include *air* and *other gases*, *water*, *kerosene*, *gasoline*, and *oils*; see also **Non-Newtonian Fluid**.

Nibbling A *Manufacturing* term related to *sheet metal working*; a shearing operation which uses a machine called a *nibbler* to move a small straight punch up and down rapidly into a die to produce several overlapping holes in the sheet when placed in the gap; used to produce intricate slots and notches.

Nichrome A *Materials Engineering* term; an alloy of *nickel* (Ni), *chromium* (Cr), and *iron* (Fe) consisting of 60% Ni, 15% Cr, and 25% Fe; used in making *electrical resistances.*

Nickel Alloys A *Materials Engineering* term; the alloys of *nickel* (Ni) which contain nickel as the main constituent besides other elements present in various amounts; important nickel alloys include *nickel steel, invar, monel metal, cupronickel, German silver, nichrome, constantan, nickel coinage*, and *silver coinage.*

Nickel Coinage A *Materials Engineering* term; an alloy of *nickel* (Ni) and *copper* (Cu) consisting of 25% Ni and 75% Cu; used in making *coinage.*

Nickel Steel A *Materials Engineering* term; an alloy of *nickel* (Ni) and *iron* (Fe) which contains 2.5 to 5% Ni; extremely hard, tough, and yet elastic; because of its rust-proof nature, it is used in laying *underground cables*, making *armour plates, aeroplanes*, and *automobile parts.*

Nickel A *Chemistry* term; a metallic element having symbol Ni with atomic number 28; a greyish white metal which melts at 1453 °C; hard, malleable and has ferromagnetic nature, though much less than iron, but has high electrical and thermal conductivities (both nearly 15% of those of silver); used as an important *alloying element* and imparts hardness when added to steel; important alloys include *nickel steel, invar, monel metal, cupronickel, German silver, nichrome, constantan, nickel coinage*, and *silver coinage.*

Nitriding A *Manufacturing* term related to *heat treatment*; a *case hardening process* which involves heating of *low alloy steel* parts in a *nitrogenous atmosphere* (usually *ammonia*) to a temperature of 900–1000 °F which allows nitrogen to be absorbed in the surfaces of the parts.

Noble Gases A *Chemistry* term; also called **Inert Gases**; the atoms of *zero group* in the periodic table having *stable electronic configuration* as their valence shell (outermost shell) is completely filled up; have very little tendency to form chemical bonds and are the most stable; *molecules of all noble gases are monoatomic* as they do not combine with themselves; examples include *helium* (He), *neon* (Ne), *argon* (Ar), *krypton* (Kr), *xenon* (Xe), and *radon* (Rn).

Noble Metals A *Chemistry* term; the metals not easily affected by *acids and corrosive agents*; also not easily tarnished by *oxidation*, when exposed to atmosphere, and hence retain their brightness; a few examples include *gold, silver*, and *platinum.*

Node An *Electrical* and a *Project Management* term; in *electrical* terms, a point where three or more connections to elements or sources are made; in project management terms, a point, where two or more project activities are meeting.

Nodular Cast Iron A *Materials Engineering* term; also called **Ductile Iron** or **Ductile Cast Iron**; a form of cast iron which has *excellent strength* ($E = 172$ GPa), *good castability*, and *machining* properties; relatively *ductile* as compared to grey cast iron due to presence of *spheroidal graphite precipitates* rather than flakes which are formed during solidification when treated with magnesium.

Nominal Strain A *Strength of Materials* term; also called **Engineering Strain**; the strain related to the *gauge length* of a test specimen, when there is no change in its cross-sectional area; see also **True Strain**.

Nonagon A *Mathematics* term; a *polygon* having nine (9) sides; see also **Polygon**.

Non-conservative Force A *Physics* term; a *non-conservative force* acts on a particle, if the particle, after going through a complete round trip, returns to its initial position with changed kinetic energy; examples include *viscous force* and *frictional force*; see also **Conservative Force**.

Nonconsumable Electrode A *Welding* term; an electrode which does not get consumed during the welding process; usually made of *graphite* or *tungsten*; finds use in *atomic hydrogen welding*, *TIG welding*, *plasma arc welding*, and others; see also **Consumable Electrode**.

Nonconventional Sources of Energy A *Power Plant* term; the sources of energy which have been focussed in the recent past but were not considered previously; inexhaustible, cheaper, and environment-friendly and have tremendous scope of exploitation; not reliable as they involve intermittent operations; a few examples include *solar*, *wind*, *sea*, *geothermal*, and *biomass*; see also **Conventional Sources of Energy** and **Renewable Sources of Energy**.

Non-critical Activity A *Project Management* term; a project activity which has variation in its starting time and has a definite slack; see also **Critical Activity**.

Noncrystalline Materials A *Materials Engineering* term; see **Amorphous Materials**.

Nondestructive Inspection A *Materials Engineering* term; see **Nondestructive Testing**.

Nondestructive Testing A *Materials Engineering* term; also called **Nondestructive Inspection**; the evaluation of mechanical properties of materials and possible defects in their structures without impairing their usefulness; a few examples include *acoustic-emission testing*, *liquid-penetrant testing*, *X-radiography*, and *magnetic-particle testing*; see also **Destructive Testing**.

Nonferrous Metals A *Materials Engineering* term; metals which are almost *free of iron*; include *copper*, *zinc*, *aluminium*, *manganese*, *lead*, *tungsten*, *titanium*, *brass*, *bronze*, *gold*, *silver*, *platinum*, *uranium*, and others; compared to ferrous metals, they are more expensive; offer *superior* and *unique* properties not obtainable in ferrous metals like *good corrosion resistance*, *low density*, *ease of fabrication*, *colour choices*, and *high thermal and electrical conductivity* (copper has approximately five times, and aluminium has approximately three times more electrical conductivity than iron); have *good formability*; lack structural strength in pure state which can be increased in the alloyed form; have *poor weldability* compared to ferrous materials; can be *cast*, *formed* or *machined* easily; mostly used in the *alloyed form*; see also **Ferrous Metals.**

Non-harmonic Oscillation A *Physics* term; the oscillation (motion) which cannot be expressed in terms of a single harmonic function like sine function or cosine function, but can be expressed as a *combination of two or more than two harmonic oscillations*; mathematically expressed as:

$$y = a\sin(\omega t) + b\sin 2\omega t \tag{1}$$

$$y = a\sin\left(\frac{2\pi}{T}t\right) + b\sin\left(\frac{4\pi}{T}t\right) \tag{2}$$

where

y	= displacement of the particle from mean position at any instant t
a or b	= maximum displacement or amplitude of displacement of the particle
ω	= angular frequency $= 2\pi n = \frac{2\pi}{T}$
n	= frequency of harmonic oscillation
T	= time period of harmonic oscillation; see also **Harmonic Oscillation.**

Non-intrusive Method A *Metrology* term; the measurement of a physical quantity which does not involve insertion of a probe into the system, rather the probe has negligible interaction with the system; a typical example is the use of a *laser Doppler velocimeter* (LDV) to measure the velocity of a flowing fluid; see also **Intrusive Method.**

Non-isotropic Material A *Materials Engineering* term; see **Anisotropic Material.**

Nonlinear Regression A *Statistics* term; a process of analysis in which a set of data points best fits a polynomial relationship of the form $y = ax^3 + bx^2 + cx + d$, where the parameters a, b, c, and d are known as the *fit parameters* to be determined as a part of regression analysis; see also **Linear Regression.**

Non-loaded Joint A *Strength of Materials* term related to a *truss*; the joint in a truss which does not carry any external load and has no support reaction.

Non-metals A *Materials Engineering* term; the materials which are *brittle* as well as *poor conductor of heat and electricity* (except *graphite*); have lustre because of their ability to reflect light (except *iodine*); exist in all the three states (e.g. *sulphur*, *phosphorus*, and *iodine* are solids; *bromine* is a liquid, and *oxygen*, *nitrogen*, *hydrogen*, *chlorine*, etc. are gases); see also **Metals.**

Non-newtonian Fluid A *Fluid Mechanics* term; a fluid which does not obey the *Newton's law of viscosity*; the shear stress for a non-newtonian fluid is not directly proportional to the velocity gradient; shows a nonlinear relationship between shear stress and velocity gradient; not relevant for engineering usage; a few examples include *blood, honey, sugar syrup, printer's ink, paste, liquid plastics, slurries*, and *colloidal suspensions, polymer solutions, cake batter*, and *tar*; see also **Newtonian Fluid**.

Nonpolar Covalent Bond A *Chemistry* term; a covalent bond formed between *two similar atoms*; the shared electron pair is attracted equally by both atoms and lies exactly midway between them; found in *hydrogen* (H_2), *fluorine* (F_2) or *chlorine* (Cl_2) molecules; see also **Polar Covalent Bond**.

Nonpolar Molecule A *Chemistry* term; a molecule having a *nonpolar covalent bond*; consists of *two similar atoms*; a few examples include *hydrogen molecule* (H_2), *fluorine molecule* (F_2), and *chlorine molecule* (Cl_2); see also **Polar Molecule**.

Non-prismatic Bar A *Strength of Materials* term; a bar of non-uniform (varying) cross section; for example, a tapered circular or square bar; see also **Prismatic Bar**.

Nonrecurring Cost An *Operations Management* term related to cost *analysis*; the nonrepetitive cost; typically involves developing or establishing a capability or capacity to operate; for example, the purchase cost of land (for making a plant) along with its cost of construction are nonrecurring costs; see also **Recurring Cost**.

Nonrenewable Sources of Energy A *Power Plant* term; the sources of energy which can be depleted after their use and cannot be reused and also cannot be immediately replenished; not environment-friendly; a few examples include *fossil fuels* such as *coal, petroleum* and natural gases, and *nuclear fuels*; see also **Renewable Sources of Energy** and **Nonconventional Sources of Energy**.

Non-rigid Truss A *Strength of Materials* term related to a *truss*; a truss which is not rigid; does not obey the equation $m = 2j - 3$ where, m = number of members and j = number of joints in the truss; classified into *under-rigid* and *over-rigid* truss; see also **Rigid Truss**.

Non-uniform Flow A *Fluid Mechanics* term; a fluid flow is said to be non-uniform, if the fluid velocity changes from point to point in the flow field; a few examples include *flow through a converging or diverging pipe, flow of water through a river bed or any other non-prismatic channel*; see also **Uniform Flow**.

Normal Cost A *Project Management* term; the lowest cost of completing an activity using normal resources in the normal manner; see also **Crash Cost**.

Normal Distribution Curve A *Statistics* term; also called **Normal Probability Distribution Curve** or **Gaussian Distribution Curve**; the most commonly occurring *probability* distribution curve in which the probability density function is expressed in terms of *mean* and *standard deviation* of the data; a *bell-shaped curve* symmetrical about the peak value of the graph where symmetry of the curve indicates that one-half the probability is associated with the values to the left of the peak and the other half with the values to the right; the area under the curve is unity which implies that probability of all the measured values lying under the curve is unity; indicates that 68% of the possible outcomes will lie between ± 1 standard deviation (± 1σ) from the expected value, 95% of all outcomes will lie between ± 2 standard deviations (± 2σ) from the expected value, and 99% of all outcomes will lie between ± 3 standard deviations (± 3σ) from the expected value.

Normal Force A *Physics* term; a force which acts along the central axis of a structural member; tension and compression are normal forces.

Normal Probability Distribution Curve A *Statistics* term; see **Normal Distribution Curve**.

Normal Strain A *Strength of Materials* term; see **Longitudinal Strain**.

Normal Stress A *Strength of Materials* term; also called **Direct Stress** or **Longitudinal Stress**; the internal restoring force which acts perpendicular to the unit area of a body when stressed within the elastic limit; produces change in length and volume of a body; classified into *tensile stress* and *compressive stress*.

Normal Time A *Time Study* and a *Project Management* term; in *time study* terms, the time normally required in the completion of a job by an *average worker*; does not include the allowance time; in *project management* terms, the time required to complete a project activity using normal resources in the normal manner.

Normalizing A *Manufacturing* term related to *heat treatment*; a process of heating a metal in *still air* and then cooling it at a faster rate than that used in annealing in order to obtain harder, stronger metal than obtained by annealing; used to improve the grain structure of a metal and return it to normal by removing stresses caused by uneven cooling after *welding, casting,* or *forging*.

Normative Analysis A *Management* term; an analysis of what ought to be; for example, if it is argued that there should be a law requiring seat belts to be worn while driving a vehicle, it is normative analysis.

Norms A *Management* term; the standards or styles of behaviour considered acceptable or typical for a group of people.

Notching A *Manufacturing* term; a sheet metal operation used to remove pieces (or various shapes) from the edges of a metal sheet.

Nozzle A *Fluid Mechanics* term; a flow passage device of varying cross-sectional area in which velocity of a fluid (gas or liquid) increases in the direction of flow at the expense of its pressure drop; see also **Diffuser**.

NPSH A *Fluid Mechanics* term related to a *pump*; see **Net Positive Suction Head**.

N-Type Semiconductor An *Electronics* term; an *extrinsic semiconductor* in which the electrical conductivity is mainly due to *negatively charged electrons* that are called the majority carriers, and the holes are the minority carriers; produced when a pure semiconductor is doped with a *pentavalent impurity atom* such as *phosphorus*; see also **P-Type Semiconductor**.

Nuclear Energy A *Chemistry* term related to *radioactivity*; the energy obtained from an atomic nucleus, either by breaking a heavy nucleus (*nuclear fission*) or by joining two lighter nuclei (*nuclear fusion*); obtained in a very huge amount, even with a very little quantity of nuclear fuel; the most neat and clean form of energy, although very dangerous, if handled carelessly.

Nuclear Fission A *Chemistry* term related to *radioactivity*; a process of splitting of a heavy nucleus into two or more lighter nuclei; involves a certain loss of mass (called *mass defect*) which is converted into huge amount of energy according to the *Einstein's mass energy relationship* $E = mc^2$ where m = mass lost and c = velocity of light in vacuum; for example, *uranium* $(92)^{235}$, when bombarded with neutrons, undergoes nuclear fission in two ways, giving *barium* (56) and *krypton* (36) in the first case, and *strontium* (38) and *xenon* (54) in the second case, as the products of fission, according to the following reactions:

$$\mathrm{U}(92)^{235} + n(0)^1 \rightarrow \mathrm{U}(92)^{236} \rightarrow \mathrm{Ba}(56)^{141} + \mathrm{Kr}(36)^{92} + 3n(0)^1$$

$$\mathrm{U}(92)^{235} + n(0)^1 \rightarrow \mathrm{U}(92)^{236} \rightarrow \mathrm{Sr}(38)^{90} + \mathrm{Xe}(54)^{143} + 3n(0)^1$$

Energy released during nuclear fission varies with the process and also with the number of neutrons released during fission process; on an average, approximately 200 meV of energy and 2.5 neutrons (n) are released in the fission of U $(92)^{235}$; the principle of nuclear fission is used in the preparation of an *atom bomb*; see also **Nuclear Fusion**.

Nuclear Force A *Chemistry* term; a force which binds the neutrons and protons together in a nucleus; does not depend on the charge on nucleons; a *non-central* and *non-conservative short-range attractive force* which operates within the nucleus only in the distance range of 10^{-14} m; the *strongest force in nature*, which is 10^{38} times stronger than the *gravitational force* and 100 times stronger than the *electrostatic force*.

Nuclear Fusion A *Chemistry* term related to *radioactivity*; a process of joining of two or more lighter nuclei to form a heavy nucleus; releases enormous amount of energy due to *mass defect*, using the Einstein's mass energy relationship $E = mc^2$ where m = mass lost and c = velocity of light in vacuum; *the energy of the sun is due to nuclear fusion*, where a large number of *hydrogen* (H) atoms combine together to form *helium* (He) atoms in the ratio of 4:1, that is, 4 hydrogen atoms unite to form one atom of helium; the formation of one mole of He $(2)^4$ from 4 mol of H $(1)^1$ releases 27.3×10^8 kJ of energy; the principle of nuclear fusion is used in the preparation of *hydrogen bomb* which is far more powerful than an atom bomb; see also **Nuclear Fission.**

Nuclear Reaction A *Chemistry* term related to *radioactivity*; a reaction which leads to atomic transformations or transmutations; involves changes in the number of nucleons present in the nucleus, forming new atomic species; nuclear reactions involve energies a million times greater than those involved in chemical reactions; the products in a nuclear reaction are generally heavy nucleus and one of the particles, which may be an *electron*, a *proton*, a *deuteron*, a *neutron* or an *alpha* (α) *particle* or merely *a gamma* (γ) *ray*.

Nucleus A *Chemistry* term; the *central part* of an *atom* of a chemical element; discovered by Ernst Rutherford (1871–1937), called the *father of nuclear physics*; has a spherical shape with a radius of 10^{-13} cm, composed of *protons* and *neutrons*; contains almost the entire mass of an atom; positively charged because of presence of protons (positively charged particles), whereas electrons move around the nucleus in different orbits.

Nugget A *Welding* term; a small quantity of the weld metal used to join the parts in *resistance spot*, *resistance seam*, or *projection welding*.

Null Vector A *Physics* term; also called **Zero Vector**; a vector which has *zero magnitude*; obtained by subtracting a vector from itself.

Numerical Control A *Manufacturing* term related to *automation,* abbreviated as **NC**; a form of programmable automation pioneered by *John T. Parsons*; a control system which digitally communicates to machine tools such as mills and lathes to produce a part; offers *flexibility* of changing the program of instructions according to the type of a job to be processed; has three basic components, namely program of instructions, a controller unit (also called machine control unit, MCU) and a machine tool.

O

Objective Function An *Operations Research* term; a *mathematical function* used to express the objectives of a problem.

Oblique Collision A *Physics* term; see **Two-Dimensional Collision**.

Oblique Projection An *Engineering Drawing* and a *Machine Drawing* term; a type of projection in which views are drawn by using parallel projectors (projection lines) *oblique to the plane of projection*, where the front face of the object represents its true size and shape, as it is placed parallel to the picture plane and the receding lines representing the other two sides are usually drawn at 45° to the horizontal; all lines which are parallel on the object are also parallel on the drawing and the vertical lines on the object are vertical in the drawing also; the *isometric projection* is an example of *oblique projection*; see also **Orthographic Projection**.

Observed Time A *Time Study* term; the time observed to be elapsed in an operation using a *stopwatch*.

Obtuse Triangle A *Mathematics* term; also called **Obtuse-Angled Triangle**; a triangle whose any one angle is obtuse, that is, greater than 90°; see also **Acute Triangle**.

Obtuse-Angled Triangle A *Mathematics* term; see **Obtuse Triangle**.

Octagon A *Mathematics* term; a *polygon* having eight (8) sides; see also **Polygon**.

Octahedral Shear Theory A *Strength of Materials* term related to *theory of material's failure*; see **Maximum Distortion Energy Theory**.

Octet Rule A *Chemistry* term; the principle of attaining a maximum of *eight electrons* (ns^2p^6) in the valence shell (the outermost orbit) of an atom which gives stable electronic configuration to the atom.

Odd Number A *Mathematics* term; any number not divisible by 2; a few examples include 3, 5, 13, etc.; the sum of any two odd numbers is always an *even number*; see also **Even Number**.

D. K. Singh, *Dictionary of Mechanical Engineering*,
https://doi.org/10.1007/978-981-99-1722-8_15

Offset Method A *Strength of Materials* term; a method used to identify the *yield stress* for a material which shows no discernible point on its stress–strain curve where yielding is evident; involves drawing a line parallel to the linear portion of the engineering stress–strain curve at offset stain value of 0.2% or even smaller values such as 0.1% or 0.02%; the stress corresponding to the point of intersection defines yield stress.

Offset Strain A *Strength of Materials* term; a strain value which offsets (prevents) failure, if material's dimensions are kept below this value; corresponds to 0.2% or even smaller values such as 0.1% or 0.02% of linear strain; used to determine yield points for brittle materials.

ohm An *Electrical* term; the SI unit of *electrical resistance*; denoted by Ω; one ohm of electrical resistance is defined as a pressure of one volt which causes a current of one ampere to flow in a conductor.

Ohm's Law An *Electrical* term; an electrical law proposed by *Georg Simon Ohm* (1789–1854), a *German* physicist in 1827; states that the current (I) passing through a conductor is proportional to the potential difference (V) across its ends; given as:

$$V = IR$$

where

R = proportionality constant, called resistance.

Oil A liquid material used extensively in the industry both as a *fuel* and as a *lubricant* for machinery and products.

Oldham Coupling A *Design* term; a *temporary fastening device* used to connect two circular rotating shafts for power transmission; a type of *flexible coupling* which takes care of axial as well as lateral misalignment of the two shafts.

One-Dimensional Collision A *Physics* term; also called **Head on Collision**; the collision in which the initial velocities as well as the final velocities of both colliding bodies are along the same straight line; for example, the collision between two small spherical bodies may be one-dimensional, if the direction of travel of one body passes through the centre of second body; see also **Two-Dimensional Collision**.

One-Dimensional Motion A *Physics* term; the motion of a body along one direction (along a straight line or a well-defined straight path); a few examples include the *motion of a train along a straight railway track* and the *falling of an object from a certain height*; see also **Two-Dimensional Motion** and **Three-Dimensional Motion**.

One-Off Production An *Operations Management* term; see **Jobbing Production**.

Onyx A *Chemistry* term; one of the three *naturally occurring amorphous forms of silica* (SiO_2), other two include *Jasper* and *Agate*.

Opal A *Chemistry* term; a coloured (white) *variety of quartz used as a gem*; see also **Jasper** and **Amethyst**.

Opaque Material A *Materials Engineering* term; a material which does not transmit any radiation through it; the transmissivity of opaque materials is zero.

Open Belt Drive A *Theory of Machines* term; a belt drive used to connect two parallel shafts moving in the same direction with a maximum centre-to-centre distance of 15 m; the total length (L) of the open belt is given as:

$$L = \pi(r_1 + r_2) + \frac{(r_1 - r_2)^2}{x} + 2x$$

where

r_1 = radius of the bigger pulley
r_2 = radius of the smaller pulley
x = centre-to-centre distance between two pulleys; see also **Crossed Belt Drive**.

Open-Channel Flow A *Fluid Mechanics* term; the flow of a liquid in a channel or conduit that is not completely filled; a free surface exists between the flowing liquid (usually water) and fluid above it (usually the atmosphere); examples include the drainage system of water through the numerous creek and river systems; the *Froude number* plays a significant role in open-channel flow; a laminar flow rarely exits in open channel.

Open Circuit Voltage A *Welding* term; the voltage at the terminals of a welding machine when it is turned on without performing any welding operation; indicates a state of *no flow of current* in the welding circuit as it is not complete; see also **Load Voltage**.

Open Circuit An *Electrical* term; a circuit which does not provide a complete path for electrical flow; see also **Closed Circuit**.

Open Curve A *Mathematics* term; see **Simple Curve**.

Open Riser A *Manufacturing* term related to *casting*; also called **Top Riser** as it sits on the top of a casting; a *conventional riser* which is *open to the atmosphere.*

Open System A *Thermodynamics* term; a *thermodynamic system* which allows both mass and energy transfer across its boundary; the mass within the system may not remain constant; most of the engineering devices are generally open systems; see also **Closed System**.

Open-Channel Flow A *Fluid Mechanics* term; the flow of a fluid in a channel or conduit which is not completely filled, and there exists a free surface between the flowing fluid (usually water) and the fluid above it (usually the atmosphere); a few examples include *flow in rivers*, *irrigation ditches*, and *aqueducts*.

Open-Coiled Helical Spring A *Strength of Materials* term related to a *spring*; a cylindrical helical spring in which the plane of the coil of spring makes a certain angle with the horizontal (called the *helix angle*); see also **Close-Coiled Helical Spring**.

Open-Die Forging A *Manufacturing* term; a *hot metal working process* used to shape a heated workpiece by placing it between two flat dies (that allow metal to flow freely) and then pressed or hammered; see also **Closed-Die Forging**.

Open-Loop System An *Electrical* term; a control circuit in which the system output has no effect on the control; see also **Closed-Loop System**.

Operating Expenditure A *Financial Management* term; an outlay of funds by a firm which is expected to produce benefits within one year; see also **Capital Expenditure**.

Operation Chart A *Motion Study* term; also called **Left Hand/Right Hand Chart**; a tool used to analyse body movement (primarily *arms* and *hands*); designed to show economy of motion by pointing out wasted motion and idle time (delay).

Operation Usually a *Manufacturing* term; a transformation process which transforms inputs into outputs of greater value.

Operations Management A *Management* term; also called **Production Management**; an important discipline of management; the management of a production system which converts *raw materials*, *labour*, *capital*, and *management inputs* into desired *outputs of products and services*; *manufacturing operations management* focuses on the elimination of wasteful efforts and achieving greater efficiency in production using Frederick Taylor's scientific management tools; activities in *manufacturing operations management* include organizing work, selecting processes, arranging layouts, locating facilities, designing jobs, measuring performance, controlling quality, scheduling work, managing inventory, and planning production; pioneers in *manufacturing operations management* include *Frank and Lillian Gilbreth*, *Frederick Winslow Taylor*, and *Henry Gantt*; *service operations management* deals with intangible or nonphysical goods (services), and thrusts upon delivering efficient services such as prompt delivery of requested items by courier firms, personal care by nursing staffs in a hospital, and quick redressal of queries by customer care staffs of a company.

Operations Manager A *Management* term; a manager connected to a manufacturing unit and deals with people, technology, and deadlines of work completion.

Operations Research Model A *mathematical model* consisting of mathematical structures which are solved by using mathematical techniques; used to solve management problems; some of the important operations research models include *linear programming*, *nonlinear programming*, *dynamic programming*, and *genetic algorithms*.

Operations Research Abbreviated as **OR**; the application of scientific methods to the study of alternatives in a problem using quantitative data with a view to arrive at a best solution; involves minimizing or maximizing the objective functions under the given constraints.

Operations Strategy A *Management* term; a strategy which determines how and where a product or service is to be manufactured.

Operatives A *Management* term; those people who work directly on a job or task and have no responsibility to oversee the work of others; for example, workers; represent the general employees of an organization.

Operator-Machine Chart A *Motion Study* term; see **Activity Chart**.

Opportunity Cost A *Financial Management* term; the cost associated with opportunity foregone (lost) or opportunity not utilized to monetary advantage as a result of selecting one alternative instead of another; the salvage value of facilities and equipment may be an opportunity cost, since assets which are abandoned or replaced may be saleable.

Opposite Sides A *Mathematics* term; any two sides not starting or originating from the same initial point; see also **Adjacent Sides**.

OPT An *Operations Management* term: see **Optimized Production Technology**.

Optical Microscopy A *Physics* term; a technique which uses a light radiation to reveal the microstructural features in a material such as grain boundaries which require less than about 2000 magnification.

Optical Torsion Meter A *Metrology* term; a device used to measure *torque* by measuring the angle of twist over a fixed length of a shaft using *optical means* (a beam of light and mirrors); see also **Mechanical Torsion Meter**.

Optics A *Physics* term; a branch of *physics* which deals with the study of various phenomena connected with *light* and *optical instruments* like *microscope* and *telescope.*

Optimal Solution An *Operations Research* term; a solution which satisfies both the conditions of a problem and its given objective; for example, a manufacturing company must determine the combination of resources available so as to enable it to manufacture products in a way which not only satisfies its production schedule, but also maximizes its profits or minimizes cost.

Optimistic Time A *Project Management* term; the shortest possible time to complete a project activity, if everything goes right; used to calculate the *expected time of completion* of an activity in *programme evaluation and review technique* (PERT); see also **Pessimistic Time.**

Optimization An *Operations Research* term; a process used to find the conditions which give the maximum or minimum value of a function.

Optimized Production Technology An *Operations Management*, abbreviated as **OPT**; a complete production planning and control information system which uses the concepts of theory of constraints for managing the schedules of bottleneck operations for complex job shop environments.

OR See **Operations Research.**

Orbit A *Chemistry* term; also called **Energy Shell** or **Energy Level**; the circular path in which electrons move around the nucleus of an atom; orbits are designated by principal quantum number (n), a positive integer which can take values such as 1, 2, 3, and 4; orbits are also denoted by shells such as K, L, M, and N taken in that order starting from the nucleus; each orbit is associated with a definite amount of energy, and the farther the orbit from the nucleus, greater is the energy associated with it; energy of an electron remains constant in a particular orbit and is emitted or absorbed by an atom only when an electron moves from one orbit to another.

Orbital Speed A *Physics* term; the minimum speed required to put a satellite into a given orbit around the earth; the orbital speed is independent of mass of the satellite, but depends upon the mass and radius of the earth around which the satellite revolves; decreases with an increase in the radius of orbit or an increase in the height of the satellite; equals to 7.92 km/s, when the satellite is orbiting very close to the surface of the earth; given as:

$$V = \sqrt{\frac{GM}{r}} = \mathrm{R}\sqrt{\frac{g}{R+h}}$$

where

G = universal gravitational constant = $6.67 \times 10^{-11}\ \mathrm{N.m^2.kg^{-2}}$
M = mass of the earth
R = radius of the earth
h = height of the satellite above the surface of the earth
r = radius of the orbit of the satellite = $R + h$.

Order Cycle An *Operations Management* term related to *inventory*; the time between placing and receiving an order in an inventory system.

Order Point An *Operations Management* term; the point of time at which an order for procuring a material is placed; determined by estimating the time between placing the order and receiving another batch (called lead time) of that material.

Ordering Cost An *Operations Management* term; also sometimes called **Setup Cost**, but is slightly different; the cost associated with placing an order for inventory; ordering costs include writing the order, processing the order through the purchasing system, postage, processing invoices, processing accounts payable, and the work of the receiving departments such as handling, testing, inspection, and transporting; a part of total annual cost of inventory; decreases with increased order size of inventory; one of the costs that finds use in determining economic order quantity (EOQ); see also **Setup Cost** and **Holding Cost**.

Ordering Lead Time An *Operations Management* term; the time involved between placement and receiving of an order for materials; see also **Manufacturing Lead Time**.

Ordinary Glass A *Chemistry* term; also called **Soda Glass** or **Window Glass** or **Soft Glass**; a type of glass and a common structural material which consists of 15% Na_2O, 10% CaO, and 75% SiO_2 (all by weight) and has the approximate chemical formula: $Na_2O.CaO.6SiO_2$; softens at a comparatively lower temperature and can be easily shaped in different forms (hence called *soft glass*); used in making cheap *window panes, bottles, dishes*, etc.

Ore A *Chemistry* term; a *naturally occurring mineral* from which a metal can be conveniently and economically extracted; usually occurs in nature in the form of big lumps which are broken into small pieces with the help of crushers or grinders (process is called *crushing*); for example, *haematite* (Fe_2O_3) and *magnetite* (Fe_3O_4) are the two important *ores of iron*; see also **Mineral**.

Organic Compound A *Chemistry* term; a chemical compound in which one or more atoms of carbon are *covalently* linked to atoms of other elements, usually *hydrogen, oxygen,* or *nitrogen*; see also **Inorganic Compound**.

Organic Materials A *Chemistry* term; contain *animal* or *vegetable* cells (*dead* or *alive*) or organic compounds which are usually dissolved in organic solvents like *alcohol* or *carbon tetrachloride* but not dissolved in water; see also **Inorganic Materials**.

Organic Organization A *Management* term; an organizational structure which is low in *specialization, formalization,* and *centralization*; a highly adaptive form of organization which changes rapidly as per requirements; involves formal rules, non-standardized jobs, and technically proficient and trained professionals to handle diverse problems.

Organization A *Management* term; a systematic arrangement of human and material resources grouped together for a specific purpose or aim; for example, *a manufacturing plant*; also a group of people with defined relationship to each other and has a specific goal to achieve; for example, *a police force, an army, a bank*, etc.; see also **Manufacturing Organization** and **Service Organization**.

Organizational Culture A *Management* term; a set of standard values and norms which control organizational members' interactions with each other and with suppliers, customers, and people outside the organization.

Organizational Design A *Management* term; a process by which managers select and manage aspects of structure and culture so that an organization can control the activities necessary to achieve its goals.

Organizational Structure A *Management* term; a formal system of *task* and *authority* relationships which controls how people coordinate their actions and uses resources to achieve organizational goals or objectives; examples include *line organization, functional organization, line and staff organization, line and functional organization,* and *matrix organization.*

Organizing A *Management* term; a *management function* which deals with determining what jobs are to be done by whom, that is, distributing the task among the people and assigning them authority and responsibility.

Orientation A *Management* term; the activities carried out to introduce new employees to the organization and their work units.

Orthographic Projection An *Engineering Drawing* and a *Machine Drawing* term; a projection method in which the rays of sight (called the *projectors*) are taken *perpendicular to the plane of projection*; assumes an observer to be positioned at infinite distance from the plane of projection such that the projectors will be parallel to each other; the most commonly employed projection method of drawing; see also **Oblique Projection**.

Orthotropic Material A *Materials Engineering* term; has different properties in the three orthogonal directions; a typical example is *wood*; see also **Isotropic Material**.

Oscillatory Motion A *Physics* term; also called **Vibratory Motion**; the motion in which a body moves back and forth (to-and-fro) over the same path repeatedly about a fixed point (called *mean position* or *equilibrium position*) in a definite interval of time; the motion is repeated between two extreme positions; a few examples include *swinging back and forth of a wall clock pendulum*, the *motion of the prongs of a tuning fork, the internal motion of atoms within molecules*, and the *motion of the bob of a simple pendulum* (on being displaced once from its mean position and left to itself); every oscillatory motion is periodic (as it is completed in a definite interval of time), but every periodic motion need not be oscillatory (as the latter involves back and forth motion, whereas the former may not); for example, the circular motion is a periodic motion, but it is not oscillatory; see also **Periodic Motion**.

Otto Cycle A *Thermodynamics* term related to *air standard cycle*; an air standard thermodynamic cycle used in *petrol engine*; named after its *German* inventor *Nicolas Otto* (1832–1891) who is recognized for having developed the first practical design for liquid-fuelled piston engines; consists of *four* reversible processes, namely reversible adiabatic (isentropic) compression, constant volume heat addition, reversible adiabatic (isentropic) expansion, and constant volume heat rejection; the thermal efficiency depends only on its *compression ratio*; for the same compression ratio and same heat rejection, the Otto cycle has higher thermal efficiency than the Diesel cycle; see also **Diesel Cycle**.

Outside Stakeholders A *Management* term; people who do not own the organization and are also not employed by it, but do have some interest in it; include *customers*, *suppliers*, the *government*, *trade union*, *local communities*, and the *general public*; see also **Inside Stakeholders**.

Outsourcing An *Operations Management* term; a process of moving some of a firm's internal activities and decision responsibility to outside providers; an important activity of *supply chain management*.

Overhanging Beam A *Strength of Materials* term related to a *beam*; a statically determinate beam whose certain length extends beyond its support ends on one or both sides.

Overhead Costs An *Operations Management* term related to *cost analysis*; the overhead costs are plant operating costs and categorized as *indirect costs*; a few examples of overhead costs include *electricity*, *general repairs*, *property taxes*, *supervision*, etc.; different types of overhead costs include *administration*, *production*, *selling*, *distribution*, and *R & D* costs; the *administration overhead cost* is related to *administration,* and its example include *office rent*, *salaries and wages of clerks*, *postage* and *telephone bills*, *Director's* and *General Manager's fees*, etc.; the *production overhead cost* (also called *factory overhead*) includes all indirect expenses incurred by a firm from the receipt of production order of a good to its dispatch to customer, and its examples include *rent*, *insurance premium*, *repair costs of building*, *depreciation cost*, *wages of maintenance men*, *general workers*, etc., *security cost*, *cost of water*, *fuel*, and *power*, etc.; the *selling overhead cost* is used to increase the volume of sales, and its examples include *advertising cost*, *rent of sales rooms*, *consumer service cost*, *service cost after sales*, etc.; the *distribution overhead cost* is related to transportation of goods to the end users, and its example includes *warehouse charges*, *loading/unloading charges*, *upkeep of delivery vehicles*, etc.; the *R & D overhead cost* is related to research activities for developing product or service, and its example includes the cost associated with improving the fuel efficiency of an engine.

Overhead Position A position of a task which is performed from the underside.

Overheating A *Manufacturing* term; a process of heating a substance beyond its critical range temperature; impairs a substance's properties.

Overprocessing An *Operations Management* term; the use of excessive resources for executing a task; one of the seven types of waste identified by *Taiichi Ohno* (1912–1990), known as the *father of Toyota Production System* (TPS).

Overproduction An *Operations Management* term; an undesirable production condition in which products are manufactured in larger quantities than they are actually required; one of the seven types of waste identified by *Taiichi Ohno* (1912–1990) who is known as the *father of Toyota Production System* (TPS).

Over-rigid Truss A *Strength of Materials* term related to a *truss*; also called **Redundant Truss**; a non-rigid truss which consists of surplus members than required to make it a rigid truss; for an over-rigid truss: $m > 2j - 3$ where m = number of members, and j = number of joints in the truss; see also **Under-rigid Truss**.

Owner's Equity A *Financial Management* term; the owner's initial investments in a company plus any earnings of the company which have not been distributed as dividends (called retained earnings).

Oxidation A *Chemistry* term; a chemical reaction in which a metal combines with the atmospheric oxygen; the removal of an electron from an atom or a molecule; see also **Reduction**.

Oxide A *Chemistry* term; a compound or a molecule formed by the reaction between a metal and oxygen; for example, *zinc oxide*, *aluminium oxide*, etc.

Oxidizing Flame A *Chemistry* term; a type of *oxyacetylene gas flame* which contains excess of oxygen as compared to fuel gas (*acetylene*) in the mixture; finds use in gas welding to join *copper alloys* and *zinc and zinc alloys*; see also **Carburizing Flame** and **Neutral Flame**.

P

Pa A *Physics* term; see **pascal**.

Parabolic Formula A *Strength of Materials* term related to a *column*; also called **J. B. Johnston Formula**; a parabolic equation-based empirical formula used to find the *crippling* (*critical*) *load* for *intermediate-length columns* with *centric loading*; finds use in the *machine*, *automotive*, *aircraft*, and *structural-steel construction fields*; expressed as:

$$P_{\mathrm{j}} = A\left[\sigma_{\mathrm{s}} - K\left(\frac{l_e}{k}\right)^2\right]$$

where

P_j = Johnson's crippling load
A = area of cross section of the column
σ_c = direct crushing stress (yield stress)
l_e = Equivalent/effective length of the column
k = radius of gyration
K = a constant for the column material
$= \frac{\sigma_c^2}{4\pi^2 E} = \frac{\sigma_c^2}{64E}$ (for pinned ends column)
E = modulus of elasticity of the column material.

Paradigm A *Management* term; a way of thinking about a given subject which defines how one views *events*, *relationships*, *ideas*, etc. within the boundaries of that subject.

Parallel Flow Heat Exchanger A *Thermodynamics* term; a type of heat exchanger in which both the hot and cold fluids flow past the heat transfer surface in the *same direction*; see also **Counter Flow Heat Exchanger** and **Cross Flow Heat Exchanger**.

D. K. Singh, *Dictionary of Mechanical Engineering*,
https://doi.org/10.1007/978-981-99-1722-8_16

Parallel Lines A *Mathematics* term; the lines (two in number) having constant distance between them, and lie in the same plane; do not intersect each other, when produced indefinitely in either direction.

Parallel Sunk Key A *Design* term related to a *key*; a sunk key having a *uniform square or rectangular cross-section* throughout its length, but ends of the key may be either squared or rounded; used in the case, when the mating part or mounting is required to slide along the shaft; normally fitted into keyway provided on the shaft with the help of set screws.

Parallel Vectors A *Physics* term; two or more vectors are said to be parallel, when they are parallel to the same line; if parallel vectors have the same direction, they are called like vectors, and if they have opposite directions, they are called unlike vectors.

Parallel-Axes Theorem A *Strength of Materials* term; used to find the *moment of inertia* of a plane area about an axis parallel to its centroidal axis; can be expressed as $I_{AB} = I_{XX} + ah^2$ where I_{AB} = moment of inertia of the cross-section about the desired axis *AB*, I_{XX} = moment of inertia of the cross section about the centroidal axis *XX*, a = cross-sectional area and h = perpendicular distance between two axes; see also **Perpendicular Axis Theorem.**

Parallelogram Law of Forces A *Physics* term; see **Law of Parallelogram of Forces.**

Parallelogram A *Mathematics* term; a quadrilateral whose both pairs of opposite sides are parallel.

Parent Metal A *Welding* term; also called **Base Metal**; the metal to be welded or cut.

Pareto Analysis A *Quality Management* term named after *Vilfredo Pareto* (1848-1923), a nineteenth-century *Italian* economist; also called **Pareto Rule** or **80/20 Rule**; states that 80% of the activity is caused by 20% of factors; implies that most of the quality problems result from only a few causes, and so by concentrating on the 20% of the factors (called the 'vital few'), 80% of the quality problems can be sorted out; a method of identifying the causes of poor quality.

Pareto Chart An *Operations Management* term; a graphical plot or chart named after *Vilfredo Pareto* (1848-1923), a nineteenth-century *Italian* economist; used in finding a few vital factors which cause most of the activity; resembles a bar chart which uses the cumulative frequency curve to identify the few vital factors which need considerations in a given situation or process; see also **Pareto Rule** or **80/20 Rule.**

Pareto Rule An *Operations Management* term; see **Pareto Analysis.**

Part Drawing A *Machine Drawing* term; a number of *pictorial representations* of a *single part of a machine* required *to facilitate its manufacture*; contains all the *dimensions, limits* and *special finishing processes* such as *heat treatment, honing, lapping, surface finish,* etc.

Partial Derivative Operator A *Mathematics* term; denoted by ∂; used to find the rate of change of one parameter with respect to another while treating other variables constant; see also **Total Derivative Operator**.

Partial Factor Productivity A *Management* term; also called **Single-Factor Productivity**; a ratio of the output to only one input, say labour or machines or materials or capital or energy; see also **Total Factor Productivity**.

Participative Leader A *Management* term; see **Democratic Leader**.

Particle Size An *Air Pollution* term; the size of a liquid or solid particle expressed as the average or equivalent diameter.

Particle A *Physics* term; an object which occupies negligible space or has zero dimensions but has some mass.

Particulate Matter An *Air Pollution* term; *the finely divided solid* or *liquid particles* which are capable of being temporarily suspended in air or other gasses.

Parting Sand A *Casting* term; a *fine grained, unbounded dry silica sand* used to prevent the bonding of sand in the cope with sand in the drag of the moulding flask.

Partnership A *Management* term; a type of business which is owned by two or more people, and is operated for profit; *finance, insurance,* and *real estate firms* are the most common types of partnership; public accounting and stock brokerage partnerships often have a large number of partners.

pascal A *Physics* term; the SI unit of *pressure* and *stress*; abbreviated as **Pa**; named after a seventeenth-century *French* scientist and philosopher *Blaise Pascal* (1623–1662) who is recognized for his contribution in *fluid mechanics*; equivalent to N/m^2; and $1 \text{ Pa} = 1 N/m^2$; see also **Kilopascal, Megapascal**, and **Gigapascal**.

Pascal's Law A *Fluid Mechanics* term; proposed by the *French* scientist *Blaise Pascal* (1623–1662); states that the pressure in a fluid at rest is same at all points, if they are at the same height; has wide engineering applications; finds use in *hydraulic lift* and *hydraulic brakes*.

Passive Pressure A *Civil Engineering* term; a *horizontal earth pressure* exerted by a soil mass to something being pushed against it; for example, pressure against the sides of a building's below-grade construction which resists the overall push of the wind against the building; see also **Active Pressure**.

Patent Drawing A *Machine Drawing* term; the pictorial and self-explanatory drawing connected to new *designs being invented*; must be systematically correct and should illustrate completely each feature of the claimed invention; *not useful for production purposes* as they are not as detailed as the shop drawings.

Pathline A *Fluid Mechanics* term; the actual path (trajectory) travelled by an individual fluid particle over a certain time period; *in steady laminar flow,* the *pathline coincides with a streamline*, but for unsteady flow, both pathline and streamline are different; see also **Streamline**.

Pattern Allowance A *Casting* term; the enlargement of a sand-mould pattern over its actual size to take care of various factors such as *shrinkage*, *machining*, and others in casting; the important pattern allowances include *shrinkage allowance*, *machining allowance*, *draft allowance*, and *rapping allowance.*

Pattern A *Casting* term; a replica of the object to be cast; used to produce a mould to cast the object; made of *wood* (mostly), *plastic* or *metal*; the selection of a pattern material depends on factors such as *size and shape of the casting*, the desired *dimensional accuracy*, the *number of castings* being produced, and the *moulding process* to be used; dimensionally more accurate castings are made using a plastic or metal pattern.

Paul Richard Heinrich Blasius (1883–1970) A *German* scientist and one of Prandtl's students who provided an analytical solution to the *boundary layer equations*; also demonstrated that pipe resistance is related to the Reynolds number (a non-dimensional group).

P-Chart A *Quality Control* term; a commonly used *control chart for attributes* which is based on counting *proportion defective* after checking every item in a randomly selected sample; involves a yes–no decision regarding the item or service to be either defective or not defective; uses *binomial distribution* for sampling distribution; the standard deviation of the binomial distribution of proportion defective is gives as:

$$\sigma_{\mathrm{p}} = \sqrt{\frac{p(1-p)}{n}}$$

where '*p*' is the average proportion defective or the target value represented by the central line on the chart and '*n*' is the sample size; the upper and lower control limits for a *p-chart* are given as:

$$\mathrm{UCL} = p + z\sigma_{\mathrm{p}}$$

$$\mathrm{LCL} = p - z\sigma_{\mathrm{p}}$$

where '*z*' is the normal deviate (number of *standard deviations* from the average); when a sample proportion defective falls outside the control limits, the analyst assumes that proportion defective generated by the process has changed, and he searches for the assignable cause; see also **C-Chart**.

PDCA Cycle An *Operations Management* term; see **Deming Wheel**.

Peat A *Chemistry* term; a type of coal which contains 60% of *carbon*.

Peaucellier Mechanism A *Theory of Machines* term; a mechanism used to *generate exact straight-line motion*; consists of *eight links*; see also **Hart Mechanism**.

Peen Rammer A *Manufacturing* term; a *foundry tool* and a *type of rammer* used to ram the sand mass in *pockets* and *corners*; see also **Hand Rammer** and **Floor Rammer**.

Peer Evaluation A *Management* term; a performance assessment in which co-workers provide input into the employee's performance.

Peltier Effect A *Physics* term; a *thermoelectric effect* which is the inverse of *Seebeck effect*; named after a *French* physicist *Jean Charles Peltier* (1785–1845) who first observed this phenomenon in 1834; states that when a current is passed through two junctions made from two dissimilar wires, it is found that one junction heats up while the other gets colder; forms the basis for *thermoelectric refrigeration*; see also **Seebeck Effect**.

Pelton Turbine A *Fluid Mechanics* term; also called **Pelton Wheel** or **Impulse Wheel** or **Pelton Wheel Turbine**; a turbomachine named after its *American* inventor *Lester A. Pelton* (1829–1908) who invented it in 1878; a tangential flow, single-stage, impulse hydraulic turbine which uses a high-velocity water jet to strike the blades of the turbine under atmospheric pressure conditions; particularly suited to high head applications (from 200 m to more than 1000 m).

Pelton Wheel Turbine A *Fluid Mechanics* term; see **Pelton Turbine**.

Pelton Wheel A *Fluid Mechanics* term; see **Pelton Turbine**.

Pentagon A *Mathematics* term; a polygon having *five* (5) sides.

Percent Strain A *Strength of Materials* term; see **Percentage Strain**.

Percentage Elongation A *Strength of Materials* term; a ratio of the increase in length to the original length of a test specimen, expressed on a percentage basis; a measure of the amount of stretching (ductility) which occurs within the gauge length of a tensile test sample; a ductile material has a higher value of the percentage elongation; see also **Percentage Reduction in Area**.

Percentage Reduction in Area A *Strength of Materials* term; another measure of ductility; a ratio of the decrease in cross-sectional area to the original cross-sectional area of a test specimen, expressed on a percentage basis; see also **Percentage Elongation.**

Percentage Strain A *Strength of Materials* term; also called **Percent Strain**; the strain expressed on percentage basis.

Perfect Friction A *Fluid Mechanics* term; see **Viscosity.**

Perfect Gas A *Chemistry* term; also called **Ideal Gas**; a theoretical (hypothetical) gas in which the intermolecular forces are negligible; obeys ideal gas equation (also called equation of state), $pv = RT$, where p is the absolute pressure, v is the specific volume, T is the absolute temperature, and R is the characteristic gas constant (a ratio of the universal gas constant to the molar mass (molecular weight), which is different for each gas; a gas whose specific heats are constant and independent of temperature, and for which $C_p - C_v = R$, where C_p and C_v are respectively the specific heats at constant pressure and constant volume; at standard temperature and pressure (STP) and also at low pressure and high temperature, most gases behave like an ideal gas with reasonable accuracy; many gases such as *nitrogen, oxygen, hydrogen, noble gases*, and some heavier gases like *carbon dioxide* can be treated like ideal gases.

Perfect Truss A *Strength of Materials* term that relates to a *truss*; see **Rigid Truss.**

Perfectly Elastic Body A *Physics* term; a body which regains its original conditions (shape and size) immediately and completely after the removal of deforming force from it; *quartz* and *phosphor bronze* are near to perfectly elastic body; however no body is perfectly elastic; see also **Perfectly Plastic Body.**

Perfectly Elastic Demand An *Economics* term; the demand is said to be perfectly elastic, when even a small change in price causes a major change in demand pattern (either becomes too more or too less or even zero); found in case of sales of products during an off season or to clear the stock; the price–demand curve for a perfectly elastic demand is a horizontal line (parallel to the demand axis), which indicates that the customer can purchase any quantity of the product at one price; see also **Perfectly Inelastic Demand.**

Perfectly Inelastic Collision A *Physics* term; the collision in which two bodies stick to each other after collision; the *coefficient of restitution, e* for perfectly inelastic collision is zero; a few examples include sticking of an arrow in a target and hitting of a target by a bullet which gets embedded in the target.

Perfectly Inelastic Demand An *Economics* term; the demand is said to be perfectly inelastic, when it does not change with the change in price, that is, the quantity demanded is independent of a price change; the price–demand curve for a perfectly inelastic demand is a vertical line (parallel to the price axis), which indicates that the customer can pay any price for a certain quantity of goods; usually the demand for *essential items* is perfectly inelastic, for example, the demand for *insulin* for *diabetic patients* is perfectly inelastic; see also **Perfectly Elastic Demand.**

Perfectly Plastic Body A *Physics* term; a body which does not regain its original conditions (shape and size) at all on the removal of deforming force, howsoever small the deforming force may be; a few examples of nearly perfectly plastic bodies include *putty*, *mud*, and *paraffin wax*; however no body is perfectly plastic; see also **Perfectly Elastic Body**.

Perforating A *Manufacturing* term; a sheet metal operation used to produce a number of closely spaced small holes in a sheet metal blank.

Performance Rating Factor A *Motion Study* term, abbreviated as **PRF**; a factor used to compare the speed of a worker with respect to the normal speed of a well-trained worker under ordinary operating conditions; used to normalize the work-time of slow and fast workers; describes how much above or below average the worker's performance on each work element is; for example, a performance rating factor of 1.00 indicates that the worker is working at normal speed; a performance rating factor of 1.20 indicates 20% faster than normal (faster worker), and a performance rating factor of 0.80 indicates 20% slower than normal (slow worker).

Perimeter A *Mathematics* term; the total length of boundary of a closed figure; equals to sum of all the sides of the closed figure; for example, the perimeter of a rectangle, a square or a triangle is given as:

$$\begin{aligned} P &= 2 \times (l + b) \quad \text{[for a rectangle]} \\ &= 4 \times a \quad \text{[for a square]} \\ &= a + b + c \quad \text{[for a triangle]} \end{aligned}$$

where

P = perimeter
l = length of the rectangle
b = breadth of the rectangle or side of the triangle
a = side of the square or the side of the triangle
c = side of the triangle.

Periodic Function A *Physics* term; any function which repeats itself regularly over a given interval of time or space; *represents a periodic motion*; a function $f(t)$ is said to be periodic, if for all values of t, there is some positive constant T such that

$$f(t) = f(t + T) = f(t + 2T)$$

where T is called the period of $f(t)$ which can be $2T$, $3T$, ... etc.; for example, sin (t) and cos (t) *are periodic functions having a period of* 2π *radian*.

Periodic Motion A *Physics* term; also called **Cyclic Motion**; the motion repeated identically after equal intervals of time; a few examples include *the rotation of the earth around the sun, the rotation of the earth about its own polar axis, the rotation of the moon around the earth*, and *the motion of hands of a clock*; see also **Oscillatory Motion**.

Periodic Order Quantity An *Operations Management* term, abbreviated as **POQ**; a lot sizing rule by which an order is placed on periodic basis based on the requirement of demand; see also **Fixed Order Quantity** and **Lot-for-Lot**.

Periodic Review System An *Operations Management* term; an inventory management system of *periodic monitoring* the status of inventory in which orders for the inventory are placed at *equal intervals of time*; the ordered quantity of inventory is not fixed, rather it varies with every order; the *production quantity model* is based on this system; see also **Continuous Review System.**

Periodic Table A *Chemistry* term; a tool for classifying the chemical elements in which *elements are arranged in increasing order of their atomic numbers* such that the elements with similar properties fall in the same vertical column, called *group*, and the elements with dissimilar properties fall in different groups; the *Mendeleev's periodic table was the first breakthrough* in the classification of elements pioneered by a Russian chemist, Dmitri Ivanovich Mendeleev; the latest and the most commonly used table is based on the electronic configuration of elements and is called the *long form of periodic table*, also called the *Bohr's table.*

Peripheral Speed A *Physics* term; the speed of an object at the periphery of a circle; given as:

$$V = \frac{\pi DN}{60} \text{m/s}$$

where

V = peripheral speed
D = diameter of the circle (m)
N = revolution per minute
$\pi = \frac{22}{7}$

Permanent Fastener A *Design* term; a fastening device used to join two parts of a machine or structure permanently and their separation is not possible without damaging the parts or the device itself; for example, *rivet*; see also **Temporary Fastener**.

Permanent Mould Casting A *Manufacturing* term related to *casting*; also called **Hard Mould Casting** as it uses a *hard metallic mould* or **Near-Net Shape Casting** as the final-shaped castings require little or no machining or **Gravity Die Casting** or **Gravity Casting** as the molten metal enters the metallic mould (*die*) under gravity without using any pressure; a casting process which uses a metallic mould made of *grey cast iron* (mostly), *bronze* or other materials like *graphite* or *refractories* to produce dimensionally more accurate castings with minute surface details and close dimensional tolerances; the typical products made by this method include *automobile pistons*, *cylinder heads*, *connecting rods*, *cooking utensils*, *refrigerator parts*, and *small gear blanks.*

Permanent Mould A *Casting* term; see **Die**.

Permissible Stress A *Strength of Materials* term; see **Working Stress**.

Perpendicular Axis Theorem A *Strength of Materials* term; used to find the *moment of inertia* (*second moment of area*) of a section about an axis perpendicular to its plane area; equals to the sum of its moments of inertia about the two rectangular axes lying in the plane of the area; given as:

$$I_z = J = I_x + I_x$$

where

$I_z = J$ = polar moment of inertia of the section about z-axis
I_x = moment of inertia of the section about x-axis
I_y = moment of inertia of the section about y-axis; see also **Parallel-Axes Theorem**.

Perpetual Motion Machine of First Kind A *Thermodynamics* term, abbreviated as **PMM 1**; a device which violates the first law of thermodynamics (by creating energy); see also **Perpetual Motion Machine of Second Kind**.

Perpetual Motion Machine of Second Kind A *Thermodynamics* term, abbreviated as **PMM 2**; a device which violates the second law of thermodynamics; see also **Perpetual Motion Machine of First Kind**.

Perpetual Motion Machine A *Thermodynamics* term, abbreviated as **PMM**; any device which violates the first or the second law of thermodynamics; see also **PMM 1** and **PMM 2**.

Personal Error A *Metrology* term; a type of *systematic error* which occurs due to *inexperience* of an observer; a few examples include *taking readings of an instrument without setting it properly*, and *taking observations without following proper precautions*.

Personal Income An *Economics* term; a part of national income received by a person from all sources; see also **National Income**.

PERT A Project *Management* term; see **Program Evaluation and Review Technique**.

Pessimistic Time A *Project Management* term; the longest possible time to complete a project activity, assuming everything goes wrong; used to calculate the *expected time of completion* of an activity in *program evaluation and review technique* (PERT); see also **Optimistic Time.**

Petrol Engine An *IC Engine* term: an internal combustion engine which works on the principle of *Otto cycle*; also called **Spark Ignition Engine** abbreviated as **SI Engine**; the working fluid is the fuel–air mixture which is ignited by a spark produced by a spark plug; uses a lower compression ratio as compared to a diesel engine; see also **CI Engine** and **Diesel Engine**.

Pewter A *Materials Engineering* term; an alloy of *tin* (Sn) and *lead* (Pb) consisting of 75% (or 80%) Sn and 25% (or 20%) Pb; used in making *cups*, *mugs*, and other *utensils*; see also **Solder**.

Phase Diagram A *Materials Engineering* and a *Thermodynamics* term; in *materials engineering* terms, a diagram which shows the *composition* and *phases* of an alloy system with respect to *temperature*; the *iron-carbon phase diagram* is the most important one, which has wider commercial applications; in *thermodynamics* terms, the plot between pressure (on the vertical axis) and temperature (on the horizontal axis) which shows all the three phases of a *pure substance.*

Phase Difference A *Physics* term; the difference in phase of two vibrating particles which indicates the lack of harmony in the vibrating states of the two particles at a given instant.

Phase A *Physics* term; a physical quantity which completely defines the position and direction of motion of an oscillating particle at any instant with respect to its mean position; if the displacement of a particle at any instant of time t is represented by the equation:

$$y = a\sin(\omega t + \phi) = a\sin\left(\frac{2\pi}{T}t + \phi\right),$$

then the quantity $(\omega t + \phi)$ or $\left(\frac{2\pi}{T}t + \phi\right)$ is called the phase of oscillation at time t.

where

Φ = phase constant
ω = angular frequency $= 2\pi n = \frac{2\pi}{T}$
n = frequency of oscillation
a = amplitude of oscillation
T = time period of oscillation.

Phenolic A *Chemistry* term; a *thermosetting polymer* produced by the polycondensation of *phenol* and *formaldehyde*; has good dielectric properties and surface hardness; widely used in making *electric iron handles, switch covers, socket boxes,* and *fan-motor housings.*

Philip B. Crosby (1926–2001) One of the quality gurus and pioneers of *quality management* who authored two famous books, '*Quality Is Free*' (1979) and '*Quality Without Tears*' (1984); also evolved two important quality terms '*cost of poor quality*' (in his book '*Quality Is Free*') and '*zero defects*' (in his book '*Quality Without Tears*').

Phosgene A *Chemistry* term; a chemical compound having the chemical formula of $COCl_2$, also called **Carbonyl Chloride**; an extremely *poisonous gas* obtained when carbon monoxide (CO) combines with chlorine (Cl_2) in the presence of *sunlight*.

Phosphor Bronze A *Chemistry* term; a *copper-base* alloy containing up to 11% *tin* (Sn) and small amounts of *phosphorus* (P) along with *copper* (Cu); has good *corrosion* and *fatigue resistance*, *high tensile strength*, and high capacity to absorb energy which makes it suitable as a spring material.

Photo Diode An *Electronics* term; a *semiconductor device* which can convert light into current or voltage; used to detect the presence or absence of minute quantities of light; finds use in consumer electronic devices such as *compact disc* (CD) *players*, *smoke detectors*, and *remote control receivers for televisions*, etc.; also used for various medical applications such as *detectors for computed tomography* or *instruments to analyse samples* (*immunoassay*).

Photocell A *Physics* term; see **Solar Cell.**

Photochemical Blanking A *Nonconventional Machining* term; a nonconventional *chemical machining* (CM) process which removes materials from a workpiece by using a *photographic* technique; can produce extremely thin metal without distortion; can easily machine *fragile parts* and *hard* and *brittle* materials; typical photochemical products include *fine screens*, *printed circuit boards*, and *masks for colour television.*

Photochemical Reaction A *Chemistry* term; any chemical reaction initiated as a result of absorption of light.

Photoelectric Effect A *Physics* term; a phenomenon discovered by *J. J. Thomson* (1856–1940) in which electrons are emitted from the surface of a metal when a light of a certain frequency strikes on it.

Photoelectrons A *Physics* term; the electrons emitted from the surface of a metal on being subjected to a light of a certain frequency.

Photon A *Physics* term; a bundle of energy which light consists of; the energy of a photon is equal to $h\upsilon$, where υ is the frequency of the light and h is a constant, called *Planck's constant* whose numerical value is equal to 6.626×10^{-34} J s.

Phototransistor A digital switch which produces an output when light falls on it.

Photovoltaic Cell A *Physics* term; see **Solar Cell.**

Physical Model The reconstruction of an actual system and environment; a few examples include *scale model of an aeroplane*, *testing of a new product in the market* or *testing of a bomb.*

Physical Property A *Materials Engineering* term; a property which is not related to the composition of materials or to the application of external forces; mostly *structure-insensitive* and hence do not change with change in the structure of the materials; a few physical properties include *specific heat*, *density*, *thermal conductivity*, *thermal expansion*, *melting point*, *boiling point*, and *electrical conductivity.*

Physical Science The science which deals with *non-living things*; include subjects like *physics, chemistry, geology, geography, astronomy, astrology*, and *oceanology*; see also **Biological Science**.

Physics The most basic branch of science which deals with the study of *nature* and *natural phenomena*; two domains of interest in physics include *macroscopic* and *microscopic*; *macroscopic domain makes up classical physics and microscopic domain makes up modern physics*; five main branches of physics include *mechanics, heat & thermodynamics, electromagnetism, theory of relativity,* and *quantum mechanics.*

Physiological Need A *Management* term related to *motivation*; one of the five levels of human needs proposed by *Abraham Harold Maslow* (1908–1970) in his '*Need-Hierarchy Theory*' of motivation; represents the basic needs of human beings such as *hunger*, *thirst*, and the *need for shelter*.

Pi Bond A *Chemistry* term abbreviated as *π bond*; a type of *covalent bond* formed by *lateral (sidewise) overlapping* of the *two half-filled orbitals*; *weaker than sigma (σ) bond*; see also **Sigma Bond**.

Piercing A *Manufacturing* term; a *metal working operation* used to produce indentation (impression) in the form of cavity on the surface of a workpiece by using a punch; also used to produce hollow regions in forgings; see also **Punching** and **Blanking**.

Piezoelectric Effect A *Physics* term; a phenomenon in which a material produces a small voltage when subjected to mechanical pressure.

Piezometer *Fluid Mechanics* term; a simple device used to measure fluid pressure; consists of a vertical glass tube to be mounted at the point of pressure measurement (shown as the height of liquid column above the point); not suitable for measuring large pressure as it will require long vertical tubes; cannot measure negative gauge pressure (vacuum).

Piezometric Head A *Fluid Mechanics* term; the sum of *pressure head* and *datum head*; a more convenient means, as the sum of the two heads is easily measurable by employing a simple piezometer.

Pig Iron A *Materials Engineering* term; a product of *blast furnace*; obtained by heating the *iron ore* in the blast furnace, and the liquid iron is poured into moulds to form what is known as pigs of iron; a base metal used in steel making; is *hard* and *brittle*; contains considerable amounts of dissolved *carbon, manganese, silicon, phosphorus*, and *sulphur*.

Pillar File A *Manufacturing* term; one of the types of a file used in the fitting shop which resembles a hand file, but has reduced width and increased thickness; used in the filing of *narrow works* such as *keyways*, *slots*, and *grooves*.

Pin Holes A *Casting* term; a type of casting defect which occurs in the form of *small gas cavities* just below the surface of a casting due to excessive moisture content and poor permeability of the moulding sand.

Pin Joint A *Design* term; a t*emporary fastening device* which uses a pin to fasten two circular rods subjected to tensile forces; permits a small amount of flexibility.

Pin Support A *Physics* and a *Strength of Materials* term; see **Hinge Support**.

Pin A *Design* term; a *fastening element* used in *pin joint*; *used to connect two circular shafts*; classified into *cylindrical pin* and *conical pin.*

Pinion A *Theory of Machines* term related to *gear*; used in conjunction with a *rack*; the *smaller gear* of the two mating gear wheels; see also **Rack**.

Piston Rod An *IC Engine* term; a round rod attached to the piston of an engine, which in turn, is connected to the connecting rod; executes *reciprocating motion* inside the cylinder of the engine.

Piston An *IC Engine* term; a *cylindrical* component of an IC engine which makes a reciprocating motion (also called *to-and-fro motion* or *up and down motion*) in the engine cylinder to compress the fuel or expel the products of combustion.

Pit Moulding A *Casting* term; the process of making a mould in a *pit* instead of a moulding flask; used for *extremely large size castings.*

Pitch Circle Diameter A *Theory of Machines* term related to *gear*; also called **Pitch Diameter**; the diameter of the pitch circle or the theoretical diameter of a screw thread lying between major and minor diameters; see also **Major Diameter** and **Minor Diameter**.

Pitch Circle A *Theory of Machines* term related to *gear*; an imaginary circle around which the teeth on a gear or any other uniformly toothed device are evenly spaced.

Pitch Diameter A *Theory of Machines* term related to *gear*; see **Pitch Circle Diameter**.

Pitch A *Gear* term; see **Circular Pitch**.

Pitot Tube A *Fluid Mechanics* term, named after *Henry Pitot* (1695–1771), a *French* physicist; a metering device used to measure the *velocity of a fluid flow* using *Bernoulli's equation*; consists of a thin glass tube with both ends open, having a right-angle bend, where one end of the tube faces the upstream flow and brings it to rest at its inlet; the velocity of the flow (V) is given as:

$$V = \sqrt{2gh}$$

where

h = height of liquid column (in case of open channel) or difference between liquid columns (in case Pitot tube lies in a pipe and is attached with a piezometer)
g = acceleration due to gravity = 9.81 m/s^2.

Pitting A *Strength of Materials* term; a surface fatigue failure which occurs due to many repetitions of high contact stresses.

Pivot Bearing See **Foot-Step Bearing.**

P-Kanban An *Operations Management* term; see **Production Kanban.**

Plain Carbon Steel A *Materials Engineering* term; also called **Carbon Steel**; an alloy of *iron* and *carbon* with small percentage of *manganese, silicon, sulphur* and *phosphorus* present as impurities but not as alloying elements; carbon content may vary in the range from a trace to 1.7%; finds use in making components where strength and other requirements are not too severe; classified into *low carbon, medium carbon,* and *high carbon* steels.

Planar Structure A *Strength of Materials* term; a structure lying in a single plane; for example, a *simple beam* or a *cantilever beam.*

Plan-Do-Check-Act Cycle An *Operations Management* term; see **Deming Wheel.**

Plane of Bending A *Strength of Materials* term; the plane in which all the loads act and all deflections occur; the plane in which a beam bends.

Plane of Symmetry A *Strength of Materials* term; an *imaginary plane* which divides the cross section of a beam into two identical parts; a plane about which the two parts of the cross section of a beam are the mirror image of each other.

Plane Stress A *Strength of Materials* term; a state of stress in which an element of a beam's cross section is subjected to normal and shear stresses in only two directions, say along x and y directions of a xy-plane; a common stress condition which exists in all ordinary structures such as in the *walls of pressure vessels*, in the *webs and flanges of beams of various shapes*, and in a wide variety of structures subjected to the combined effects of axial, shear and bending loads as well as internal pressure.

Plane Truss A *Strength of Materials* term related to a *truss*; a truss in which forces through the members as well as the external loads acting at the joints lie in a plane; see also **Space Truss.**

Plane A *Mathematics* term; a flat smooth surface which extends indefinitely in all directions.

Planimeter An *IC Engine* term; an instrument used to measure the area of an *indicator diagram.*

Planned Order Schedule An *Operations Management* term related to **MRP;** the primary output of MRP; a schedule or plan of the quantity of each material to be ordered in each time period over the entire planning horizon in a production system; used by *purchasing* to place orders with *suppliers*, and by *production* to order *parts, subassemblies* or *assemblies* from *upstream production departments*.

Planning Horizon An *Operations Management* term; the length of time covered by an aggregate plan; it is typically of one year.

Planning A *Management* term; one of the main functions of a manager which focuses on setting the organizational goals and objectives as well as the actions to achieve them; requires decision-making, that is, choosing a course of action from among alternatives.

Plant Layout An *Operations Management* term related to *facilities planning*; see **Facility Layout**.

Plant Location An *Operations Management* term related to *facilities planning*; the finding or locating a place for establishing a plant; should be near to a source of raw materials, market and workforce (skilled and unskilled); supported by infrastructural facilities such as road, power, water and transportation; a suburban area which combines the advantages of both urban and rural areas is the best choice for establishing a plant as it offers many benefits such as skilled workforce, increased scope of expansion because of abundant availability of cheap land, pollution-free environment and improved working atmosphere.

Plaster Mould Casting A *Manufacturing* term related to *casting*; a casting process which uses a *plaster of paris mould* to produce castings of high dimensional accuracy and good surface finish; typical parts produced by this method include *gears, valves*, and *ornaments*.

Plastic Deformation A *Strength of Materials* term; a process of permanently deforming a body with the application of an external force, and the body does not recover its original shape and size on the removal of the force; finds use in giving useful and desired shapes to products used for various applications; see also **Elastic Deformation**.

Plastic Material A *Materials Engineering* term; a material which does not fully recover its unstrained (original) state, but retains some permanent strain on the removal of the load applied on it, when stressed within elastic limit; the unloading path does not follow the loading path and the process is *irreversible*; for example, *rubber* is a plastic material; see also **Elastic Material**.

Plastic A *Chemistry* term; a chemically produced material which can be moulded into shape while soft and then set into a hard or slightly flexible form.

Plasticity A *Physics* term; a property of a metallic body which enables it to get permanently deformed on the application of a deforming force; used in producing useful products; see also **Elasticity**.

Plasticizer A *Chemistry* term; a complex, low molecular weight organic compound; used as an *additive* to impart flexibility and softness to a polymer during the polymerization process.

Plate A form of material (metal, plastic and others) having thickness greater than 6 mm; used for making *structural bodies* like *machine structures, boilers, bridges, ship hulls, nuclear vessels*, etc.; see also **Sheet**.

Platinum A *Chemistry* term; a *heavy, soft, ductile*, and *malleable noble metal* which has the specific gravity of 21.4 and the melting point of 1773 °C; a costly metal; used in making *platinum thermometers, scientific apparatus*, and *jewellery*.

Plug-in Hybrid Electric Vehicle An *Electric Vehicle* term, abbreviated as **PHEV**; a hybrid electric vehicle which uses a larger battery that uses plug-in charging facility for fast charging using Level 2 charging; can travel more distances (10–40 miles) with battery assistance before gasoline engine can be used to drive the vehicle further; a few examples of PHEV include Audi A3 E-Tron, Ford Escape PHEV, Kia Optima, Mitsubishi Outlander PHEV, Toyota Prius, Volvo XC90 TB; see also **Hybrid Electric Vehicle**.

Plywood A *Materials Engineering* term; a composite material formed by thin sheets of wood that are glued together using high-strength thermosetting plastic adhesives; has a greater strength-to-weight ratio than a board of equal geometry made from a single species of wood.

PMM 1 A *Thermodynamics* term; see **Perpetual Motion Machine of First Kind**.

PMM 2 A *Thermodynamics* term; see **Perpetual Motion Machine of Second Kind**.

PMM A *Thermodynamics* term; see **Perpetual Motion Machine**.

Pneumatic Control System A *Physics* term; a control system which employs the changes of air pressure for the functioning of an operating device.

Pneumatics The physical science and technologies associated with the mechanics of pressurized gases; mechatronic devices work on the principle of pneumatics.

Point Defect A *Materials Engineering* term; a *zero-dimensional disorder* in a crystalline structure caused either by heating during the processing of a material or by the introduction of impurity atoms or ions in the material; the two common types of point defects include *vacancy* and *interstitial*; see also **Dislocation**.

Point Load A *Physics* and a *Strength of Materials* term; see **Concentrated Load**.

Point of Contraflexure A *Strength of Materials* term; also called **Inflection Point**; a point of zero moment in the bending moment diagram (BMD) where the bending moment changes sign from positive to negative and vice versa; corresponds to zero curvature of the deflected beam.

Point A *Mathematics* term; a geometrical figure which has no length or breadth, but has only a position that can be specified by coordinates.

Point-to-Point System A *Manufacturing* term related to *automation*; also called **Positioning System**; a motion control system used in *numerical control* (NC) to move the cutting tool to a predefined location and then start the work; the speed or path by which this movement is accomplished is not important; denotes the *lowest level of motion control* between the tool and workpiece; highly suitable for *drilling operations* and *spot welding*.

Poise A *Fluid Mechanics* term; the CGS unit of *dynamic viscosity*, named after *Jean Poiseuille* (1797–1869), a *French* physician and scientist who studied the flow of blood through capillaries in the human body; equivalent to dyne-s-cm^{-2} or gm/(cm-s); one poise is the coefficient of viscosity of a liquid in which a viscous force of 1 dyne acts per cm^2 of a liquid layer having a velocity-gradient of 1 cm per sec per cm perpendicular to the direction of flow; 1 poise = 0.1 kg/(m-s) = 0.1 Pa-s; see also **Centipoise**.

Poiseuille Flow A *Fluid Mechanics* term; see **Hagen–Poiseuille Flow**.

Poisson's Ratio A *Strength of Materials* term; a ratio of the lateral strain to the longitudinal strain; a useful parameter to connect elastic constants; denoted by υ (nu); theoretically its value should lie between -1 and 0.5, but it is never negative, and practically lies between 0 and 0.5; for most metals its value is 0.33; other values of Poisson's ratio include 0 for cork, 0.1 for cement concrete, and 0.5 for rubber and incompressible materials.

Poka Yoke An *Operations Management* term; a *mistake-proofing technique* used in *JIT manufacturing*; requires a process or product to be redesigned in a manner so that it is difficult to make a mistake in the process or product; the *Taguchi approach* of robust design eliminates quality defects; a jig which holds pieces for processing might be designed in a manner to allow pieces to be held only in the correct orientation.

Poka-Yoke An *Operations Management* term; also called **Mistake-Proofing** or **Full Proofing**; a *Japanese* technique to design a process in a manner so that it is difficult to make a mistake in the process; any foolproof device or mechanism which prevents defects from occurring; for example, the Taguchi approach of '*robust design*' eliminates quality defects; similarly, a jig used to hold a workpiece during its processing must be designed in a manner to allow the workpiece to be held only in the correct orientation.

Polar Covalent Bond A *Chemistry* term; a covalent bond formed between two *dissimilar atoms*; one of the two atoms in the bond generally has a greater tendency to attract the electrons towards itself, hence electron pair is pulled closer to that atom, resulting in *unsymmetrical distribution* of electrons leading to charge separation; found in molecules such as *water* (H_2O) and *hydrogen fluoride* (HF); see also **Nonpolar Covalent Bond**.

Polar Modulus A *Strength of Materials* term; a ratio of the polar moment of inertia of a shaft's cross-section to its radius; also equals to a ratio of the torque applied to the maximum shear stress on the shaft; polar modulus of solid and hollow shafts are given as:

$$\text{Polar modulus for a solid circular shaft} = \pi d^3/16$$

$$\text{Polar modulus for a hollow circular shaft} = \pi\left(D^4 - d^4\right)/16d$$

where

d = diameter of a solid shaft or inside diameter of a hollow shaft
D = outside diameter of a hollow shaft.

Polar Molecule A *Chemistry* term; a molecule having a *polar covalent bond*; consist of *two dissimilar atoms*; *has a permanent dipole moment*; examples include *water molecule* (H_2O) and *hydrogen fluoride molecule* (HF); see also **Nonpolar Molecule.**

Polar Moment of Inertia A *Strength of Materials* term; the moment of inertia of a plane area about an axis perpendicular to the plane of the area; denoted by J or I_z; given as:

$$\begin{aligned} J &= I_z = I_x + I_y \\ &= 2I_x \quad (\text{for a circular shaft, as } I_x = I_y) \\ &= \frac{\pi d^4}{32} \quad (\text{for a solid circular shaft}) \\ &= \frac{\pi\left(D^4 - d^4\right)}{32} \quad (\text{for a hollow circular shaft}) \end{aligned}$$

where

d = diameter of the solid shaft or inside diameter of the hollow shaft
D = outside diameter of the hollow shaft; see also **Second Moment of Area.**

Polar Orbit A *Physics* term; the orbit whose angle of inclination with the equatorial plane of the earth is 90°.

Polar Satellite A *Physics* term; a low altitude satellite which revolves in a *polar orbit* around the earth, and is placed at about 500–800 km height from the surface of the earth; moves around the poles of the earth from north to south direction, whereas the earth rotates around its own axis from west to east direction; its time period of revolution is about 100 min; used for collecting information about atmospheric conditions (such as *temperature* and *pressure* required for cloud formation), ozone layer in the atmosphere, and to detect the *ozone hole over Antarctica*, which are extremely useful for remote sensing, meteorology as well as for environmental studies of the earth; see also **Geostationary Satellite.**

Polarity An *Electrical* term; the condition of being electrically positive or negative.

Policy A *Management* term; a general statement or understanding which guides a manager's thinking in decision-making; ensures that decisions fall within certain boundaries.

Political Skill A *Management* term; a management skill which relates to a manager's ability to build a power base and to establish influential connections; helps a manager to get more promotions and compete for resources.

Polyester A *Chemistry* term; a thermosetting polymer produced by the polycondensation of *polycarboxylic acid* and *polyhydric alcohol*; has excellent dielectric properties and surface hardness; widely used in making *cloth*, *paper*, *mat*, and *insulation for wire and cables.*

Polyethylene A *Chemistry* term; also called **Polythene**; a *thermoplastic polymer* formed by polymerization of a large number of *ethylene molecules* covalently bonded to one another; used in the form of *sheets*, *film*, and *wire insulation*; a few typical products include *bottles*, *pipes*, and *housewares.*

Polygon Law of Forces A *Physics* term; see **Law of Polygon of Forces**.

Polygon A *Mathematics* term; a *closed plane figure* formed by three or more sides; *triangle* is the simplest polygon having three sides; other examples of polygon include *rectangle*, *square*, *pentagon*, *hexagon*, etc.

Polymer A *Chemistry* term; a *high molecular mass macromolecule* which consists of repeating structural units derived from monomers; can resist corrosion and chemical attacks; often used in composites both as fibres and as a matrix; has wide range applications; most of the polymers are electrical insulators.

Polymerization A *Chemistry* term; the process by which monomers join together to form a polymer; for example, the *formation of polyethylene from ethylene*; classified into *addition polymerization* and *condensation polymerization.*

Polystyrene A *Chemistry* term; a commonly used *plastic material* which may be formed into a variety of shapes by the simple application of heat; a low-cost, easy-to-process, rigid, crystal clear, brittle material which displays exceptional resistance to wear and corrosive attacks by chemicals.

Polythene A *Chemistry* term; see **Polyethylene**.

Polytropic Process A *Thermodynamics* term; a process which follows the equation $PV^n =$ constant, where P and V respectively represent pressure and volume, and 'n' is an exponent which equals to 1 for an isothermal process, 0 for a constant pressure process, ∞ for a constant volume process and γ $(= C_p/C_v)$ for an adiabatic process; the work done during a polytropic process undergoing between two states 1 (initial state condition) and 2 (final state condition) is given as:

$$W = (P_1V_1 - P_2V_2)/(n - 1)$$

POQ An *Operations Management* term; see **Periodic Order Quantity**.

Porosity A *Manufacturing* term related to *casting*; see **Shrinkage**.

Positioning System A *Manufacturing* term related to *automation*; see **Point-to-Point System**.

Positive Displacement Machine A *Fluid Mechanics* term; a fluid device in which energy transfer is accompanied by volume changes which occur due to movement of the boundary (positive displacement) in which the fluid is confined; a few examples include *piston-cylinder arrangements*, *gear pumps*, and *lobe pumps*.

Positive Displacement Pump A *Fluid Mechanics* term; a pump in which a fluid is sucked and pushed out alternatively in a cylinder using a piston or plunger involving displacement of the fluid; develops positive (high) pressure used to lift the fluid to a required height; suitable for *high pressure applications* (ranging up to 40 MPa) like *pumping viscous liquids* or *thick slurries*, and for applications where precise amounts of liquids are to be handled as in case of medical applications; a few examples include *flexible-tube peristaltic pump*, *three-lobe rotary pump*, *gear pump*, and *double screw pump*; *a human heart is also a good example of positive displacement pump*; see also **Rotodynamic Pump**.

Positive Integers A *Mathematics* term; any positive number such as 1, 2, 3, 4, etc.; see also **Negative Integers**.

Potash Glass A *Chemistry* term; also called **Hard Glass** because of its hardness; a type of glass having high melting point which makes it to withstand higher temperatures; contains potassium carbonate (K_2CO_3), calcium carbonate ($CaCO_3$), and sand or quartz (silica, SiO_2) fused together in proper proportions.

Potential Energy A *Physics* term; also called **Datum Energy**; the capacity of a body to do work by virtue of its *position*, *configuration* or *state of strain*; for example, the water at the top of a waterfall can rotate a turbine when allowed to fall on it, and its energy is called the *gravitational potential energy*; a wound clock spring may keep the clock running by virtue of its state of strain and its energy is called the *elastic potential energy*; when a body is kept on the earth's surface, its potential energy is zero, but once it is raised to a certain height, it acquires a certain value of potential energy, depends on mass and height of a body; equals to the product of mass (m), acceleration due to gravity ($\boldsymbol{g}$), and height (h), that is, $\boldsymbol{mgh}$.

Potential Flow A *Fluid Mechanics* term; see **Irrotational Flow**.

Potential Function A *Fluid Mechanics* term; see **Velocity Potential**.

Potential Head A *Fluid Mechanics* term; see **Elevation Head**.

Potential Vortex A *Fluid Mechanics* term; see **Free Vortex**.

Pound Force A *Physics* term; the *British* unit (the FPS unit) of force; denoted by lbf; 1 pound force is that force which acts on a body of one pound mass and produces an acceleration equal to the standard acceleration of gravity on earth, that is, 32.2 ft/s^2.

Pound Mass A *Physics* term; the British unit (the FPS unit) of *mass*; 1 pound mass is equal to 0.4536 kg.

Pound Weight A *Physics* term; a unit of force in the *British* system of units (the *FPS unit*); the pull of gravity on the unit pound mass with an acceleration due to gravity of 32.2 ft/s^2; 1 pound weight is equal to 32.2 poundal; see also **Poundal**.

Poundal A *Physics* term; a unit of force in the British system of units (the FPS unit); the force which acts over a body of unit pound mass and produces an acceleration of 1 ft/s^2.

Pouring Basin A *Manufacturing* term related to *casting*; a part of the *gating system*; a funnel-shaped cavity which lies at the *top* of the mould which contains molten metal used for making the castings.

Power Hacksaw A *Manufacturing* term; a hacksaw operated by power.

Power Spinning A *Manufacturing* term; see **Shear Spinning**.

Power Stroke A *Thermodynamics* term related to an *IC engine*; also called **Expansion Stroke** or **Working Stroke**; one of the strokes of a piston in an (internal combustion) engine which develops power during an expansion process.

Power Transmission A *Strength of Materials* term; the power transmitted by a solid as well as a hollow shaft is given as:

$$P = 2\pi NT/60\,(\text{W}) \tag{1}$$

$$= \pi NT/30{,}000\,(\text{kW}) \tag{2}$$

$$= 2\pi NT/4500\,(\text{HP}) \tag{3}$$

where

T = torque acting on the shaft in N-m in Eqs. (1) and (2) and in Kgf-m in Eq. (3)
N = revolutions per minute (rpm) of the shaft
π = a constant.

Power A *Physics* and a *Management* term; in *physics* terms, the time rate of doing work; a ratio of the work done to the time taken; the product of *force* and *velocity* in *linear motion*; the product of *torque* and *angular velocity* in *rotational motion*; the

SI unit is *watt* (W) which is equivalent to joule/sec (J/s); its another popular unit is *horsepower* (hp) which is used to describe the output of *automobiles*, *motorbikes*, etc.; in *management* terms, the ability of individuals or groups to induce or influence the beliefs or actions of other persons or groups; see also **Authority**.

ppm An *Air Pollution* term; the short form of *parts per million*; a unit of measure of the concentration of gases in air; expressed as parts of the gas per million (10^6) parts of the air–gas mixture, both by volume.

Precedence Relationship A *Project Management* term; the sequential relationship of project activities to each other.

Precious Metals A *Materials Engineering* term; relatively scarce, highly corrosion-resistant metals which are valued for their colour, lustre and malleability; a few examples include *gold*, *silver*, and *platinum*.

Precision Casting A *Manufacturing* term related to *casting*; see **Investment Casting**.

Precision Forging A *Manufacturing* term; also called **Net-Shape Forging** or **Near-Net-Shape Forging**; a *hot metal working process* used to produce highly finished parts requiring no further machining.

Precision A *Metrology* term; the repeatability of a measurement process; also indicates how closely the individual measured values of a physical quantity agree with each other when measurements are carried out under identical operating conditions at a short interval of time; a precision instrument always gives the same reading; used in conjunction with *accuracy*; determined by the *least count* of the measuring instrument; smaller the least count, greater is the precision; see also **Accuracy**.

Predictive Maintenance An *Operations Management* term; also called **Condition Monitoring**; a method of *preventive maintenance* in which the condition of the equipment is constantly monitored, and the decision to carry out maintenance is taken based on the analysis of the equipment's behaviour; see also **Preventive Maintenance** and **Breakdown Maintenance**.

Present Worth Method A *Financial Management* term; see **Present-Value Analysis**.

Present-Value Analysis A *Financial Management* term; also called **Present Worth Method**; a process of determining the amount of money to be invested now at a specific rate of interest to accumulate to a certain amount in the future.

President A *Management* term; a high level manager (a corporate official) who is the next most powerful position after the chairman and chief executive officer (CEO) of a company, and reports directly to the CEO; shares the principal responsibility for managing the company's business with the CEO.

Press Forging A *Manufacturing* term; another method of applying forces in a forging operation, other than drop forging, in which a workpiece (heated or cold) is shaped by *gradually applying pressure* through a ram; effectively utilizes the deformation energy in shaping the workpiece, the major part of which goes as a waste in drop forging; also offers higher dimensional accuracy to the workpiece even in a single operation; see also **Drop Forging**.

Press Work A *Design* term; an operation which involves shaping of a workpiece by means of a *press* (*mechanical* or *hydraulic*) which applies the required pressure.

Pressed Casting A *Manufacturing* term related to *casting*; also called **Corthias Casting** named after *Corthias of France*; a *variation of the permanent mould casting* in which the hot molten metal is pressurized to enter the mould cavity of an open-ended metal mould (die); limited to the production of ornamental castings of open design.

Pressure Angle A *Theory of Machines* term related to a *gear*; also called **Angle of Obliquity**; an angle between the common normal at the point of contact and the common tangent at the pitch point; varies between $14\frac{1}{2}^\circ$ and $22\frac{1}{2}^\circ$ for involute gears, but the normal value is 20°.

Pressure Die Casting A *Manufacturing* term related to *casting*; see **Die Casting**.

Pressure Drop A *Fluid Mechanics* term; the differential pressure which exists between two points in a fluid flow system; the resistance to flow between the two points.

Pressure Gauge A *Power Plant* term related to a *boiler*; a *boiler mounting* used to measure steam pressure in the boiler; also called **Steam Gauge**; fitted in the front and at the top of the boiler for clear inspection.

Pressure Gradient A *Fluid Mechanics* term; the rate of change of pressure with distance; represents the negative of the surface force per unit volume due to pressure; denoted by ∇p; expressed as:

$$\begin{aligned} \text{grad}\, p = \nabla p &= \left(\mathbf{i}\frac{\partial p}{\partial x} + \mathbf{j}\frac{\partial p}{\partial y} + \mathbf{k}\frac{\partial p}{\partial z}\right) \\ &= \left(\mathbf{i}\frac{\partial}{\partial x} + \mathbf{j}\frac{\partial}{\partial y} + \mathbf{k}\frac{\partial}{\partial z}\right) p \end{aligned}$$

Pressure Head A *Fluid Mechanics* term; an energy parameter which appears in the *Bernoulli equation*; represents the height of a fluid column which produces the static pressure; expressed as $P/\rho g$ or P/γ; (P = pressure of the fluid, ρ = density of the fluid, g = acceleration due to gravity = 9.81 m/s^2 and γ = specific weight of the fluid); see also **Velocity Head** and **Potential Head**.

Pressure Pouring A *Manufacturing* term related to *casting*; see **Low Pressure Casting**.

Pressure Ratio An *IC Engine* term; a ratio of the pressure of a fluid after compression to the pressure of the fluid before compression; used to find the efficiency of an *IC engine* or a *pump*; see also **Compression Ratio.**

Pressure Transducer An *Electronics* term; a sensor-based pressure measuring device which converts a *pressure effect* into an *electrical effect* like *change in voltage, resistance,* or *capacitance.*

Pressure Vessel A *Strength of Materials* term; a *cylindrical* or *spherical* container used to store liquids or gases under pressure; a few examples include *compressed air tanks, boilers, water-supply pipes, fire extinguishers, spray cans, penstocks,* and *reservoirs*; used in thermal and nuclear power plants, chemical industries, water, steam and air supply system and in space and ocean depths; classified into *thin-walled* or *thick-walled* types depending upon diameter-thickness (d/t) ratio; the thin-walled pressure vessels have (d/t) ratio greater than 20, and are subjected to two types of stresses: the longitudinal stress acting along the axis and the hoop stress (circumferential stress) acting along the circumference of the vessels; the thick-walled pressure vessels have (d/t) ratio less than 20 and are subjected to three types of stresses: the radial stress acting on the thickness (annular ring), the hoop stress, and the longitudinal stress.

Pressure A *Fluid Mechanics* and a *Thermodynamics* term; the external normal force which acts on unit area of a gas or a liquid; the SI unit is *pascal* (Pa) which is equivalent to N/m^2; the unit pascal is named after the French mathematician *Blaise Pascal* (1623–1662) who contributed to early principles of hydrostatics; the other units of pressure include kgf/cm^2 or kgf/m^2 (the MKS unit), $dyne/cm^2$, mm of Hg, cm of Hg, bar (1 bar $= 10^5$ Pa), or torr (1 mm of Hg); has dimension of $[ML^{-1}\,T^{-2}]$.

Prevention Cost A *Quality Management* term; a part of *cost of quality* associated with preventing the occurrence of defects in a product during its manufacturing; includes cost of redesigning the product to make it easily reproducible, cost of redesigning the process to remove the causes of poor quality, cost of improved raw materials, and cost for training employees; see also **Appraisal Cost**.

Preventive Maintenance An *Operations Management* term; a type of maintenance carried out at regular time intervals (during regular operation of facilities or machines) in an attempt to prevent their breakdown (by conducting regularly scheduled inspections, machine adjustments, lubrication, cleaning, parts replacement, painting, and the needed repairs and overhauls); see also **Breakdown Maintenance** and **Predictive Maintenance**.

PRF A *Work Study* term; see **Performance Rating Factor**.

Price Elasticity of Demand An *Economics* term; see **Elasticity of Demand.**

Price An *Economics* term; the amount at which a product is sold per unit; reflects the amount what a customer is willing to pay.

Primary Bond A *Chemistry* term; an atomic bond formed by the transfer or sharing of outer orbital electrons; relatively a stronger bond than secondary bond; examples include *ionic bond* and *covalent bond*; see also **Secondary Bond**.

Primary Sources of Energy A *Power Plant* term; include *water* (by storing water to a certain height by constructing a dam), *sun*, *nuclear fuels*, and *fossil fuels* such as *coal*, *oil*, and *natural gases*; see also **Secondary Sources of Energy**.

Prime Cost An *Operations Management* term related to *cost analysis*; the sum of *direct material cost*, *direct labour cost*, and *direct expenses*; it is very related to *manufacturing activities*.

Prime Mover A *Power Plant* term; a mechanical device used to convert energy obtained from a natural resource into mechanical energy; a few examples include *gas turbines*, *hydraulic turbines*, and *internal combustion engines*.

Prime Number A *Mathematics* term; a number which has no factor other than 1and the number itself, that is, a prime number has only two factors: 1 and the number itself; a few examples include 2, 3, 5, and 7, but not 4, 6, 8, and 9; 1 is not a prime number, as it has exactly one factor; 2 is the only even prime number; every prime number other than 2 is odd, but every odd number is not necessarily a prime number; every even number greater than 4 can be expressed as the sum of two odd prime numbers; see also **Twin Primes** and **Composite Number**.

Priming A *Thermodynamics* term related to *steam*; the wetness fraction of steam expressed in percentage; see also **Quality of Steam**.

Principal Axis A *Strength of Materials* term; an axis about which the product moment of inertia of a plane area is zero; all planes have two principal axes about which the product moment of inertia is zero.

Principal Plane A *Strength of Materials* term; a plane of zero shear stress within a material; there can be two mutually perpendicular principal planes at any arbitrary point within a material under plane stress condition, where one plane carries the maximum (major) principal stress and the other plane carries the minimum (minor) principal stress; see also **Principal Stress**.

Principal Stress A *Strength of Materials* term; a normal stress which acts on a principal plane; can be tensile or compressive; its maximum and minimum values are respectively referred to as the maximum (major) principal stress (σ_1) and the minimum (minor) principal stress (σ_2); see also **Principal Plane**.

Principle of Conservation of Energy A *Physics* term; states that the total energy of an *isolated system* remains conserved (does not change); implies that energy can neither be created nor be destroyed, but can be transformed from one form to another; since the universe as a whole can be treated as an isolated system, the sum total of energy of all kind in the universe remains constant at all times; if one part of the universe loses energy, its other part would gain an equal amount of energy; forms the basis of the *first law of thermodynamics*.

Principle of Conservation of Linear Momentum A *Physics* term; also called **Law of Conservation of Linear Momentum**; states that when the vector sum of the external forces acting on a system is zero, the total linear momentum of the system remains constant (although the momenta of individual particles may change); the law is universal, and holds true even in atomic and nuclear physics, where the Newtonian mechanics fails; equivalent to Newton's third law of motion.

Principle of Conservation of Mass A *Physics* term; states that the amount of mass in a system is constant; the time rate (t) of change of a system mass (M) is equal to zero; expressed as:

$$\left(\frac{\mathrm{d}M}{\mathrm{d}t}\right)_{\text{system}} = 0$$

Principle of Continuity A *Fluid Mechanics* term; also called **Equation of Continuity**; states that for an incompressible non-viscous fluid flowing through a tube of non-uniform cross section, the product of the area of cross-section (A) and the velocity of flow (V) is same at every point in the tube; based on the principle of *conservation of mass*; the mathematical expression of the principle of continuity is known as *equation of continuity*; for any two points 1 and 2 in the fluid flow, the principle of continuity can be mathematically expressed as:

$$A_1 V_1 = A_2 V_2 = \text{constant} \tag{1}$$

The above equation shows that for a *steady incompressible fluid flow*, the fluid velocity varies inversely proportional to the cross-sectional area, being larger in the narrower parts of the tube; for a *compressible fluid flow* (where the density of the fluid changes with pressure), the principle of continuity can be expressed as:

$$\rho_1 A_1 V_1 = \rho_2 A_2 V_2 = \text{constant} \tag{2}$$

where A_1, V_1 and ρ_1 are respectively the cross-section, velocity and density of the fluid at section 1, and A_2, V_2 and ρ_2 are respectively the cross-section, velocity and density of the fluid at section 2 of the tube.

Principle of Dimensional Homogeneity A *Physics* term; also called **Law of Dimensional Homogeneity** or **Principle of Homogeneity of Dimensions** or **Fourier's Law of Dimensional Homogeneity**, named after a *French* mathematician and physicist *Baron Joseph Fourier* (1768-1830) who proposed it in 1822; states that the dimensions of physical quantities on both sides of a dimensionally homogeneous equation are same; the basic equation which represents a physical phenomena must be valid for all systems of units and the form of equation does not depend on the units of measurement; only that mathematical equation is correct, in which the dimensions of the various terms on one side of the equation are equal to the respective dimensions of these terms on the other side of the equation; powers of all the three quantities, namely *M* (*mass*), *L* (*length*), and *T* (*time*) must match on either side of the formula, and in case, power of even one of them fails to match, the equation is invalid; used for checking the correctness of an equation; also used in deducing relation amongst different physical quantities; mere dimensional correctness of an equation does not ensure its physical correctness, for example, dimensionally, both *work* and *torque* are same, but physically they differ.

Principle of Homogeneity of Dimensions A *Physics* term; see **Principle of Dimensional Homogeneity**.

Principle of Minimum Chain of Command A *Management* term; emphasizes on the minimum number of hierarchical levels in an organization consistent with its goals and the environment in which it operates; implies that an organization should be kept as flat as possible, and managers should be evaluated for their ability to control organizational activities with the smallest number of managers possible.

Principle of Moment A *Physics* term; the algebraic sum of moments of coplanar forces about a point is zero; concludes that the sum of counterclockwise moments about a fixed point is equal to the sum of clockwise moments about the same point; provides the basis for rotational equilibrium of a body.

Principle of Superposition A *Strength of Materials* term; states that the effect of a given combined loading on a structure may be obtained by determining separately the effects of various loads and combining the results obtained, provided that the deformation produced be each load is small and varies linearly proportional to the load applied and does not affect the conditions of application of other load.

Principle of Thermometry A *Thermodynamics* term; states that two bodies in contact and in thermal equilibrium must have the same temperature; forms the basis of temperature measurement.

Principle of Transmissibility A *Physics* term; also called simply **Transmissibility of Forces**; states that a force can be applied at any point on its given line of action provided its effect remains unaltered.

Principles of Motion Economy An *Operations Management* term related to *work study*; also called **Principles of Motion Study**; a broad set of guidelines proposed by *Frank Gilbreth* (1868–1934) for the efficient design of work and to economize motions involved in a task; focuses on *work arrangements*, the *use of human hands and body*, and the *use of tools and equipment* for improving and conserving the motions involved in a task.

Principles of Motion Study An *Operations Management* term related to *work study*; see **Principles of Motion Economy**.

Principles of Scientific Management A *Management* term; a famous book authored by *F. W. Taylor* (1856–1915) and published in 1911; discusses scientific methods, principles, and tools for improving the methods of doing a task in the solution of managerial problems.

Prismatic Bar A *Strength of Materials* term; a straight structural member of uniform cross-section throughout its length; finds use in a *tension test*; see also **Non-Prismatic Bar**.

Prismatic Channel A *Fluid Mechanics* term; a channel with a constant slope and cross- section.

Probability Density A *Statistics* parameter; the number of occurrences of an event divided by the total sample number.

Probability Distribution A *Statistics* term; the number of ways in which the occurrence of a given event can take place; has different characteristics depending on the types of processes in the operation and nature of variances of the parameters; the probability distribution modes include the *binomial distribution*, the *Poisson distribution*, and *the normal distribution*; a typical example of probability distribution may be a graphical representation of the average death rate against different age groups.

Probability A *Statistics* term; the likelihood of an event occurring measured on a scale between 0 and 1, where zero value corresponds to impossibility and the unit value corresponds to the certainty of an event taking place.

Probe Method A *Metrology* term; see **Intrusive Method**.

Problem Statement A *Design* term; a part of a design brief which clearly and concisely identifies a client's or target consumer's problem, need, or want.

Process Capability A *Quality Control* term; denoted by Cp; the ability of a conversion process to produce products within design specifications (tolerances); a measure of spread of the process; for example, for a process with standard deviation σ, the latest process capability is 6σ.

Process Control A *Quality Control* term; a process which involves monitoring of a production process (using statistical quality-control methods) so that a consistent uniform product will be produced; see also **Production Control**.

Process Flowchart An *Operations Management* term; also called simply **Flowchart**; a chart used in motion study to show the various steps in a job, operation or process using symbols within the flow of the production process.

Process Innovation An *Operations Management* term; the innovation which leads to improvement in various operational processes such as improving manufacturing facilities, increasing product quality, and faster distribution; see also **Product Innovation**.

Process Layout An *Operations Management* term related to *facilities planning*; also called **Functional Layout**; a *process-oriented* plant layout in which groups of similar multifunctional machines are positioned at well-defined separate locations; used to produce a variety of nonstandard products in relatively small batches (batch production) as in a custom machine shop; failure of one or two machines does not stop production activities as additional machines are in standby for support; offers maximum manufacturing flexibility; involves more in-process inventory; suitable for producing *customized* products; see also **Product Layout**.

Process Planning An *Operations Management* term; a process of determining the sequence (using route sheet) of individual manufacturing operations needed to produce a given part or product.

Process Production An *Operations Management* term; see **Continuous Production.**

Process Technology An *Operations Management* term; a collection of *equipment, people,* and *systems* used to produce a firm's products and services.

Procurement Cost An *Operations Management* term; the cost related to the *procurement of supplies* (goods and services) from a supplier.

Procurement An *Operations Management* term; a process of purchasing goods and services from a *supplier*; an important activity in supply chain management (SCM).

Producer Gas A *Chemistry* term; an important *industrial fuel* which is a mixture of *carbon monoxide* (CO) and *nitrogen* (N_2), that is, ($CO + N_2$); produced when *air* is passed over *hot coke*; see also **Water Gas**.

Producer's Risk A *Quality Control* term; the probability of rejecting a good lot; see also **Consumer's Risk**.

Produce-to-Order An *Operations Management* term; see **Make-to-Order**.

Produce-to-Stock An *Operations Management* term; see **Make-to-Stock**.

Product Flexibility An *Operations Management* term; the ability of a production system to quickly change from producing one product or service to producing another; see also **Volume Flexibility**.

Product Innovation A *Management* term; the innovation which leads to the enhancement of a product's physical *attributes* and *capabilities* in order to make it look more attractive and customer-friendly; see also **Process Innovation**.

Product Layout An *Operations Management* related to *facilities planning*; also called **Line Layout** or **Assembly Line** or **Production Line**; a *product-oriented* plant layout in which highly expensive, standardized machines are arranged in a linear way in sequence of operations being performed on the machines; used to produce *standardized* products of limited product designs; failure of even one machine in the layout can stop the entire production, but eliminates the scope of in-process inventory; suitable for *mass production* of similar products as well as for a JIT production system, but is not suitable for manufacturing customized products; finds use in *automobile industry*; see also **Process Layout**.

Product Life Cycle An *Operations Management* term; the life cycle of a new product during which it moves through several demand stages, from introduction to growth to maturity and then to decline; forms the basis of demand forecasting.

Product Moment of Inertia A *Strength of Materials* term; see **Product of Inertia**.

Product of Inertia A *Strength of Materials* term; also called **Product Moment of Inertia** or **Product Second Moment of Area**; with respect to two perpendicular axes *x* and *y*, the product of inertia, I_{xy} of an area is defined as: $I_{xy} = \int xy.dA$, where *x* and *y* are the distances of the area *dA* from the two axes *y* and *x* respectively; can be *positive*, *negative* or *zero*; for any axis of symmetry of the area, $I_{xy} = 0$; the SI unit is m^4.

Product Reliability A *Statistics* term; the likelihood that a product will perform its intended function for a prescribed lifetime under the specified operating conditions; see also **Machine Reliability**.

Product Second Moment of Area A *Strength of Materials* term; see **Product of Inertia**.

Product Structure Tree An *Operations Management* term; see **Bill of Materials**.

Production Activity Control An *Operations Management term*; see **Shop Floor Control**.

Production Capacity An *Operations Management* term; the maximum rate of production of an organization.

Production Control An *Operations Management* term; a process of controlling production activities to bring them under control; includes activities such as *inspection, measuring*, and *suggesting corrective measures;* see also **Process Control**.

Production Drawing An *Engineering Drawing*; also called **Working Drawing**; the most important and extensively used drawing used for the production of various components in the industry.

Production Function An *Operations Management* term; also called **Total Product Function**; a numerical or mathematical expression of a relationship between the quantities of inputs used and the maximum quantity of outputs which can be produced.

Production Kanban An *Operations Management* term; also called **P-Kanban**; a card physically attached to a container or cart, which authorizes *production of goods* in a production system; a distinctive feature of JIT production system; see also **Withdrawal Kanban**.

Production Line An *Operations Management* term related to *facilities planning*; see **Product Layout**.

Production Lot An *Operations Management* term; also called simply **Lot**; a certain quantity of material produced at one stage of production, stored in inventory, and then sent to the next stage in production or shipped to customers.

Production Management A *Management* term; see **Operations Management**.

Production Planning An *Operations Management* term; the planning of production activities related to products to be manufactured, its volume required, the methods of production, the costs of production, the machines and equipment needed, and the workforce requirements.

Production Quantity Model An *Operations Management* term; also called **Production Rate Model** or **Economic Production Quantity Model** or **Gradual Usage and Non-Instantaneous Receipt Model**; an inventory model in which the order for inventory is received or delivered gradually over time as per the requirements, rather than all at once as in the case of basic EOQ model; involves less inventory holding as compared to basic EOQ model, as it uses part of the inventory while it is being produced; this model is commonly employed when the inventory user is also the producer, where a part is produced to be used in a large assembly; see also **Basic EOQ Model**.

Production Rate Model An *Operations Management* term; see **Production Quantity Model**.

Production Rate A *Operations Management* term; the number of units of a product produced per unit time (such as per hour or per day) by a manufacturing unit.

Production System An *Operations Management* term; a system which deals with production activities for producing finished goods from raw materials; inputs for a production system include *raw materials*, *equipment* (*machine tools*), *tooling* and *fixtures*, *energy* (*electrical energy*) and *labour*.

Production Technology A *Manufacturing* term; a quantitative relationship between inputs and outputs.

Production Also called **Manufacturing**; a transformation process used in a production system to convert *raw materials* into *finished products* which are finally consumed by the customers.

Productivity A *Management* term; a ratio of the output produced to the amount of input; a measure of utilizing the resources effectively in producing goods and services; the productivity of a worker relates to his efficiency and the productivity of an organization relates to its growth; considered as a determinant and an indicator of national competitiveness; classified into *labour*, *capital*, and *material* productivity.

Products of Combustion An *IC Engine* term; the components leaving the combustion chamber of an engine; include H_2O and CO_2 in case of *complete combustion*, and C, CO, H_2, or OH in case of *incomplete combustion*.

Program Evaluation and Review Technique A *Project Management* term, abbreviated as **PERT**; one of the two project network approaches developed in the 1956s by the *U.S. Navy* during the development of the *Polaris Missile Project* (a ballistic missile-firing nuclear submarine); a *probabilistic approach* of project management which uses three time estimates, namely the *optimistic time* (a), the *pessimistic time* (b) and the *most likely time* (m) for every activity; used when emphasis is on shortening and monitoring the project execution time without too much concern for cost implications; useful especially for large projects consisting of many activities and involving uncertainties, for example, research and development (*R & D*) projects where activity time is not certain and can only be predictable such as developing new drugs, planning and introducing new products, developing new and more powerful computer systems, etc.; the expected time of completion (t_e) for each activity in PERT is calculated as:

$$t_e = \frac{a + 4m + b}{6}$$; see also **Critical Path Method**.

Programmable Controller A *Computer* term; see **Programmable Logic Controller**.

Programmable Logic Controller A *Computer* term; also called **Programmable Controller**; a specialized heavy-duty computer system used for process control in factories, chemical plants, and warehouses; closely similar to traditional relay logic.

Progressing A *Management* term; see **Controlling**.

Project Crashing A *Project Management* term; also called simply **Crashing**; a method for shortening the completion time of a project by reducing the time of one or more of the critical activities to less than its normal activity time through the use of additional resources (labour and materials); used when overall reduction in the project cost is expected to be realized or when time is an important factor of consideration.

Project Management A *Management* term; a technique to schedule, manage, and complete the activities of a project on time and in estimated budgets.

Projectile Motion A *Physics* term related to a *projectile*; a projectile has motion along two mutually perpendicular directions, one in the *vertical direction* and another in the *horizontal direction*; in projectile motion, *the body returns to the ground at the same angle and with the same speed with which it was projected*; the kinetic energy of the projectile is maximum at the point of projection and also at the point of reaching to the ground, and is minimum at its highest point; *a missile follows a projectile motion.*

Projectile A *Physics* term; a body thrown in space from the ground with some initial velocity at a certain angle to the horizontal (called *angle of projection*), and then is allowed to return under the action of gravity alone; has two velocity components during its flight, one in the *vertical direction* and another in the *horizontal direction*; a few examples include *a missile* or *a rocket.*

Projection Welding A *Welding* term; a type of *resistance welding*; the joint is produced when one or two projection impressions (locations of high heat concentration) produced on one of the two work parts (thinner one) collapse during application of pressure after being heated to plastic state; *produces a stronger joint than spot welding*; see also **Spot Welding**.

Projection An *Engineering Drawing* and a *Machine Drawing* term; a view imagined to be projected on a projection plane.

Proof Load A *Strength of Materials* term related to a *spring*; the *maximum* load which a spring can take up without undergoing permanent deformation.

Proof Stress A *Strength of Materials* term related to a *spring*; the stress which corresponds to proof load; denotes the maximum stress which a spring can be subjected to without undergoing permanent deformation.

Propeller Jet An *IC Engine* term; see **Propjet**.

Propeller Turbine A *Fluid Mechanics* term; also called simply **Propeller**; a hydraulic turbine which is nearly identical to a *Kaplan turbine* except that *blades are fixed*; the flow rate is regulated only by the *wicket gates*; an *axial-flow, single-stage, reaction hydraulic turbine* which produces thrust by imparting linear momentum to a fluid, and operates without an outer housing; most suited for *low head* and *high volume flow rate* conditions; see also **Kaplan Turbine**.

Propeller A *Fluid Mechanics* term; see **Propeller Turbine**.

Proper Fraction A *Mathematics* term; a fraction whose numerator is less than the denominator; a few examples include, $\frac{1}{2}$, $\frac{2}{3}$, $\frac{3}{4}$ etc.; see also **Improper Fraction**.

Property The observable characteristics of a material; for example, *internal energy*, *enthalpy*, and *entropy* are *thermodynamic properties* of a material.

Propjet An *IC Engine* term; the short form of **Propeller Jet**; a variation of a *turbojet engine* in which the turbine drives not only the compressor but also a propeller which provides most of the propulsive force.

Propped Cantilever Beam A *Strength of Materials* term related to a *beam*; also called **Fixed-Simple Beam**; a statically indeterminate beam which is originally a cantilever beam of increased length that has a roller support at anywhere over its length providing a vertical reaction to prevent its deflection.

Proton A *Chemistry* term; one of the three fundamental particles of an atom of a chemical element discovered by *Eugen Goldstein* (1850–1930), a German physicist in 1886; a *positively charged* particle which resides in the nucleus of the atom and makes it positively charged; has charge equal and opposite to that of an electron (1.60206×10^{-19} C) and mass equal to 1.6725×10^{-27} kg; see also **Electron** and **Neutron**.

Prototype Design A *Design* term; the initial product design which exhibits the basic characteristics of a product's *form*, *fit*, and *function* which will be required in the final design.

Prototype A *Design* term; a one-of-a-kind working model of a solution developed for testing purposes; the physical system for which the predictions are to be made using a model; see also **Model**.

Protractor A *Mathematics* term; a geometrical instrument used for measuring a given angle or constructing an angle of given magnitude; usually made out of a thin transparent plastic sheet in the shape of a semi-circular arc; has degree marks on the curved edge, which is divided into 180 equal parts, each part indicates 1°.

Pseudoplastic Fluid A *Fluid Mechanics* term; also called **Shear Thinning Fluid**; a *non-Newtonian fluid* whose viscosity decreases with increase in applied shear stress; for example, *paint* (a plastic fluid) which is very viscous when it is poured from the can or when it is picked up by a paintbrush, but later becomes thin (very less viscous) when applied on the wall; see also **Dilatant Fluid**.

Psychrometric Chart A *Thermodynamics* term related to *humidity*; a chart drawn between specific humidity (ω) on the vertical axis and dry-bulb temperature on the horizontal axis for a standard atmospheric pressure of 760 mm of Hg; also includes the *saturation line*, the *lines of constant wet-bulb temperature*, the *lines of constant enthalpy* (nearly parallel to the lines of constant wet-bulb temperature), the *lines of constant relative humidity*, and the *lines of constant specific volume*; a useful tool to know the state of the atmospheric air at a specified pressure; serves as a valuable aid in visualizing the *air-conditioning processes*.

Psychrometric Properties A *Thermodynamics* term; the properties of the moist air.

Psychrometry A *Thermodynamics* term; the study of the behaviour of moist air.

P-Type Semiconductor An *Electronics* term; an extrinsic semiconductor in which the electrical conductivity is mainly due to *positively charged holes* that are called the majority carriers, and the electrons are the minority carriers; produced when a pure semiconductor is doped with a *trivalent impurity atom* such as *boron*; see also **N-Type Semiconductor**.

Pull Manufacturing System An *Operations Management* term; see **Pull Production System**.

Pull Production System An *Operations Management* term; also called **Pull Manufacturing System** or simply **Pull System**; a *customer-driven system of production* in which goods are produced only when they are ordered by the customers; used in modern manufacturing; involves least inventories and thrusts on customized products; a JIT manufacturing system uses this concept; see also **Push Production System**.

Pull System An *Operations Management* term; see **Pull Production System**.

Pulley A *Design* term; a free-spinning wheel used to transmit motion or power from one shaft to another by means of a rope or belt which runs over it; mostly made of cast iron.

Pulverization A *Materials Engineering* term related to metallurgy; a process of converting small pieces of an ore (obtained from breaking of big lumps) into *powdered form* with the help of a *ball mill* or *stamp mill*; see also **Benefaction**.

Pump Efficiency A *Pump* term; a ratio of the *increase in the mechanical energy* of a fluid to the *mechanical energy input*; see also **Turbine Efficiency**.

Pump A *Fluid Mechanics* term; a *fluid machine* (turbomachine) and a *work-consuming device* which adds mechanical energy to a fluid by performing work on it, thereby increasing pressure of the fluid; suffers from irreversible losses due to friction, internal leakage, flow separation on blade surfaces, turbulent dissipation, etc.

Punch A *Manufacturing* term; a *fitting tool* used to produce marking or locating centres on a workpiece.

Punching A *Manufacturing* term; a *sheet metal operation* used to produce a cavity (hole) in a sheet metal blank using a die-punch assembly; also used to produce seamless tubing from a cylindrical billet; punching differs from blanking in the end result part; in *punching*, the punched out part is considered waste and the remaining part is the useful product, whereas in *blanking*, the punched out part is the useful product and the remaining part is treated as waste; see also **Blanking**.

Purchasing An *Operations Management* term; the activities related to procurement of materials and supplies consumed during the production of products; see also **Procurement**.

Pure Bending A *Strength of Materials* term; also called **Simple Bending**; a bending condition in which a beam is subjected to *only bending moment* and the effect of shear force is neglected.

Pure Rotational Motion A *Physics* term; the motion of a rigid body in which every particle of the body moves in a circle (lying in a plane perpendicular to the axis, and has its centre on the axis); a few examples include motion of a *ceiling fan* and *a potter's wheel*; see also **Pure Translational Motion**.

Pure Shear A *Strength of Materials* term; a state of stress in which an element of a beam's cross section is subjected to *shear stress* only.

Pure Substance A *Thermodynamics* term**;** a substance of homogeneous nature which has invariable chemical composition throughout; exists in three phases, namely *solid*, *liquid*, and *gas*; a few examples include the *atmospheric air*, the *steam-water mixture*, and the *combustion products* of a fuel.

Pure Translational Motion A *Physics* term; the motion of a rigid body in which every particle of the body has the same velocity at a particular instant of time; for example, sliding of a rectangular block down an inclined plane produces pure translational motion; see also **Pure Rotational Motion**.

Push Manufacturing System An *Operations Management* term; see **Push Production System**.

Push Production System An *Operations Management* term; also called **Push Manufacturing System** or simply **Push System**; a system of production in which goods are produced according to a schedule based on anticipated demand; used in *material requirements planning* (MRP) and *economic order quantity* (EOQ); see also **Pull Production System**.

Push System An *Operations Management* term; see **Push Production System**.

PV Cell A *Physics* term; see **Solar Cell**.

Pyramid Structure A *Management* term; an organizational structure which resembles a pyramid; consists of fewer and fewer managers at each level as we move up from the base; most of the organizations have a pyramid structure.

Pyramid A *Mathematics* term; a solid whose base is a plane rectilinear figure and whose side faces are triangles having a common vertex outside the plane of the base; see also **Regular Pyramid**.

Pyrex Glass A *Chemistry* term; a type of glass which has very low coefficient of expansion because of presence of oxides of boron and aluminium (B_2O_3 and Al_2O_3); can withstand *high temperatures* as well as *sudden changes in temperatures* effectively, making it *highly suitable for laboratory glasswares* such as *flasks*, *beakers*, etc.; approximately contains silica 80% (SiO_2), 4% sodium oxide (Na_2O), 0.5% calcium oxide (CaO), 0.5% potassium oxide (K_2O), 12% boric anhydride (B_2O_3), and 3% aluminium oxide (Al_2O_3).

Pyrometry A *Measurement* term; the science of measurement of *high temperatures* usually above the gold point (1064.43 °C).

Pythagoras Theorem A *Mathematics* term; a theorem related to a *right angle triangle*; states that the square of the hypotenuse (h) is equal to the sum of the squares of base (b) and perpendicular (p); given as:

$$h^2 = b^2 + p^2$$

Q

QC An *Operations Management* term; see **Quality Circle**.

QFD A *Quality Management* term; see **Quality Function Deployment**.

QLF A *Quality Management* term; see **Quality Loss Function**.

QS 9000 A *Quality Management* term; an international quality standard based on the ISO 9000 standards but contains additional requirements which are particular to the *automotive industry*; the big three automakers of the US, namely *General Motors*, *Ford*, and *Daimler-Chrysler*, are QS 9000 certified companies.

Quadrilateral Pyramid A *Mathematics* term; a pyramid whose base is a quadrilateral; see also **Pyramid**.

Quadrilateral A *Mathematics* term; a closed plane figure made of four line segments which do not intersect except at their end points; a few examples include *trapezium*, *rhombus*, *rectangle*, and *square*.

Qualitative Factors A *Statistics* term; also called **Intangible Factors**; the *unquantifiable factors*; the factors which are difficult to measure numerically; for example, *service characteristics are qualitative factors*; see also **Quantitative Factors**.

Qualitative Methods An *Operations Management* term related to *forecasting*; also called **Judgement Methods**; these methods use *management judgement*, *expertise*, and *opinion* to make the demand forecast; useful when no past demand data are available or if available they are not in sufficient amount; most frequently used for *long-range* forecasts; a few qualitative methods include *Delphi*, *historical analogy*, and *product life cycle*: see also **Quantitative Methods**.

Quality at the Source A *Quality Management* term; the defect-free quality of the starting materials, that is, raw materials, to produce defect-free final products.

D. K. Singh, *Dictionary of Mechanical Engineering*,
https://doi.org/10.1007/978-981-99-1722-8_17

Quality Circle An *Operations Management* term, abbreviated as **QC**; also called **Quality Control Circle**; a small, voluntary group of people or a team of preferably 6–8 workers and supervisors of an organization who meet frequently to discuss and solve problems; a quality improving technique as well as an important concept in JIT manufacturing system which was initially aimed at addressing quality problems in Japan; also used in addressing other organizational problems such as cost reduction, facilities improvement, etc.

Quality Control Circle An *Operations Management* term; see **Quality Circle**.

Quality Control A *Quality Management* term; a quality-related process which ensures that quality of a product or a service meets the standards specified by the designer; exercised at every level of a product being manufactured, right from the raw material stage to production stage to finished product stage, where final inspection is performed to test its overall functional and appearance quality; see also **Total Quality Management**.

Quality Function Deployment A *Quality Management* term, abbreviated as **QFD**; a quality improving technique introduced in Japan by *Yoji Akao* (1928–2016) in 1966, which was extensively used by *Toyota*; a concept of formalizing customer inputs into product design by identifying customer wants and eliminating wasteful product features and activities which do not contribute; translates the customer requirements (*voice of the customer*) into appropriate technical design requirements at every stage of design and manufacture of a product or service; used to enhance communication between different functions such as *marketing*, *operations*, and *engineering*, thereby producing better product designs in less time than the conventional design processes.

Quality Gurus An *Operations Management* term related to *quality*; people who greatly contributed to the improvement of quality; prominent quality gurus include *Joseph M. Juran*, *W. Edwards Deming* (1900–1993), *Armand V. Feigenbaum* (1920–2014), *Kaoru Ishikawa* (1915–1989), *Walter Shewhart* (1891–1967), and *Philip B. Crosby* (1926–2001).

Quality Loss Function A *Quality Management* term, abbreviated as **QLF**; a *qualitative measure* of the effectiveness of quality control, often in terms of the economic losses a customer suffers on account of a defective product; see also **Quality Function Deployment**.

Quality Management A *Management* term; a process of improving the *design*, *features*, *appearance*, *on time delivery*, and other related parameters of a product or service in order to *ensure customer satisfaction*.

Quality of Conformance A *Quality Management* term; the degree to which the product or service meets the design specifications during the production process; ensures that product or service is produced according to designed parameters, for example, if new tyres do not conform to specifications, they wobble.

Quality of Steam A *Thermodynamics* term related to *steam*; the *dryness fraction* of steam expressed in percentage; see also **Priming.**

Quality A *Management* term; the degree to which a product or service conforms to its design specifications and functional requirements; quality measuring parameters include *performance, features, reliability, conformance, durability, serviceability, aesthetics*, and *safety*; a quality product or a quality service offers the highest degree of customer satisfaction.

Quantitative Factors A *Statistics* term; the *quantifiable factors*; the factors which can be measured in numerical terms; for example, the physical dimensions of a product such as *length, breadt*h, and *height* are quantitative factors; see also **Qualitative Factors.**

Quantitative Methods An *Operations Management* term related to *forecasting*; also called **Time Series Methods** as a set of observed values (usually *sales*) are measured over successive periods of time; these methods use past demand data to make the future forecast of demand; most frequently used in *short-range* forecasts of products in manufacturing and service companies; a few quantitative methods include *moving average, weighted moving average, exponential smoothing*, and *regression analysis*; see also **Qualitative Methods.**

Quantity Discount An *Inventory Management* term; a policy of allowing unit item cost to vary with the volume ordered; usually the item cost decreases as volume increases due to economies of scale in production and distribution; used as a means to clear the old stocks.

Quantum Numbers A *Chemistry* term; the numbers in an atom which assign electrons to discrete energy levels; the four quantum numbers include *principal quantum number* (n), *azimuthal quantum number* (l), *magnetic quantum number* (m), and *spin quantum number* (s).

Quarter-Circle A *Mathematics* term; the *one-fourth* part of a circle.

Quarter-Elliptic Leaf Spring A *Strength of Materials* term related to a *spring*; a type of leaf spring which is also called **Cantilever Laminated Spring** because of its similarity to a cantilever beam loaded with a point load at its free end.

Quartz A *Chemistry* term; the most common *naturally occurring crystalline form of silica* (SiO_2) which has a specific gravity of 2.648; offers certain advantages over glass such as acid and alkalis resistant, very small coefficient of expansion, and can stand sudden change in temperature which prevents its breakage on sudden heating and cooling; can be shaped into vessels such as flasks and beakers; extensively used as a piezoelectric material; its use has made possible to develop extremely accurate clocks, modern radio and television broadcasting and mobile radio communications; transparent quartz is used in making optical instruments and lenses; the coloured varieties of quartz are used as gems, for example, *amethyst* (purple), *jasper* (red), and *opal* (white); sand is also a crystalline form of quartz formed as result of 'weathering' or disintegration of rocks, and further breaking down by violent movements of water.

Quasi-Static Process A *Thermodynamics* term; a *reversible* process in which a system deviates from one equilibrium state by only an infinitesimal amount throughout the process; closely approximates to a succession of equilibrium states; infinite slowness is the characteristic feature of a quasi-static process.

Quenching A *Manufacturing* term related to *heat treatment*; a heat treatment process in which a heated metal workpiece is *cooled rapidly* at a controlled rate in *water* or *oil*; *the oil quench is slow* but prevents cracks formation caused by rapid expansion of the workpiece being treated; *the water quench is fast* and is used for *medium-carbon* and *low-alloy steels.*

Quick Response Logistics An *Operations Management* term related to a *supply chain*; the fast movement of goods in a supply chain using various tools of information technology such as *electronic data interchange* (EDI), *bar coding*, the use of *electronic point of sale* (EPOS), and *laser scanners.*

R

R11 A *Refrigeration* term; see **Freon-11**.

R12 A *Refrigeration* term; see **Freon-12**.

R22 A *Refrigeration* term; see **Freon-22**.

Rack and Pinion A *Theory of Machines* term related to a *gear*; the limiting case of a gear set consisting of a *rack* (*a gear of infinite radius*) and a *pinion* (the *smaller gear* of the two mating gear wheels); often used in the *steering mechanism of an automobile's front wheels*, where the centre point of the pinion remains fixed and the rack moves horizontally as the pinion rotates; the rack can be supported by rollers, or it can slide on a smooth (lubricated) surface.

Rack A *Theory of Machines* term related to a *gear*; used in conjunction with a *pinion*; a portion of a gear, which has *infinite radius* or zero curvature and has an *infinitely large number of teeth*; its pitch surface is a plane surface, and the pitch circle circumference is a straight line; moves horizontally as the pinion rotates; mostly used in a *lathe* to transmit motion to the *saddle*; see also **Pinion**.

Radial Drawing A *Manufacturing* term; see **Deep Drawing**.

Radial Flow Turbine A *Fluid Mechanics* term; a turbine in which a fluid flows along the radius of the shaft of the turbine, either inward or outward; for example, *Francis turbine* is a radial flow turbine; see also **Axial Flow Turbine** and **Mixed Flow Turbine**.

Radial Flow A *Fluid Mechanics* term; a fluid flow which occurs along the radius of the shaft of a turbine or pump, either inward or outward; occurs in *Francis turbine*; see also **Axial Flow** and **Mixed Flow**.

Radial Forging A *Manufacturing* term; see **Swaging**.

Radial Flow Pump A *Fluid Mechanics* term; see **Centrifugal Pump**.

D. K. Singh, *Dictionary of Mechanical Engineering*,
https://doi.org/10.1007/978-981-99-1722-8_18

Radiation A *Heat Transfer* term; the third basic mode of heat transfer after *conduction* and *convection* in which heat is emitted or absorbed in the form of *electromagnetic waves* (or photons); occurs as a result of changes in electronic configurations of atoms or molecules (without direct physical contact); does not require any material medium; the fastest mode of heat transfer as it travels at the speed of light and can travel in *vacuum*; for example, the energy from the sun reaches the earth through the process of radiation; not related to radiation in the context of nuclear processes or power generation; see also **Conduction and Convection**.

Radio Frequency Identification An *information technology* tool, abbreviated as **RFID**; a *wireless technology* to obtain real-time information about an object; consists of a microchip tag to be attached to the object which transmit identification and location information; it has provided excellent product traceability; for example, using RFID, the progress of a part through the production process can be monitored, and when finished, its location in the warehouse can be tracked; also extensively used in *logistics* to trace the locations of vehicles carrying goods for distribution.

Radioactivity A *Physics* term; a *nuclear property* of *radioactive elements* such as *uranium* (*U*), *plutonium* (*Pu*), *radium* (*Ra*), or *thorium* (*Th*) discovered by the *French* physicist *Henri Becquerel* (1852–1908) in 1895 in which elements emit radiations such as *alpha* (α), *beta* (β), and *gamma* (γ) rays; its measuring unit include *curie* (ci), *millicurie* (mci), and *microcurie* (μci), of which *curie* is the bigger unit of *radioactivity*.

Radiographic Inspection A *Materials Engineering* term; see **X-Radiography**.

Radiographic Testing A *Materials Engineering* term; see **X-Radiography**.

Radius of Curvature A *Strength of Materials* term; the radius of a curved shape produced by bending of a structural member (beam).

Radius of Gyration A *Physics* term; the radial distance from the axis of rotation of a body, the square of which, when multiplied by the total mass of the body, gives the moment of inertia of the body about that axis; also defined as the distance from a given axis, the square of which, when multiplied by the total area of a plane figure, gives the second moment of area about that axis; denoted by k; expressed as: $k_x = \sqrt{\frac{I_x}{M}}$, $k_y = \sqrt{\frac{I_y}{M}}$ or $k_x = \sqrt{\frac{I_x}{A}}$, $k_y = \sqrt{\frac{I_y}{A}}$ where M = mass of the body, A = area of the plane figure, and I_x and I_y = moment of inertia of the body or a plane figure about x-axis and y-axis, respectively.

Radius A *Mathematics* term; a line segment which starts from the centre of a circle and ends on its circumference; a circle has an infinite number of radius of length equal to one-half of the diameter of the circle; expressed in the units of length such as *mm, cm, inch,* or *m*.

Rammer A *Manufacturing* term; a *foundry tool* used to distribute the sand mass uniformly around the pattern during making of a mould; classified into peen, hand, and floor rammers.

Random Error A *Metrology* term; also called **Chance Error**; an error which occurs due to non-specific causes like *natural disturbances*; occurs randomly and is *erratic in nature*; varies in magnitude and can be positive or negative; arises due to random and unpredictable variations in the experimental conditions such as *temperature, pressure, voltage supply,* and *mechanical vibration* or due to personal errors of the observer doing the experiment; cannot be eliminated, but may be characterized by *statistical analysis*; *follows the Gaussian law of normal distribution*; can be minimized by repeating the observation several times and taking the arithmetic mean of all the observations, as the mean value would be very close to the true value; see also **Systematic Error**.

Random Numbers A *Mathematics* term; the numbers selected in such a way that every number has an equal chance or probability of selection.

Range Anxiety An *Electric Vehicle* term; the concern that an electric vehicle's batteries will die before it reaches its destination.

Range Chart A *Quality Control* term; see **R-Chart**.

Range An *Electric Vehicle* and a *Measurement* term; in *electric vehicle* terms, the total distance covered by an *electric vehicle* (EV) in one full charge of battery; in *measurement* terms, the spread between the maximum and minimum values of measurement.

Rankine Cycle A *Thermodynamics* term; an *idealized thermodynamic cycle* for a *heat engine*, named after a *Scottish* engineer *William John MacQuorn Rankine* (1820–1872) who made several outstanding contributions in the field of thermodynamics; a *vapour power cycle* used in thermal power plants for generating electricity; for a given temperature limits, the Rankine cycle efficiency is always lower than the Carnot cycle efficiency because not all the heat is added at the highest temperature in the *Rankine cycle.*

Rankine Scale A *Thermodynamics* term; a thermodynamic temperature scale used in the *English* system as well as an absolute temperature scale; named after its inventor *William Rankine* (1820–1872); uses the unit *rankine* (R) for temperature measurement; its absolute zero coincides with the absolute zero of the *Kelvin* scale; the *Rankine* (R) and the *Kelvin* (K) scales are related as: T (R) = 1.8 T (K); see also **Celsius Scale, Fahrenheit Scale,** and **Kelvin Scale**.

rankine A *Thermodynamics* term; the unit of temperature in the *English* system, named after *William Rankine* (1820–1872); denoted by R.

Rankine's Formula A *Strength of Materials* term related to a *column*; see **Rankine-Gordon Formula**.

Rankine's Theory A *Strength of Materials* term; see **Maximum Normal Stress Theory**.

Rankine-Gordon Formula A *Strength of Materials* term related to a *column*; also called simply **Rankine's Formula**; an empirical formula used to find the *crippling* (*critical*) *load* for both *short* and *long* columns; expressed as:

$$P_r = \frac{\sigma_C A}{1 + K\left(\frac{l_e}{k}\right)^2}$$

where

P_r = Rankine's crippling load
σ_c = direct crushing stress (yield stress)
A = area of cross section of the column
l_e = equivalent or effective length of the column
k = radius of gyration
K = a constant for the column material
$= \frac{\sigma_C}{\pi^2 E}$
E = modulus of elasticity of the column material

Rapid Prototyping A *Manufacturing* term, abbreviated as **RP**; a collection of CAD data-driven physical model construction technologies which use additive manufacturing processes.

Rapping Allowance A *Casting* term; see **Shake Allowance**.

Rat Tail A *Casting* term; a type of *casting defect* which occurs in the form of a rat tail; produced mainly on the flat surfaces of a casting *due to thermal expansion of the moulding sand.*

Raw Material A *Materials Engineering* and an *Inventory Management* term; the basic or starting material which is used in making finished products; the quality of the raw materials is important for the production of defect-free products.

R-Chart A *Quality Control* term; also called **Range Chart**; a *control chart for variables* which uses the *range of sample* to monitor the process variability; to calculate the range of a set of sample data, the smallest is subtracted from the largest measurement in each sample; in case of the data falling outside the control limits, the process variability is considered not in control; the control limits for the *R*-chart are obtained as: $UCL = D_4\ R$ and $LCL = D_3\ R$, where R = average of the range values of each sample and the central line of the control chart (the target value); D_3 and D_4 are constants which provide 3 standard deviation (3σ) limits for a given sample size and are obtained from a table; see also **X-Bar Chart**.

Reactants A *Chemistry* term; the components which take part in a chemical reaction.

Reaction Turbine A *Fluid Mechanics* term; a *hydraulic turbine* in which a part of the pressure change occurs in the rotor and remaining in the stator; may be radially inward, axial, or mixed (partially radial and partially axial); produces more power than an impulse turbine of the same diameter, net head, and volume flow rate; also has higher efficiency than an impulse turbine; not suitable for *very high head*, because of leakage and frictional head losses; *Francis turbin*e and *Kaplan turbine* are examples of a reaction turbine; see also **Impulse Turbine**.

Reactive Force A *Physics* term; a force which results from a reaction; for example, the force produced at the positions of fixed supports of a structure (beam or column).

Real Fluid A *Fluid Mechanics* term; a fluid which involves friction and has viscosity; see also **Ideal Fluid**.

Reaming A *Manufacturing* term; a process of finishing a drilled hole by means of a reamer; see also **Drilling** and **Boring**.

Reciprocating Engine An *IC Engine* term; an engine in which the piston reciprocates; that is, it has *to-and-fro motion* in the cylinder of the engine; classified into *spark-ignition* (SI) and *compression-ignition* (CI) engines.

Reciprocating Pump A *Fluid Mechanics* term; a *positive displacement pump* which works on the principle of sucking a fluid and pushing (compressing) it out alternatively by the reciprocating motion (to-and-fro motion) of a plunger or piston in a cylinder, whose thrust lifts the fluid to a certain height; operates at low speeds and involves intermittent operation resulting in *pulsating and fluctuating delivery*; involves very high initial cost (approximately 4 times that of centrifugal pump for the same capacity); suitable for *small discharge* and *high heads*; see also **Centrifugal Pump**.

Recrystallization Temperature A *Materials Engineering* term; the temperature at which the strength of a particular metal decreases and is more easily formed.

Recrystallization A *Materials Engineering* term; a process in which *nucleation* and *growth* of a *new stress-free microstructure* results from a cold-worked microstructure.

Rectangle A *Mathematics* term; a parallelogram whose each angle is a right angle, that is, equal to 90°; see also **Parallelogram**.

Rectangular Tube A tube having a rectangular cross section.

Rectangular Weir A *Fluid Mechanics* term; the simplest type of weir; a vertical thin plate having a *straight top edge*; named so because the cross section of the liquid which passes over it is rectangular when placed in an *open channel*; used to measure the liquid flow rate.

Rectilinear Motion A *Physics* term; also called **Linear Motion**; the one-dimensional motion of an object along a straight line; a few examples include the motion of a train along a straight railway track and the falling of an object from a certain height; the three important equations of rectilinear motion under constant acceleration are given as:

$$v = u + at \tag{1}$$

$$s = ut + \frac{1}{2}at^2 \tag{2}$$

$$v^2 = u^2 + 2as \tag{3}$$

where

u = initial velocity of the particle
v = final velocity of the particle
t = time of motion
a = acceleration of the particle
s = displacement of the particle.

In case of a particle under *constant retardation*, the positive sign (+) in all the above three equations is replaced by the negative sign (–); *a common example of rectilinear motion is that of a body falling towards the earth.*

Recurring Cost An *Operations Management* term related to cost *analysis*; the repetitive costs that occur when an organization produces *similar goods or services on a continuing basis*; a fixed cost paid on a repeatable basis is also a recurring cost; for example, in an organization providing architectural and engineering services, office space rental (a fixed cost) is also a recurring cost; see also **Nonrecurring Cost**.

Red Hardness A *Manufacturing* term; also called **Hot Hardness**; the ability of a material to maintain its hardness at increased temperatures; permits higher operating speed to a cutting tool.

Red Iron A *Chemistry* term; see **Haematite**.

Redrawing A *Manufacturing* term; a cold metal working process used to increase the depth of the drawn cup obtained in deep drawing by subjecting it again to deep drawing.

Reducing Flame A *Chemistry* term; see **Carburising Flame**.

Reduction A *Chemistry* term; a chemical reaction in which a metal combines with hydrogen; the addition of an electron to an atom or a molecule; see also **Oxidation**.

Redundant Truss A *Strength of Materials* term related to a *truss*; see **Over-Rigid Truss**.

Redundant A *Physics* term; any reaction in excess of those which can be determined by statics alone; the number of redundant is the same as the degree of indeterminacy.

Reed Switch An *Electrical* term; an electromagnetically operated digital switching device.

Refractories A *Materials Engineering* term; see **Refractory Materials**.

Refractory Materials A *Materials Engineering* term; also called **Refractories**; the *non-metallic materials* which can withstand *high temperatures*; the stable oxides of elements such as *silicon, aluminium, magnesium, calcium*, and *zirconium*; a few examples include *quartz, sand, silica brick, magnesite, dolomite*, and *carbon*; classified into *acidic, basic*, and *neutral refractories.*

Refractory Metals A *Materials Engineering* term; metals and alloys which are *resistant to high temperatures*; for example, *molybdenum.*

Refrigerant A *Thermodynamics* term; a chemical substance used as a working fluid in a refrigeration system to produce cooling by removing heat; a few refrigerants include *ammonia* (NH_3), *carbon dioxide* (CO_2), and *Freons.*

Refrigeration Cycle A *Thermodynamics* term; a *thermodynamic cycle* used to cool or heat a space.

Refrigeration A *Thermodynamics* term; a process of removing heat from a body; a process used to produce a *cooling effect* using a *refrigerant.*

Refrigerator A *Thermodynamics* term; a device which removes heat from a body at a lower-than-ambient temperature thus producing cooling using a *refrigerant*; see also **Heat Pump**.

Regenerator A *Thermodynamics* term; a heat exchanger which preheats the air before it enters the burner by using the hotter air which leaves the turbine.

Regression Method An *Operations Management* term related to *forecasting*; also called **Causal Model Method**; a *quantitative forecasting* method based on *regression analysis* used to establish a relationship which may exist between two or more variables; aims to formulate a model of *cause*-and-*effect* based on the past collected data, and then using this model to forecast the future demand; the most commonly used form of causal model is a *single-variable linear model* of the form $y = ax + b$, but the multiple regression models (nonlinear) which follow a *polynomial relationship* of the form $y = ax^3 + bx^2 + cx + d$ are more useful than a single-variable ones, since forecasts often relate to more than one factor.

Regular Polygon A *Mathematics* term; a polygon whose all the sides as well as all the angles are equal; see also **Polygon**.

Regular Pyramid A *Mathematics* term; a pyramid whose all the sides of its base are equal; see also **Pyramid**.

Regular Tetrahedron A *Mathematics* term; a tetrahedron whose all the edges are of equal length; see also **Tetrahedron**.

Reheat Cycle A *Thermodynamics* term; a *thermodynamic cycle* used to increase the *work output* of a steam turbine; works on the principle of extracting heat from the steam (after expansion in the turbine) and reheating it by passing it again through the boiler, and then returning it to the turbine.

Reinforcement A *Materials Engineering* term; the structural component of a composite material in the form of fibres, whiskers, or particles, which are encased by a matrix material that exists in a different phase.

Relative Error A *Metrology* term; also called **Fractional Error**; a ratio of the mean absolute error to the mean value of the measured values of a physical quantity.

Relative Humidity A *Thermodynamics* term related to *psychrometry*; a psychrometric property; a ratio of the mass of water vapour in a certain volume of the moist air at a given temperature to the mass of water vapour in the same volume of the saturated air at the same temperature; a measure of air's ability to absorb moisture; the saturated air has 100% relative humidity; the high relative humidity slows down the heat rejection by evaporation, whereas low relative humidity speeds it up; denoted by ϕ, given as:

$$\phi = \frac{v_s}{v_v} = \frac{p_v}{p_s} = \frac{\omega}{0.622}\frac{p_a}{p_s}$$

where

v_s = specific volume of saturated air
v_v = specific volume of water vapour
p_v = partial pressure of water vapour
p_a = partial pressure of dry air
p_s = partial pressure of saturated air; see also **Specific Humidity**.

Reliability A *Statistics* term; the probability that a product will not fail and will work successfully when used under specified operating conditions for a specified period of time; denoted as *R*; given as: $R = 1 -$ probability of failure, where $0 \leq R \leq 1$; if $R = 0.90$, it means there is a 90% chance that the product will perform its proper function without failure.

Removable Pattern A *Manufacturing* term related to *casting*; a pattern which can be removed from the mould after being used in making the mould cavity; used in *sand casting*; made of *wood, plastic,* or *metal*; see also **Disposable Pattern**.

Renewable Energy The energy obtained from *nonconventional sources* such as *sun, wind, tidal wave*, and *biomass* which can never be depleted; a few examples include *solar energy, wind energy*, and *tidal energy*.

Renewable Sources of Energy A *Power Plant* term; the sources of energy which can be replenished in short order; freely available in nature and can be used again and again without exhausting them; that is, they are renewable in nature and are environment-friendly; a few examples include *solar*, *wind*, *sea*, and *geothermal*; see also **Nonrenewable Sources of Energy** and **Nonconventional Sources of Energy**.

Reorder Point An *Operations Management* term related to *inventory*; the inventory level at which a stock should be reordered.

Repeatability A *Work Study* term; the degree to which the same work can be done again.

Repetitive Manufacturing A type of manufacturing system in which plants assemble similar products; *automobiles*, *watches*, *microcomputers*, *pharmaceuticals*, and *televisions* are produced using repetitive manufacturing; *material requirements planning* (MRP) and *just-in-time* (JIT) techniques are useful for repetitive manufacturing.

Residual Stress A *Strength of materials* term; the stress which remains within a structural material after all the applied loads are removed.

Resilience A *Strength of Materials* term related to a *spring*; the capacity of a material to absorb energy within its elastic limit; see also **Modulus of Resilience**.

Resin A *Materials Engineering* term; a gum-like solid or semisolid viscous substance which contains polymer macromolecules that cross-link when combined with a hardener.

Resistance Butt Welding A *Welding* term; see **Upset Welding**.

Resistance Seam Welding A *Welding* term; see **Seam Welding**.

Resistance Spot Welding A *Welding* term; see **Spot Welding**.

Resistance Welding A *Welding* term; a *plastic welding process* which uses heat produced by electrical resistance of the work parts held together under pressure by two *copper electrodes*; the heat produced by electrical resistance is given as:

$$H = I^2 \cdot R \cdot t$$

where

H = heat produced in joule
I = current in ampere
R = electrical resistance in ohm
t = time of current flow in second.

Resistance welding uses *high current* and *low voltage*; does not require any *flux*, *filler metal* or *shielding atmosphere*; finds extensive applications in *automobile industries*; classified into *spot welding*, *seam welding*, *projection welding*, and *flash welding*.

Resistivity An *Electrical* term; also called **Specific Resistance**; a material's property which indicates its resistance (R) to current flow in the material in its unit length (L) having unit cross-sectional area (A); denoted by ρ; given as:

$$\rho = R/(L/A)$$

Resistor An *Electrical* term; an electrical device designed primarily to produce heat by obstructing the flow of current; commonly made of *iron* or *nichrome* wires; used in *electrical networks* and *electronic circuits.*

Resolution A *Metrology* term; the smallest change in input signal or the measured value which can be detected by an instrument; *the least count of an instrument is usually the resolution of the instrument.*

Resonance A *Physics* term; the phenomenon of attaining a very high amplitude (of an oscillation) when a body is set to oscillate by an *external periodic force* with a frequency being equal to the *natural frequency* of the body.

Resources An *Economics* term; see **Factors of Production**.

Response A *Metrology* term; an evaluation of a system's ability to effectively *sense, transmit,* and *present* all the pertinent information included in the measured quantity; classified into *amplitude response, frequency response,* or *phase response.*

Rest A *Physics* term; the state of a body with zero velocity; the state of a body in which it does not change its position with respect to time; a few examples include *a book lying on a table* and *a person sitting on a chair*; see also **Motion**.

Restrained Beam A *Strength of Materials* term related to a *beam*; see **Fixed Beam**.

Retaining Wall A *Civil Engineering* term; an engineering structure in the form of a wall used to retain *earth, soil, sand,* etc. and sustain their lateral pressures; a few examples include *basement walls* and *dams*; see also **Dam**.

Retardation A *Physics* term; also called **Deceleration**; a negative acceleration which occurs when the velocity of a body decreases.

Revenue Assets A *Financial Management* term; see **Current Assets**.

Reverberatory Furnace A *Materials Engineering* term related to *metallurgy*; a furnace in which a fuel does not come in direct contact with the charge; flames are directed from the roof of the furnace on to the charge, and air supply is controlled by vents; used for both *oxidation* and *reduction* processes.

Reverse Extrusion A *Manufacturing* term; see **Indirect Extrusion**.

Reverse Logistics An *Operations Management* term related to a *supply chain*; the process of moving products upstream from the customer back towards manufacturers and suppliers; used in case of returnable items rejected by the customers on account of quality defects; see also **Logistics**.

Reverse Polarity A *Physics* term; a polarity setup in which workpiece is connected to the negative terminal and electrode to the positive terminal of the DC electrical circuit; usually used with *consumable* electrodes; see also **Straight Polarity**.

Reversed Carnot Cycle A *Refrigeration* term; a *reversible Carnot cycle* employed as a *reversible refrigeration cycle* to get maximum possible coefficient of performance (COP) of a refrigerating machine operating between two temperature limits; does not depend on the working substance (*refrigerant*) used for refrigeration.

Reversible Cycle A *Thermodynamics* term; a cycle consisting entirely of *reversible processes*; a few examples include *Otto cycle*, *Diesel cycle*, and *Brayton cycle*; see also **Irreversible Cycle**.

Reversible Engine A *Thermodynamics* term; an engine which operates on reversible thermodynamic processes and has no losses; see also **Irreversible Engine**.

Reversible Process A *Thermodynamics* term; an *ideal process* in which a system passes through a continuous series of equilibrium states and can take place in both directions as its initial and final states coincide; a few examples include *polytropic expansion* and *compression* of a fluid, *isothermal expansion* and *compression* of a gas, *frictionless adiabatic expansion* and *compression* of a fluid, *electrolysis*, and *evaporation*; processes used in a thermodynamic cycle are reversible; see also **Irreversible Process**.

Reverted Gear Train A *Theory of Machines* term related to a *gear*; a special type of compound gear train in which the first and the last gears have the same axis; a few examples include *back gear* of a *lathe* and *clock*.

Revolution Per Minute A *Design* term, abbreviated as *rpm*; the number of rotations an object makes per minute.

Reynolds Number A *Fluid Mechanics* term; the most important *dimensionless* (nondimensional) group used in fluid mechanics; named after *Osborne Reynolds* (1842–1912), a noted *English* physicist and a pioneer in the study of pipe flow and turbulence; a ratio of the *inertia force* to the *viscous force* of a fluid; a parameter to differentiate between laminar and turbulent flow through pipes; classified into *critical*, *lower critical*, and *upper critical Reynolds number* depending upon its value; for example, the critical Reynolds number indicates a mixture of *laminar* and *turbulent* flows, whereas the lower critical Reynolds number is that Reynolds number below which the entire flow is laminar, and the upper critical Reynolds number is that above which the entire flow is turbulent; denoted by R_e; given as:

$$R_e = \frac{\rho V L}{\mu}$$

where

ρ = density of the fluid
V = velocity of fluid flow
L = characteristic length (usually the diameter of the pipe)
μ = viscosity (dynamic viscosity) of the fluid.

RFID An *information technology* tool; see **Radio Frequency Identification.**

Rheopectic Fluid A *Fluid Mechanics* term; a fluid whose apparent viscosity (that depends on shear rate) increases with time under a constant applied shear stress; see also **Thixotropic Fluid.**

Rhombus A *Mathematics* term; a parallelogram whose all the sides are equal; see also **Parallelogram.**

Ribbed Coupling See **Clamp Coupling.**

Right Angle A *Mathematics* term; an angle equal to 90°.

Right Hand Rule A *Physics* term; represents the direction of *angular displacement vector*; states that if the curvature of the fingers of the right hand represents the sense of rotation of an object, then the thumb, held normal to the curvature of the fingers, represents the direction of angular displacement vector.

Right Triangle A *Mathematics* term; also called **Right-Angled Triangle**; a triangle whose any one angle is a right angle, that is, equal to 90°; see also **Acute Triangle** and **Obtuse Triangle.**

Right-Angled Triangle A *Mathematics* term; see **Right Triangle.**

Rigid Body A *Physics* and a *Strength of Materials* term; a body which does not deform on the application of an external force or the distance between the two arbitrarily selected points in the body always remains constant; engineering mechanics deals with rigid bodies; see also **Elastic Body.**

Rigid Coupling A *Design* term; a temporary *fastening device* used *to connect two coaxial shafts*; requires the shafts to be in good alignment, both laterally and angularity; *does not allow any misalignment of the shaft axes, hence the name rigid coupling*; classified into *muff* or *sleeve* coupling, *clamp coupling, flange coupling,* and *solid* or *forged flange coupling.*

Rigid Truss A *Strength of Materials* term related to a *truss*; also called **Perfect Truss**; a truss which is rigid and non-collapsible; obeys the relation $m = 2j - 3$, where m = number of members and j = number of joints in the truss; see also **Non-rigid Truss, Under-rigid Truss,** and **Over-rigid Truss.**

Riser A *Manufacturing* term related to *casting*; a *reservoir of molten metal* which supplies additional metal to castings to compensate for shrinkage during solidification of castings.

Rivet A *Design* term; a fastening element usually used to join plates permanently; made of *wrought iron, mild steel, copper,* or *aluminium alloys*; fails due to *shear, bending,* or their combined action.

Riveting A *Design* term; a permanent method of fastening which uses rivets to join the plates that cannot be separated without damaging the rivets.

Roasting A *Materials Engineering* term related to *metallurgy*; a chemical process of converting an *ore* into its *metallic oxide* by heating it strongly *in excess of air* at a temperature below its melting point; carried out in a *reverberatory furnace*; see also **Calcination.**

Robert Manning (1816–1897) An *Irish* engineer who is known for the *Manning equation* used to obtain the fluid velocity or the flow rate in *an open channel*; proposed several formulae for *open-channel resistance.*

Robot A *programmable machine* capable of performing functions according to computer programmed instructions; used to replace the human beings for some very heavy weight, dirty, hazardous, unpleasant, or monotonous tasks; the typical applications of a robot include *loading and unloading of machine tools in a production system* and *painting and spot welding in automotive and appliance industries*; see also **Robotics.**

Robotics A science which deals with the use of robots and their mechanisms for various applications.

Robust Design A *Quality Management* term; a quality concept evolved by *Genichi Taguchi* (1924–2012), a famous *Japanese* quality guru; emphasizes that products or services should be designed in a manner so that they are robust enough to function satisfactorily despite variations in the environmental and operating conditions; used as an approach in the planning of new products and services; quality of such products or services is called robust quality.

Robust Product A *Quality Management* term; a quality concept proposed by Genichi Taguchi (1924–2012); a product which can perform under a wide range of environmental conditions without failing.

Rocker A *Physics* term; a bar which has one fixed pivot point around which it may rotate through a fixed angle range.

Rocket A device in which a solid or liquid fuel and an oxidizer are made to react in a combustion chamber to produce high-pressure combustion gases, which are expanded in a nozzle to give them very high velocity while leaving the rocket and thus producing the required thrust needed to propel the rocket.

Rockwell Hardness Test A *Strength of Materials* term; see **Rockwell Test.**

Rockwell Test A *Strength of Materials* term; also called **Rockwell Hardness Test**; a destructive method of hardness testing of a material developed by *S.P. Rockwell* (1886–1940) in 1922; commonly employed in the USA; uses either a *diamond indenter* having 120° included angle or a hardened steel ball that needs to be pressed against the test specimen; based on measuring the depth of penetration rather than the diameter of indentation as in case of Brinell test; uses three scales for hardness measurement, namely *Rockwell A* for *case hardened materials* that uses a 60 kg load, *Rockwell B* for soft materials (such as *mild steels*, *brass*, and *aluminium*) that uses a 100 kg load, and *Rockwell C* for hard materials (such as *high-speed steels*, *high carbon steel*, and *tool steels*) that uses a 150 kg load; see also **Brinell Test**.

Roll Bending A *Manufacturing* term; a cold metal working process used to give the required curvature to metal sheets or plates by means of rotating rollers so as to convert them into useful products; typical roll bending products include *channels*, *gutters*, *panels*, and *frames*; see also **Roll Forging**.

Roll Forging A *Manufacturing* term; a hot metal working process used to reduce the cross section of round or flat bar stocks by subjecting them to rotating rollers; typical roll forged products include *tapered shafts*, *leaf springs*, *knives*, and *hand tools*; see also **Roll Bending**.

Roll Threading A *Manufacturing* term related to *forming*; a cold metal working process which involves *squeezing* and *rolling* to produce threads on a blank by placing it between two serrated dies.

Roller Support A *Physics* and a *Strength of Materials* term; a type of simple support which prevents vertical movement but allows movement in the horizontal plane; also does not prevent rotation; has one vertical force reaction and no moment reaction; see also **Hinge Support**.

Rolling Contact Bearing A *Design* term; see **Anti-friction Bearing**.

Rolling Friction A *Physics* term; the friction between two surfaces in contact, of which one surface is rolling over another; caused by the deformation of a rolling object such as a wheel or a cylinder rolling over a flat surface; has point contact on the surface and has very small values; see also **Sliding Friction**.

Rolling A *Manufacturing* term; the most widely used metal working process used to reduce the thickness or change the cross section of a workpiece by placing it between two rotating rolls; can be performed in both *hot* and *cold* states (called respectively *hot rolling* and *cold rolling*); typical rolled products include *flat plates*, *sheets*, *foils* (using cold rolling), and bars of various shapes and sizes including *round*, *square*, *hexagon*, *I-section*, *channel section*, *angle section*, *T-section*, and *Z-section* (using hot rolling).

Root Diameter A *Theory of Machines* term related to a *gear*; the diameter of the dedendum circle; see also **Blank Diameter**.

Rope Drive A *Theory of Machines* term; a mechanism used to transmit *motion* or *power* between two shafts which are *considerable distance apart*, using *a rope* that runs on *grooved pulleys* attached to the shafts; useful for *high-power transmission*; a *negative drive* of power transmission as some power is lost due to *friction* and *slippage*; see also **Belt Drive** and **Chain Drive**.

Rose Metal A *Materials Engineering* term; an alloy of *tin* (Sn), *lead* (Pb), and *bismuth* (Bi) consisting of 28% Sn, 22% Pb, and 50% Bi; used in making *electrical fuses*.

Rotameter A *Fluid Mechanics* term; a *flow measuring device* consisting of a tapered vertical glass tube in which a metering float moves upward or downward by the upward motion of the fluid around it; the vertical position of the float is linearly proportional to the discharge, which can be read directly from the graduated tube.

Rotary Pump A *Fluid Mechanics* term; a type of *positive displacement pump* used for pumping *highly viscous fluids* and *slurries* such as pumping *engine oil in automobiles* and for pumping *heavy liquids* like *syrup, tomato paste, chocolate*, and *slurries* like *soups in the food industry*.

Rotary Swaging A *Manufacturing* term; see **Swaging**.

Rotational Equilibrium A *Physics* term; the equilibrium which ensures that a rigid body does not rotate or rotates with a *constant angular velocity* about an axis; requires the net external torque or the vector sum of all the torques acting on the body is zero; a body under rotational equilibrium has *constant angular momentum* and *zero angular acceleration*; mathematically expressed as:

$$\Sigma M_X = 0,\ \Sigma M_Y = 0,\ \text{and } \Sigma M_Z = 0;$$ see also **Translational Equilibrium**.

Rotational Flow A *Fluid Mechanics* term; a fluid flow in which the fluid particles rotate about their mass axes, perpendicular to the plane of motion; for a rotational flow, the vorticity is nonzero; see also **Irrotational Flow**.

Rotational Inertia A *Physics* term; see **Mass Moment of Inertia**.

Rotational Momentum A *Physics* term; see **Angular Momentum**.

Rotational Speed A *Physics* term; the speed connected with the rotation of an object; for example, a shaft has a rotational speed; a measure of how fast an object rotates about an axis; the rotational speed is measured in *revolution per minute*.

Rotodynamic Pump A *Fluid Mechanics* term; also called **Dynamic Pump**; a pump which increases the dynamic pressure of a fluid by the rotation of its impeller blades producing centrifugal action and imparting momentum to the fluid; some examples include *centrifugal pump, axial-flow pump*, and *mixed-flow pump*; see also **Positive Displacement Pump**.

Round File A *Manufacturing* term; one of the types of a file used in the *fitting shop* which has a *circular cross section*; used in the filing of *holes, round corners,* or *round slots.*

Round Key A *Design* term related to a *key*; a type of key used for light duty applications; has either uniform circular cross section throughout its length or is tapered (1:50) along its length which can fit in the hole drilled partly in the shaft and partly in the hub of the mounting; its diameter is usually equal to $0.25D$, where D is the diameter of the shaft.

Route Sheet An *Operations Management* term; a document which shows the routing of a component or assembly including the *workstations, machine tools,* and *operations times* through its production processes.

Routing An *Operations Management* term; the processing steps or stages needed to create a product or to do a job; a component of *shop floor control.*

rpm A *Design* term; see **Revolution Per Minute**.

Rubber A *Chemistry* term; also called **Natural Rubber**; an *elastic polymeric material* prepared from *latex* (a white viscous liquid), which is obtained from a latex tree; consists of long chain molecules interlocked with each other; lacks desirable properties which can be improved; the addition of compounding materials, such as *sulphur, accelerators* (*lime, magnesia, litharge,* etc.), *antioxidants, reinforcing agents* (*carbon black, zinc oxide,* etc.), *colouring agents* (*zinc sulphide, lead chromate,* etc.), and *plasticizers* (*vegetable oils, stearic acids,* etc.) improve its properties like *elasticity, tensile strength, abrasion resistance,* and *chemical resistance* and ensure its lower electrical and thermal conductivity; obtained in the form of sheets using *calendering machine*; extensively used in the manufacture of *tyres* and *tubes* for the *vehicles*; also used in making *protective coatings, adhesives, papers,* and *packing materials*; see also **Artificial Rubber**.

Rudolf Diesel (1858–1913) The *German* inventor of the *diesel engine,* a type of *internal combustion engine.*

Ruler A *Metrology* term; a length measuring instrument smaller than a meter scale; a straight edged thin strip of steel or plastic in which centimetre and millimetre are marked (called *graduations*) separately on its two edges.

Rules A *Management* term; the specified required actions or inactions (usually in the form of written statements) which do not allow any discretion; act as a means for reaching the desired goals.

Runner Blade A *Fluid Mechanics* term related to a *turbine*; see **Bucket**.

Runner A *Manufacturing* and a *Fluid Mechanics* term; in *manufacturing* terms, it relates to *casting*; a part of the *gating system*; a *horizontal channel* in the *parting line* which carries molten metal from the sprue to the mould cavity; see also **Sprue;** in *fluid mechanics* terms, it relates to a *turbine*; the moving component of a *turbine* which consists of *vanes* or *buckets* attached to a rotating shaft;.

Rusting A *Chemistry* term; a common corrosion process for *ferrous alloys* which deteriorates them; a phenomenon in which, when iron is exposed to air and moisture, it gets oxidized to a hydrous oxide layer forming over its surface which consists mainly of hydrated iron oxide ($2Fe_2O_3.3H_2O$) together with a small quantity of Fe (II) carbonate; the pure iron, if kept in contact with pure water does not rust, but even small traces of other metal such as zinc or copper in iron promote the formation of rust; the presence of chromium and nickel slows down or even prevents rusting; iron is protected from rusting by coating it with *zinc* (the process is called *galvanization*) or by painting or white-washing it with lime (CaO).

S

Saddle Key A *Design* term related to a *key*; a key of uniform width but has tapered thickness on one side where the magnitude of taper is usually 1:100; suitable for *light load applications* only as it tends to slip around the shaft, if used under heavy loads; classified into *hollow saddle key* and *flat saddle key*.

Saddle A *Lathe* term; a principal part of a lathe in the form of an H-shaped casting which moves along the bedways and carries *cross slide*, *compound slide*, and *tool post*.

SAE Society of Automotive Engineers; an *American* professional body.

Safe Load A *Strength of Materials* term; the actual load which can be placed on a beam or column without producing any buckling; equals to the buckling load divided by the factor of safety.

Safety Glass A *Chemistry* term; also called **Unbreakable Glass**; a type of glass which does not break easily under ordinary impact, hence ensures safety but can break only under heavy impacts; prepared by placing a layer of transparent plastic (usually *a sheet of vinyl acetate resin*) between two layers of glass using a suitable adhesive and joined tightly by applying heat and pressure; finds use in making *screens of automobiles*, *aeroplanes*, and *railways*.

Safety Need A *Management* term related to *motivation*; one of the five levels of human needs proposed by *Abraham Harold Maslow* (1908–1970) in his '*Need-Hierarchy Theory*' of motivation; represents the need for protection from *threats* and *danger* (also includes safety from verbal attack on a person at a meeting).

Safety Stock Inventory An *Operations Management* term; see **Buffer Stock**.

Safety Stock An *Operations Management* term; see **Buffer Stock**.

Safety Valve A *Power Plant* term related to a *boiler*; a *boiler mounting* used to maintain a safe pressure inside a boiler; fitted on the top of a boiler; helps to release the excess pressure developed in the boiler, thus prevents its bursting; at least *two* safety valves are used so that if one valve fails, the other can be used.

D. K. Singh, *Dictionary of Mechanical Engineering*,
https://doi.org/10.1007/978-981-99-1722-8_19

Sagging A *Strength of Materials* term related to *bending*; a *positive bending moment* which produces the upward concavity in a beam, as in case of a loaded simple beam; see also **Hogging**.

Saint Venant's Criterion A *Strength of Materials* term; see **Maximum Normal Strain Theory**.

Sales Index An *Operations Management* term; a ratio of the quality cost to the sales; see also **Cost Index** and **Labour Index**.

Salvage Value A *Financial Management* term; the income obtained from the sale or disposal of an asset (such as plant and equipment) at the end of a project's life.

Sample Proportion Defective A *Quality Control* term; the number of defective units in a randomly selected sample divided by the sample size; used in *p-chart*.

Sample Range A *Quality Control* term; the arithmetic difference between the highest and lowest measurement for a sample.

Sample Size A *Quality Control* term; a certain quantity of randomly selected observations of process outputs (product or service) from a lot which are subjected to quality testing measures to know the status of the lot for its acceptance or rejection.

Sampling Inspection A *Quality Control* term; a *statistical method* of inspection which takes a random sample of certain items from the entire lot, tests every item of that sample; the approval or rejection of the sample will decide the acceptance or refusal of the entire lot.

Sand Casting A *Manufacturing* term related to *casting*; a casting process which uses a *sand mould* to produce cast parts.

Sand A *Chemistry* term; a *crystalline form of quartz* formed as a result of '*weathering*' or disintegration of rocks and further breaking down by violent movements of water; used in the manufacture of *glass*, *mortar*, and *porcelain*; forms an important construction material to be mixed with *lime* or *cement*; used in making *vitreosil* (a translucent variety of *glass*); finely ground sand is used in the extraction of phosphorus from phosphates.

Sandwich Beam A *Strength of Materials* term; a recently developed *composite beam* fabricated in a sandwich manner, primarily to save material and reduce weight; typically consists of two thin faces of relatively high-strength material (such as *aluminium*) separated by a thick core of lightweight low-strength material (*plastics* and *foams*); its faces function somewhat like the flanges of an I-beam, which can withstand highest bending stresses for being located at the greatest distance from the neutral axis, whereas its core serves as a filler and provides support for the faces, stabilizing them against wrinkling or buckling; see also **Composite Beam**.

Satellite A *Physics* term; a material body revolving continuously in an orbit around a comparatively much larger material body (earth or any other planet or sun); for example, the moon is the satellite of earth and the earth is the satellite of sun; the total energy (E) of a satellite consists of *potential energy* and *kinetic energy*; when a satellite is orbiting close to the earth, its total energy is given as:

$$E = -\frac{GMm}{2R}$$

where

G = universal gravitational constant = $6.67 \times 10^{-11} \text{N.m}^2.\text{kg}^{-2}$
M = mass of the earth
m = mass of the satellite
R = radius of the earth.

For a satellite orbiting very close to the surface of the earth, the orbital speed is equal to 7.92 km/s, and the time period of revolution is 84.6 min; classified into *natural satellite* and *artificial satellite.*

Satisfiers A *Management* term related to *motivation*; see **Intrinsic Factors.**

Saturated Air A *Thermodynamics* term related to *psychrometry*; the air containing maximum water vapour at the *saturation pressure*; cannot hold any more moisture; see also **Unsaturated Air**.

Saturated Liquid A *Thermodynamics* term; the state of a liquid where both the liquid and the vapour phases coexist in equilibrium.

Saturated Vapour Line A *Thermodynamics* term; the line made up of the saturated vapour states.

Saturated Vapour A *Thermodynamics* term; the state at which all liquid is just vaporized.

Saturated Liquid Line A *Thermodynamics* term; the line made up of the saturated liquid states.

Saturation Line A *Thermodynamics* term related to *psychrometry*; a curved line in the psychrometric chart which corresponds to 100 per cent relative humidity of air.

Saturation Pressure A *Thermodynamics* term; the pressure corresponding to saturation temperature at which a substance changes its phase; see also **Saturation Temperature**.

Saturation Temperature A *Thermodynamics* term; the temperature corresponding to saturation pressure at which a substance changes its phase; see also **Saturation Pressure**.

Saw A *Manufacturing* term; a *cutting tool* consisting of a blade (usually made of high carbon and high-speed steels) having a number of small teeth, which take part in the to-and-fro cutting operation; classified into *hacksaw, band saw, circular saw,* and *abrasive disc saw.*

SAW A *Welding* term; see **Submerged Arc Welding.**

Sawing A *Manufacturing* term; a bulk-removal cutting process used to cut the correct-sized workpiece from a large raw material stock with the help of a saw.

SBU A *Management* term; see **Strategic Business Unit.**

Scab A *Casting* term; a type of *casting defect* which occurs due to thermal expansion of the moulding sand which raises a certain part of a casting over its flat surface.

Scalar Chain A *Management* term; one of the 14 principles of management proposed by *Henri Fayol* (1841–1925); a chain of command which ensures communication from top management; ranges from the ultimate authority to the lowest rank in an organization.

Scalar Principle A *Management* term related to *line authority*; states that the clearer the line of authority, the clearer will be the responsibility for decision-making and the more effective will be the organizational communication; see also **Unity of Command.**

Scalar Product of Vectors A *Physics* term; also called **Dot Product of Vectors**; the product of the magnitude of the two vectors and the cosine of the smallest angle between them; the product of the magnitude of one vector and the component of the other vector in the direction of the first vector; given as:

$$\boldsymbol{A} \cdot \boldsymbol{B} = ab \cos \theta$$

where

a = magnitude of vector $\boldsymbol{A}$
b = magnitude of vector $\boldsymbol{B}$
θ = angle between the two vectors $\boldsymbol{A}$ and $\boldsymbol{B}$

The physical scalar quantities like *work, energy, electric potential* or *gravitational potential* can be described as the scalar product of two vectors; a *scalar quantity* which *follows the commutative* and *distributive laws of algebra*; if either of the two vectors is multiplied by a scalar, the scalar product also gets multiplied by the scalar; becomes zero when two vectors are at right angles (90°) to each other; the scalar product of two similar unit orthogonal vectors is 1 (as $\cos 0° = 1$), and that of two dissimilar unit orthogonal vectors is 0 (as $\cos 90° = 0$); see also **Vector Product of Vectors.**

Scalar Quantity A *Physics* term; a physical quantity which has only magnitude and no direction; a few examples include *mass*, *distance*, *area*, and *volume*; see also **Vector Quantity**.

Scale Effect A *Fluid Mechanics* term related to *similarity laws*; the deviations from the dynamic similarity rule due to involvement of Reynolds number, Mach number, relative roughness of machines or lack of strict geometrical similarity; tends to improve the performance of larger machines.

Scalene Triangle A *Mathematics* term; a triangle whose all the three sides are of different (unequal) lengths; see also **Equilateral Triangle**.

Scar A *Casting* term; a type of casting defect which occurs in the form of a *shallow blow* mainly on the flat surface of a casting; see also **Blow**.

Scarcity An *Economics* term; the shortage which exists, when less of something is available than is wanted.

Scatter Diagram An *Operations Management* term; a graphical component of regression analysis; a graph which shows how two process variables (measurements) relate to each other; often used to verify possible causes and effects obtained from cause-and-effect diagrams;

Schematic A diagram which uses symbols to represent components of a system.

Schottky Defect A *Materials Engineering* term; a type of *point defect* in solid materials; a pair of oppositely charged ion vacancies created due to atoms or molecules moving out from the interior to the surface of the crystal.

Science A subject which proposes hypotheses or theories based on observations validated with carefully performed experiments which use many measurements; a subject which can be argued upon with definite reasoning.

Scientific Management A *Management* term; also called **Taylor's Scientific Management**; a scientific approach of management introduced by an American mechanical engineer *F. W. Taylor* (1856–1915) who is called the "*father of scientific management*"; deals with improving the methods of doing a task using scientific principles and tools in the solution of managerial problems; objective was to design jobs better, train workers, and provide financial incentives to increase workers' productivity.

SCM An *Operations Management* term; see **Supply Chain Management**.

SCOR An *Operations Management* term; see **Supply Chain Operations Reference**.

Scoring A *Design term*; a *surface failure* which occurs due to *ineffective lubrication*.

Scott-Russell Mechanism A *Theory of Machines* term; a mechanism used to produce *exact straight-line motion* merely by *copying*, but not by generating as it happens in case of *Peaucellier* and *Hart mechanisms*; similar to that of a reciprocating engine; not very useful, as it involves *sliding pair* which results in more *friction* and *wear*.

Screw Dislocation A *Materials Engineering* term; a type of *linear defect* in a material which derives its name from the *spiral stacking* of crystal planes around the dislocation line; the Burgers vector is parallel to the dislocation line; see also **Edge Dislocation** and **Mixed Dislocation**.

Screw Jack A *load lifting device* which consists of a square-threaded central rod (called a screw) fitted into the internally-threaded collar of a jack.

Screw A *Design term*; a mechanical device having threads over its certain length; used for fastening purposes; classified into *single-threaded* or *multiple-threaded* screws, where the latter has two or three threads; the standardized *screws*, *bolts*, and *nuts* all have single threads; for a double-threaded screw, lead equals to twice the pitch, whereas for a triple-threaded screw, lead equals to 3 times the pitch.

Scriber A *Manufacturing* term; a sharp pointed needle like *fitting tool* used to produce marking on a workpiece to cut it along the marked line.

Scroll A *Fluid Mechanics* term related to a *pump*; the snail-shaped casing of a centrifugal pump.

SDE Analysis An *Operations Management* term related to *inventory*; a *selective inventory control technique* which classifies the inventories based on the degree of difficulty in its availability into three categories: *S* (*scarce*) items, *D* (*difficult*) items, and *E* (*easily available*) items; *S items* may require source development (e.g. *patented items*); *D items* are not easily manufacturable and may involve one or two manufacturers only (e.g. *computers* and *printers*); *E* items are readily available in the market (e.g. *clothing*, *food*, and *jewellery*); this analysis helps in determining purchasing strategies for inventory items and to control their lead time.

Seal A *Design* term; a device used to *prevent leakage of fluids* like *gas*, *steam* or *liquids* between moving and stationary parts or between two machine members having no relative motions; also used to maintain the applied pressure or vacuum atmosphere within a sealed volume, for example, in a *refrigeration system* and *turbines*.

Seam Welding A *Welding* term; also called **Resistance Seam Welding**; a type of *resistance welding*; the joint is produced along a line as a series of *overlapping spot welds* made between the work parts held together under pressure by rotating circular electrodes; used to produce a leak-proof joint; finds applications in the joining of *gasoline tanks*, *boilers*, *automobile mufflers*, and *pressure vessels*; see also **Spot Welding**.

Secant Column Formula A *Strength of Materials* term related to a *column*; also called simply **Secant Formula**; a formula which involves a *sec* term; used to find the *crippling (critical) load* for all types of columns subjected to *eccentric loading*, where the line of action of the column forces is separated from the centroidal axis of the column by a certain distance, called *eccentricity*; expressed as:

$$\sigma = \frac{P}{A}\left[1 + \frac{eC}{k^2}\sec\left(\frac{1}{2}\sqrt{\frac{P}{AE}}\frac{l_e}{k}\right)\right]$$

where

σ = maximum stress induced in the column
P = eccentric load (crippling) on the column
A = area of cross section of the column
E = modulus of elasticity of the column material
e = eccentricity of the load
l_e = equivalent/effective length of the column
k = radius of gyration
C = distance of the outermost fibre of the column from its neutral axis.

Secant Formula A *Strength of Materials* term related to a *column*; see **Secant Column Formula**.

Second Conjugate Beam Theorem A *Strength of Materials* term; a conjugate beam-based theorem used to find *deflection* at any section of a loaded beam; states that the deflection at any section of a loaded beam is equal to the bending moment at the corresponding section of the conjugate beam; see also **First Conjugate Beam Theorem**.

Second Law of Thermodynamics A *Thermodynamics* term; a law which cannot be proved like any other physical law, as it is based on experimental observations; consists of two classical statements—the Kelvin-Planck statement, which is related to heat engines, and the Clausius statement, which is related to *refrigerators* or *heat pumps* (see them separately discussed).

Second Moment of Area A *Strength of Materials* term; also called **Geometrical Moment of Inertia** or **Area Moment of Inertia** or simply **Moment of Inertia** of a plane area; denoted by *I*; defined as:

$$I_x = \int y^2 da \text{ (about } x\text{-axis)}$$

$$I_x = \int x^2 da \text{ (about } y\text{-axis)}$$

where x and y are the distances of the area da from y and x axes, respectively; the SI unit is m^4; for a circular shaft (solid or hollow), its values are given as:

$$I_{NA} = I_x = I_j = \frac{\pi d^4}{64} \quad \text{(for a solid circular shaft)}$$

$$= \frac{\pi \left(D^4 - d^4\right)}{64} \quad \text{(for a hollow circular shaft)}$$

where

d = diameter of the solid shaft or inside diameter of hollow shaft
D = outside diameter of the hollow shaft
NA = neutral axis of the shaft's cross section; see also **Polar Moment of Inertia.**

Secondary Bond A *Chemistry* term; an atomic bond formed without any electron transfer or sharing of outer orbital electrons; examples include *metallic bond* and *coordinate bond*; see also **Primary Bond.**

Secondary Sources of Energy A *Power Plant* term; the sources of energy derived from the primary sources; for example, *hydrogen* produced from *water*; see also **Primary Sources of Energy**.

Second-Class Lever A *Physics* term; a lever in which the load is positioned between the fulcrum and the effort and both act in opposite directions; moving the load closer to the fulcrum will increase the mechanical advantage, whereas moving the load closer to effort decreases the mechanical advantage; A few examples of second-class lever include a *wheelbarrow* (used to lift a load of dirt) and a *nutcracker* (used to crack open the tough shell of a nut).

Section Modulus A *Strength of Materials* term; a ratio of the second moment of area (I) to the distance (y) of outer layer of a beam from the neutral axis; equals to the ratio of the bending moment to the maximum bending stress; a measure of strength of a beam's cross section; the SI unit is m^3.

Section An *Engineering Drawing* and a *Machine Drawing* term; the inner details of an object or a machine component on being imagined to be cut by a cutting plane and viewed after the removal of the cut portion.

Seebeck Effect A *Physics* term; a *thermoelectric effect*, named in honour of a *German* physicist *Thomas Johann Seebeck* (1770–1831) who first made this discovery in 1821; states that when the two junctions formed by two dissimilar materials are maintained at different temperatures, an electromotive force (emf) is observed across the ends of the materials which is responsible for the flow of current in the circuit; used in temperature measurement and power generation; also forms the basis for thermoelectric power generation (thermoelectric generator); see also **Peltier Effect**.

Selective Inventory Control An *Operations Management* term related to *inventory*; a process of controlling inventory on the basis of degree of control or intensity of management efforts required as all the inventories are not equally important in respect of sales, profits, availability, etc.; a few selective inventory control techniques include *ABC analysis*, *VED analysis*, *FSN analysis*, *HML analysis*, *S-OS analysis*, *SDE analysis*, *GOLF analysis*, and *XYZ analysis*.

Self-Actualization Need A *Management* term related to *motivation*; one of the five levels of human needs proposed by *Abraham Harold Maslow* (1908–1970) in his '*Need-Hierarchy Theory*' of motivation; represents the highest level of human needs which includes *continual self-development*, *trust*, *integrity*, *responsibility*, and *naturalness*.

Self-Locking Device A *Design* term; a machine which gets locked and does not move to its original position in the reversible manner; has efficiency less than 50 per cent; *a screw jack is a self-locking device.*

Self-Priming Pump A *Fluid Mechanics* term; a pump which can create a significant vacuum pressure at its inlet even in its dry state; able to lift a liquid from several metres below the pump without undergoing any priming activity.

Selling Price An *Operations Management* term related to *cost analysis*; the selling price of a product is equal to *total cost* plus (or minus) profit (or loss); *profit* occurs when the selling price of a product exceeds its total cost; *loss* occurs when the selling price of a product is less than its total cost.

Semi-Centrifugal Casting A *Manufacturing* term related to *casting*; *a type of centrifugal casting* used to produce parts with *rotational symmetry* such as a *wheel with spokes* using a rotating mould (about its vertical axis); the mould rotates at lower speed as compared to semi-centrifugal casting; see also **True Centrifugal Casting** and **Centrifuging**.

Semi-Circle A *Mathematics* term; the one-half part of a circle.

Semiconductor An *Electronics* term; a solid material whose electrical conductivity lies between metallic conductors (good electrical conductors) and ceramics insulators (good electrical insulators); the current in semiconductors flows through two charge carriers, namely *electrons* and *holes*; a few examples of *elemental semiconductors* include *silicon* (Si), *germanium* (Ge), and *tin* (Sn); a few examples of *compound semiconductors* include *gallium arsenide* (GaAs) and *cadmium sulphide* (CdS); classified into *intrinsic* and *extrinsic* semiconductors; used in making electronic devices such as *transistors* and *diodes* used in *integrated circuits*.

Semi-Diesel Engine An *IC Engine* term; an internal combustion engine which works on a dual cycle in which heat is supplied partly at constant volume and partly at constant pressure; see also **Diesel Engine**.

Semi-Metal A *Chemistry* term; see **Metalloid**.

Sensible Heat A *Physics* term; the heat which produces a temperature rise or fall, when added to or removed from a substance; associated with the kinetic energies of the molecules of the substance; when added, temperature rises as molecules of the substance absorb heat energy and move faster; when removed, temperature drops as molecules of the substance become slower on account of loss of heat energy; given as:

$$Q = m.C.\Delta T$$

where

Q = heat added or removed
m = mass of the substance
C = specific heat of the substance
ΔT = temperature difference; see also **Latent Heat**.

Sensitivity Analysis A *Management* term; a technique for systematically changing parameters in a model to determine the effects of such changes; helps to assess what the results would be like if parameters such as costs, time-scales, and market share were different from the values assumed for a plan.

Sensitivity A *Metrology* term; a ratio of the magnitude of output signal to the magnitude of input signal.

Sensor A device which responds to a physical stimulus (such as heat, light, sound, pressure, magnetism, or a particular motion) and transmits a resulting signal required for measurement or operating a control.

Service Organization A *Management* term; the organization which deals with services (intangible/nonphysical products) which are being offered to customers; a few examples include *railway, airline companies, hospitals, financial institutions,* and *telephone* companies; see also **Manufacturing Organization**.

Servomechanism An *Electrical* term; a feedback system which consists of a sensing element, amplifier, and a servomotor.

Set-Squares A set of two *triangular shaped* geometrical instruments; usually made out of a thin transparent plastic sheet whose two perpendicular edges are graduated, one in centimetres and other in inches; classified into 30° and 45° set-squares of which 30° set-square has angles of 30°, 60°, and 90°, and 45° set-square has angles of 45°, 45°, and 90°; used to draw *parallel* and *perpendicular* lines; also used to construct angles of 30°, 45°, and 90°.

Setup Cost An *Operations Management* term; the cost related to configuring tools, equipment, and machines within a factory to produce an item; setup costs include writing orders and processing for the production system, time, and labour to clean and change tools or holders, and costs associated with employees' training; see also **Ordering Cost** and **Holding Cost**.

Setup Time An *Operations Management* term; the time required to adjust the machine settings, replace materials, change tools, and do everything it takes to change over from producing one type of product to another at a workstation; the setup time reduction is a continuous process in a JIT environment to reduce the inventory levels and increase production capacity; the concept of *single minute exchange of die* (SMED) is an useful concept to reduce the setup time.

SFD A *Strength of Materials* term; see **Shear Force Diagram**.

Shaft Coupling A *Design* term; also called simply **Coupling**; a *temporary fastening device* used to connect one shaft to another (because of limitation of length of shafts) or to couple a driver shaft to a driven shaft, in such a way that when both rotate, they act as single unit; used in transmission of power from one shaft to another; the shafts may have *collinear axes*, *intersecting axes* or *parallel axes* at a small distance; generally made of *cast iron*, but may be made of *cast steel for heavy-duty work*; classified into *rigid coupling* and *flexible coupling*.

Shaft Power An *IC Engine* term; also called **Brake Power** or **Brake Horse Power** or **BHP**; the power available at the shaft of an engine; given as:

$$\begin{aligned} \text{BP or BHP} &= T\omega \\ &= \frac{2\pi NT}{60,000} \quad \text{kW}(T \text{ in N-m}) \\ &= \frac{2\pi NT}{4500} \quad \text{HP}(T \text{ in Kgf-m}) \end{aligned}$$

where

T = torque transmitted to the crankshaft (N-m or Kgf-m)
ω = angular velocity of the crankshaft (rad/sec)
N = number of revolutions per minute (rpm)
π = a constant = $\frac{22}{7}$.

Shaft Work A *Thermodynamics* term; also called **Mechanical Work**; the work produced due to rotation of a shaft; the energy transmission with a rotating shaft is very common in engineering practice; for example, lifting of a weight through a pulley attached to a rotating shaft; associated with devices such as *pump*, *turbine*, *fan*, and *compressor*.

Shaft A *Design* term; a usually circular rotating part used to transmit motion or power between two members such as from the prime mover (*steam turbine*, *gas turbine*, etc.) to a machine or from a motor to a machine tool or from an engine to the rear axle of an automobile.

Shake Allowance A *Casting* term; also called **Rapping Allowance**; a *negative pattern allowance* which is provided by *decreasing the size of the pattern* so that the increased size during the shake of pattern in the mould cavity (during its withdrawal) can be accommodated.

Shaping A *Manufacturing* term; a *machining process* used to create plane surfaces using a single-point cutting tool; usually used to produce rectangular workpieces to be used for further machining.

Shareholder A *Management* term; the owner of an organization or the persons who have invested their money in an organization by buying the organization's stock; their claim on the organizational resources is often considered superior to the claims of other inside stakeholders; they are faceless as they are not directly involved in the management activities of the organization; see also **Stockholder**.

Sharp-Crested Weir A *Fluid Mechanics* term; a vertical thin plate having a sharp edge at its top which allows a liquid to pass over it when placed in an *open channel*; used to measure the liquid flow rate; see also **Broad-Crested Weir**.

Shaw Process A *Manufacturing* term related to *casting*; see **Ceramic Mould Casting**.

Shear Centre A *Strength of Materials* term; also called **Centre of Twist** or **Flexural Centre** or **Centre of Flexure**; a point in the cross section of a beam where if a lateral load is applied, it will bend the beam without twisting along the longitudinal axis; the point of intersection of the bending axis with the plane of transverse section of a beam; the section gets twisted about this point, if the load is not applied at the shear centre; for a doubly symmetric section, the shear centre coincides with the centroid; for a singly symmetric section, the shear centre lies on the axis of symmetry.

Shear Deformation A *Strength of Materials* term; see **Shearing Strain**.

Shear Flow A *Strength of Materials* term; the shear force which acts in a longitudinal section per unit length of a beam; denoted by q.

Shear Force Diagram A *Strength of Materials* term related to *bending,* abbreviated as **SFD**; a pictorial representation which shows the distribution of shear forces at various sections of a loaded beam; the positive values of shear forces are shown above the axis of the beam (horizontal axis) and the negative values below the axis (both values as vertical lines), which are connected by a horizontal straight line in case of point loads, by an inclined straight line in case of uniformly distributed loads (*udl*) and by a parabolic curve in case of uniformly varying loads acting on the beam; see also **Bending Moment Diagram**.

Shear Force A *Physics* and a *Strength of Materials* term; a force which acts tangentially or is parallel to the surface of a body and produces change in shape of the body.

Shear Formula A *Strength of Materials* term; proposed by a *Russian* bridge and railway engineer *D. J. Jourawski* (1821–1891); used to find the shear stress at the cross sections of various shapes of beams; given as:

$$\tau = \frac{V}{Ib}Ay$$

where

τ = shear stress
V = vertical shear force
I = moment of inertia of the cross section about the neutral axis
b = width of the cross section
A = area of cross section
y = distance of the centroid of the section where shear stress is τ.

Shear Modulus A *Strength of Materials* term; see **Modulus of Rigidity**.

Shear Spinning A *Manufacturing* term; also called **Power Spinning**; a cold metal working process used to shape a flat sheet metal blank into conical or cylindrical shapes by pressing it against a power driven shaped-mandrel using rollers; see also **Tube Spinning**.

Shear Strain Energy due to Principal Stresses A *Strength of Materials* term; the shear strain energy stored in a body per unit its volume when subjected to three mutually perpendicular principal stresses; given as:

$$U = \frac{1}{12G}\left[(\sigma_1 - \sigma_2)^2 + (\sigma_2 - \sigma_3)^2 + (\sigma_3 - \sigma_1)^2\right] \tag{1}$$

For a uniaxial stress system ($\sigma_2 = 0$, $\sigma_3 = 0$, and assuming $\sigma_1 = \sigma$)

$$U = \frac{1}{6G}\sigma^2 \tag{2}$$

where

$\sigma_1, \sigma_2, \sigma_3$ = principal stresses acting along three mutually perpendicular axes x, y and z respectively
G = modulus of rigidity of the body material.

Shear Strain A *Physics* and a *Strength of Materials* term; see **Shearing Strain**.

Shear Stress A *Physics* and a *Strength of Materials* term; see **Shearing Stress**.

Shear Thickening Fluid A *Fluid Mechanics* term; see **Dilatant Fluid**.

Shear Thinning Fluid A *Fluid Mechanics* term; see **Pseudoplastic Fluid**.

Shear A *Physics* term; when a body is acted upon by an external force tangential or parallel to its surface, it undergoes a change in its shape, and the body is said to have sheared.

Shearing Strain A *Strength of Materials* term; also called **Shear Strain** or **Angular Distortion** or **Shear Deformation**; a strain produced on account of change in shape of a body without any change in its volume by a shear force; an angle (in radian) through which a plane perpendicular to the fixed surface of a body gets turned under the effect of a shear force; calculated as a ratio of the displacement of a surface under a shear force to the perpendicular distance of the displaced surface form the fixed surface of a body; denoted by φ (phi).

Shearing Stress A *Physics* and a *Strength of Materials* term; also called **Shear Stress** or **Tangential Stress**; the stress produced by a shearing force that acts on the surface of a body; a ratio of the shearing force to the shearing area; produces change in shape of a body.

Shearing A *Manufacturing* term; a process which involves mechanical cutting along a straight line completely across a *strip, sheet* or *bar* without the formation of any chip or using any heating process; used to indicate most of the sheet metal cutting operations; a few shearing operations include *piercing, notching*, and *blanking*.

Sheet Metal Operations A *Manufacturing* term; the operations used to make useful products out of a metal sheet using shearing action by means of die-punch assembly; a few examples include *blanking, punching, lancing, bending, notching*, and slitting.

Sheet A form of material (metal, plastic, and others) having thickness less than 6 mm; used extensively in making *automobile* and *aircraft bodies*, as well as *appliances, food and beverages containers,* and *kitchen equipment*; see also **Plate**.

Shell Mould Casting A *Manufacturing* term related to *casting*; also called **Croning Process** or **C Process** named after its *German* inventor *Johannes Croning* (1886–1957); a variation of *sand casting* which uses a dimensionally more accurate *thin shell mould* made of a mixture of dry silica sand free from clay and phenolic resin; used to produce more accurate castings with improved surface finish as compared to sand casting; the typical products made by this method include *gear housings, cylinder heads*, and *connecting rods*.

Shewhart Cycle An *Operations Management* term; see **Deming Wheel**.

Shielded Metal Arc Welding A *Welding* term, abbreviated as **SMAW**; also called **Stick Electrode Welding** as the electrode used in this process looks like a *stick*; an *extensively used arc welding process* which uses a consumable coated electrode that also acts as a filler metal, and the coating material provides shielding atmosphere around the joint; finds major applications in the *fabrication of ships, bridges, pressure vessels, and structural members*.

SHM A *Physics* term; see **Simple Harmonic Motion**.

Shock Load A *Physics* term; see **Impact Load**.

Shock Wave A *Physics* term; a wave generated from loud noises similar to the produced from a *gun* or *explosion*; travels faster than the speed of sound.

Shoe Brake A *Theory of Machines* term; see **Block Brake**.

Shop Drawing A *Machine Drawing* term; the complete drawing of an object comprising a number of drawings required to facilitate the fabrication of all the parts of the object and their subsequent assembly into a complete product; usually includes both the *assembly drawing* and the *part drawing*.

Shop Floor Control An *Operations Management* term; also called **Production Activity Control**; a production control method of *scheduling* and *monitoring* of day-to-day production activities in a *job shop* to know the progress of the work and to take the appropriate corrective measures in case of any deviation; the components of shop floor control include *loading*, *sequencing*, *scheduling*, and *expediting* the jobs.

Shop Loading An *Operations Management* term; see **Loading**.

Shopping Goods An *Economics* term; a type of consumer goods which are purchased after giving serious thought and after making careful comparison between alternatives with respect to their attributes and features, rather than their brand; include infrequently bought items such as *cameras* and *refrigerators*; usually sold through fewer specialized outlets such as camera shops and mobile shops (as compared to convenience goods which are sold through many retail outlets); see also **Convenience Goods** and **Speciality Goods**.

Short Column A *Strength of Materials* term related to a *column*; a column of short length having its slenderness ratio less than 30; short columns can be subjected to maximum permissible compressive stress without buckling, but they fail by yielding; see also **Medium Column** and **Long Column**.

Shortage Cost An *Operations Management* term; see **Stockout Cost**.

Shortest Processing Time Rule An *Operations Management* term; see **SPT Rule**.

Short-Term Assets A *Financial Management* term; the assets which are to be considered for a short-term, say, one year or less; include the *current assets*; see also **Short-Term Liabilities**.

Short-Term Financial Management A *Financial Management* term; a process of managing the current assets and current liabilities; one of the financial manager's most important and time-consuming activities.

Short-Term Forecasts An *Operations Management* term related to *forecasting*; have a time frame of a *few days to three months*; these forecasts tend to be for narrow or specific entities, such as forecasts of demand for individual products or services; a few examples include decisions such as replenishing inventory or scheduling employees in the near term; the *time series analysis* is most frequently used for short-term forecasts; the *management judgement method* is also used for short-term forecasts, when past data regarding a product is not available; see also **Medium-Term Forecasts** and **Long-Term Forecasts**.

Short-Term Liabilities A *Financial Management* term; the liabilities which are to be considered for a short-term, say, one year or less; include the *current liabilities*; see also **Long-Term Liabilities.**

Shovel A *Manufacturing* term; a *foundry tool* used to prepare the moulding sand or to bring it from its site of preparation to the place of its use.

Shrill A *Physics* term; a *high frequency sound* which has a high pitch.

Shrinkage Allowance A *Casting* term; a type of pattern allowance provided in a sand mould pattern in the form of increased size over its actual size to take care of reduction in volume of the castings during their solidification; it is typically 2% for *steels*, 1% for *cast iron*, and 1.25% for *aluminium.*

Shrinkage A *Manufacturing* term related to *casting*; also called **Porosity**; a type of *casting defect* in which the size of casings reduces upon solidification as molten metal shrinks on cooling; occurs when the density of the metal in its solid state is greater than its density in liquid state.

Shroud A *Fluid Mechanics* term related to a *pump*; also called **Volute**; the pump casing which surrounds the impeller blades to increase blade stiffness.

SI Engine An *IC Engines* term; see **Petrol Engine**.

SI Unit A *Physics* term; the universally accepted international system of units; includes six basic units such as *metre* (m) for length, *kilogram* (kg) for mass, *second* (s) for time, *kelvin* (K) for temperature, *ampere* (A) for electric current, and *candela* (cd) for luminous intensity.

Side Riser A *Manufacturing* term related to *casting*; see **Blind Riser**.

Siderite A *Chemistry* term; an *iron ore* having the chemical formula $FeCO_3$; contains about 48% *iron.*

Sigma Bond A *Chemistry* term, abbreviated as **σ bond**; a type of *covalent bond* formed by the *overlapping of orbitals* along the same axis (like overlapping between *s–s*, *s-p* or *p-p* orbitals); *stronger than pi* (π) *bond*; see also **Pi Bond**.

Sigma A *Greek* letter denoted by σ; defines a *standard deviation* from a bell curve; also used to denote *normal stress.*

Silica Gel A *Chemistry* term; a white amorphous form (powder) of silica (SiO_2) having the chemical formula of $SiO_2.2H_2O$; prepared by the action of water on silicon tetrachloride or tetrafluoride ($SiCl_4$ or SiF_4); used as a *catalyst in petroleum industry* and as *an adsorbent in column chromatography*; also *used to protect merchandize from moisture during storage by placing small packets of dry silica gel (blue) in the packing boxes (turn pink after absorbing moisture).*

Silica A *Chemistry* term; also called **Silicon Dioxide (SiO_2)**; a *hard* and *inert* substance chemically obtained by heating silicon in oxygen; has a very high melting point due to its highly stable covalent-bonded three-dimensional network structure; occurs in nature in several amorphous and crystalline forms; *three important crystalline forms of silica* include *quartz, tridymite,* and *cristobalite,* of which *quartz is the most common form*; the *amorphous forms of silica* include *agate, jasper,* and *onyx*; also found in certain plants and animals, for example, *the hard portions of straw and bamboo contain silica,* the skeletons of certain marine organisms (e.g. *sponges*) consist largely of silica, and *claws of animals and finger nails also contain silica*; *the most widely used ceramic material* and an *essential ingredient in glasses* and many glass ceramics; silica-based materials are used in *thermal insulation, refractories, abrasives, fibre-reinforced composites, laboratory glassware,* etc.; also used to make *optical fibres for communications*; *powdered silica is used in tyres, paints, and many other applications.*

Silicate A *Chemistry* term; a *ceramic compound* with silica (SiO_2) as a major constituent.

Silicon Bronze A *Chemistry* term; a *copper-base* alloy containing 3% *silicon* (Si) and 1% *manganese* (Mn) along with *copper* (Cu); has *good corrosion resistance* and mechanical properties.

Silicon Dioxide (SiO_2) A *Chemistry* term; see **Silica**.

Silicon A *Chemistry* term; a *chemical element* having symbol *Si* with atomic number 14 and atomic weight 28; an *essential constituent of most of the non-living materials* including *rocks, hills, clays, soils,* and *ores*; occurs most abundantly, next only to oxygen, up to about 27.7% in the earth's crust as *silica* (sand and quartz of various forms such as opal and flint) and *silicates* [*felspar* ($KAlSi_2O_8$) and *Mica* ($KH_2Al(SiO_4)_3$]; also *the most abundant element, after oxygen, in human body, to the extent of about 18%, in combined state*; a *hard* and *brittle non-metallic solid* with a very high melting point (1410 °C); has same appearance (grey colour) as graphite and also *the electrical conductivity is of the same order as that of graphite*; exists in two allotropic forms: a brown amorphous powder and a grey crystalline mass; amorphous silicon is chemically more active than the crystalline variety; used as an alloying element for steel or iron to increase its resistance to attack by acids; its alloys such as silicon bronze and manganese-silicon bronze possess strength and hardness even greater than steel; *pure silicon is used as an important semiconductor to make transistors.*

Silver Coinage A *Materials Engineering* term; an alloy of *nickel* (Ni) and *silver* (Ag) containing 5% Ni; used in making *coinage.*

Silver A *Chemistry* term; a *pure-white, soft, highly ductile,* and *malleable noble metal* which has the specific gravity of 10.5 and the melting point of 690 °C; a good conductor of *heat* and *electricity*; finds use in making *coinage* and *jewellery* and in *electroplating.*

Similitude A *Fluid Mechanics* term; the study of predicting prototype conditions from model observations; useful when an analytical or numerical solution is not possible or when testing is not practical on a full-scale prototype.

SIMO Chart An *Operations Management* term related to *work study*; a *simultaneous motion* chart developed by *Frank Gilbreth* (1868–1934); finds use in *time study* to record simultaneously the motion of each hand of an operator along with their timings while performing a work in order to identify *unwanted, inefficient*, and *ineffective* motions which could be eliminated; used in micromotion study of *short cycle repetitive jobs* and high-order skill jobs like *component assembly, packaging, inspection,* and *repetitive use of jigs and fixtures.*

Simple System A *Thermodynamics* term; a system defined by only two properties.

Simple Beam A *Strength of Materials* term related to a *beam*; also called **Simply Supported Beam** or **Hinged Beam**; a statically determinate beam which is supported at its end points of which one end is hinge supported while the other end is roller supported; one of the most widely used beams in civil structures.

Simple Bending A *Strength of Materials* term; see **Pure Bending**.

Simple Curve A *Mathematics* term; also called **Open Curve**; a curve which does not cut or cross itself; a straight line is an example of a simple curve; see also **Closed Curve**.

Simple Gear Train A *Theory of Machines* term related to a *gear*; a gear train in which each gear shaft between the follower and the driver carries one gear only; see also **Compound Gear Train**.

Simple Harmonic Motion A *Physics* term, abbreviated as **SHM**; also called **Simple Harmonic Oscillation**; a harmonic motion of a *single frequency* and *constant amplitude*; the particle executing SHM moves back and forth (to-and-fro) between two extreme positions about an equilibrium position (the mean position) through a (restoring) force (directly proportional to the displacement, but opposite to it in direction); the acceleration of the particle is maximum at the two extreme positions and zero at the mean position; a few examples of SHM include the *backward and forward swing of a simple pendulum, the up and down motion of a weight hanging on a spring*, and *the twisting and untwisting of a body suspended by a wire*; can be mathematically expressed either by Eq. (1) or (2) as:

$$y = a\sin(\omega t) = a\sin\left(\frac{2\pi}{T}t\right) \tag{1}$$

$$\text{or}\quad y = a\cos(\omega t) = a\cos\left(\frac{2\pi}{T}t\right) \tag{2}$$

where

y = displacement of the particle from mean position at any instant t
a = maximum displacement or amplitude of displacement of the particle
ω = angular frequency $= 2\pi n = \frac{2\pi}{T}$
n = frequency of harmonic oscillation
T = time period of harmonic oscillation.

Simple Harmonic Oscillation A *Physics* term; see **Simple Harmonic Motion**.

Simple Pendulum A *Physics* term; a device consisting of a *heavy mass of negligible dimensions* (called *bob*) which is suspended vertically by means of an *inextensible* and *massless* string; when moved to sideways from its resting (defined by an angle $\theta \leq 4°$ with the vertical) and allowed to swing, it executes SHM; its time period (T) is given as:

$$T = \frac{1}{n} = 2\pi\sqrt{\frac{l}{g}}$$

where

n = frequency of oscillation
l = length of the string
g = acceleration due to gravity
π = a constant.

Simple Support A *Physics* and a *Strength of Materials* term; a support provided by a hinge (also called a pin) and a roller; see also **Hinge Support** and **Roller Support**.

Simple Truss A *Strength of Materials* term related to a *truss*; a truss supported at its end points where one end is hinge supported and the other end roller supported.

Simple Vortex A *Fluid Mechanics* term; see **Free Vortex**.

Simplex Method An *Operations Research* term; developed in 1963 by *George Bernard Dantzig* (1914–2005), an *American* mathematical scientist; a *computational technique* used to solve a *linear programming problem* involving two or more variables; simplex method is the most general and widely used method.

Simply Supported Beam A *Strength of Materials* term related to a *beam*; see **Simple Beam**.

Simulation A *Computer* term; an artificial environment that attempts to closely mirror an actual condition; used for decision-making under conditions of uncertainty like where mathematical simplification is not feasible or it is difficult to observe the actual environment or actual observation of a system is too expensive; with the advent of high-speed digital computers, it has become possible to find the economical and quick solutions of a number of real-life business problems using computer simulation.

Sine Bar A *Metrology* term; a high precision and the most accurate angle-measuring instrument used in conjunction with a set of *angle gauges*.

Single Minute Exchange of Die An *Operations Management* term, abbreviated as **SMED**; a technique evolved by *Shiego Shingo* (1909–1990); used in JIT to reduce the setup time of machine during changeover from one product to another to less than a minute; the phrase single minute is actually *single-digit minute* to indicate the changeover of less than ten minutes.

Single Piece Pattern A *Casting* term; see **Solid Pattern**.

Single Sampling A *Quality Control* term; the acceptance sampling of a lot based on a single sample; see also **Multiple Sampling**.

Single-Acting Cylinder A *Mechatronics* and an *IC Engine* term; a common type of linear actuator controlled by fluid pressure in one direction only and the return stroke occurs automatically; see also **Double-Acting Cylinder**.

Single-Acting Engine A *Thermodynamics* term related to an *IC engine*; a heat engine in which the working fluid acts on only one side of the piston; internal combustion (IC) engines for road transport are always single-acting; see also **Double-Acting Engine**.

Single-Factor Productivity A *Management* term; see **Partial Factor Productivity**.

Sink Temperature: A *Thermodynamics* term; see **Ambient Temperature**.

Sintered Carbide A *Materials Engineering* term; see **Cemented Carbide**.

Siphon A *Fluid Mechanics* term; a long bent pipe used to carry water from a reservoir at a higher elevation to a reservoir at a lower elevation using air pressure to maintain the flow, when the two reservoirs are separated by some obstruction, say a hill or any high level ground.

Six Sigma Approach An *Operations Management* term related to *quality*; abbreviated as **6σ**, where σ is *standard deviation* (a measure of variability around the mean of a *normal distribution*); a benchmarked standard for quality evolved by *Bill Smith* at Motorola in 1986; uses rigorous measurement to reduce process variation and eliminate defects; allows only 3.4 defects per million (ten lakh) parts or services produced; products or services are 99.99966% perfect; the previous quality standard *three sigma* (3σ) allows 2600 defects per million products or services produced.

Skeleton Pattern A *Casting* term; a type of pattern, usually made of *wood*; used to produce moulds which form the *outer frame of the castings*; useful for *large size castings* in small numbers.

Skew Bending A *Strength of Materials* term; see **Unsymmetrical Bending**.

Skew Gear A *Theory of Machines* term related to a *gear*; see **Spiral Gear**.

Skin Hardening A *Manufacturing* term related to *heat treatment*; see **Case Hardening**.

Skin-Dried Mould A *Casting* term; a sand mould whose upper surface (*skin*) is dried to a depth of about 25 mm or more using a *gas torch* or *heater*; used for making *large size castings* because of its *high strength.*

Slab A *Manufacturing* term; a type of *steel ingot* (oblong) whose thickness varies between 2 and 6 inches and width between 5 and 6 feet; see also **Billet** and **Bloom**.

Slack A *Project Management* term; see **Total Slack**.

Slag A *Chemistry* term; a product formed when the impurities present in an ore (called *gangue*) combine with a flux; being lighter in weight it floats over the surface of the molten metal from where it can be skimmed off from time to time.

Sleeve Coupling A *Design* term; see **Muff Coupling**.

Slender Member A *Strength of Materials* term; any structural member whose length is at least 5 times greater than its largest cross-sectional dimension.

Slenderness Ratio A *Strength of Materials* term related to a *column*; a ratio of the effective length of a column to its least radius of gyration; a parameter used as a classification criteria for columns; for example, short columns have slenderness ratio less than 30, long columns have slenderness ratio greater than 120 and medium columns have slenderness ratio lying between 30 and 120; smaller value of the slenderness ratio indicates higher crippling stress.

Sliding Friction A *Physics* term; the friction acting between two sliding surfaces of the bodies; see also **Rolling Friction**.

Slip A *Theory of Machines* and a *Materials Engineering* term; in *theory of machines* terms, an *undesirable phenomenon* which occurs in an *open belt drive* because of large difference in tensions between the tight and slack side of the belt; allows the belt to slide over the pulley causing a relative motion between belt and pulley; decreases the *velocity ratio*; see also **Whip**; in *materials engineering* terms, a process by which a *dislocation* (a linear defect) moves and causes a metallic material to deform; useful in understanding the mechanical behaviour of metals.

Slitting A *Manufacturing* term; a sheet metal operation which involves lengthwise cutting of coil or a metal sheet into narrower width.

Slotting A *Manufacturing* term; a sheet metal operation used to cut elongated holes or slots.

Slug A *Fluid Mechanics* term; the *unit of mass* in the FPS system of units; equivalent to $lbf.s^2/ft$ and 1 slug is equal to 32.2 lb.

Sluice Gate A *Fluid Mechanics* term; a common type of *underflow gate* which is in the form of a vertical thin plate having a sharp edge at its bottom; used to control the liquid flow rate in *a wide open channel like a dam*; see also **Drum Gate**.

Slush Casting A *Manufacturing* term related to *casting*; a *special form of permanent mould casting*; used to produce *hollow thin-walled castings* (without using any core) which are usually painted or finished to give them good looks of expensive metals; frequently used to produce *ornamental* and *decorative* objects such as *lamp bases* and *stems*, *statuettes*, and *toys* from low melting point materials like *zinc*, *tin*, and *lead* alloys.

SMAW A *Welding* term; see **Shielded Metal Arc Welding**.

SMED An *Operations Management* term; see **Single Minute Exchange of Die**.

Smith Forging A *Manufacturing* term; also called **Hammer Forging**; the most basic and the oldest forging process practised from the ancient times by a village blacksmith; used to shape a heated workpiece by striking it repeatedly with a hammer in a closed or open die.

Smog An *Air Pollution* term; a mixture of *smoke* and *fog*.

Smoke An *Air Pollution* term; the small gas-borne particles which result from incomplete combustion of fuels; predominantly consist of carbon and other combustible material; see also **Smog** and **Smust**.

Smoothing Constant An *Operations Management* term related to *forecasting*; a numerical parameter which finds use in the *exponential smoothing method* of demand forecasting; reflects the weightage assigned to the most recent demand data; denoted by α (alpha) and assumes a value between 1 and 0; zero (0) value is a case of '*infinite smoothing*' which does not reflect the most recent demand at all, and it indicates that current period forecast equals to the last period forecast and the forecast once set never changes; as α approaches zero, the forecast will react and adjust more slowly to differences between the actual demand and the forecasted demand; the unit (1) value is a case of '*no smoothing*' which implies that the current period forecast equals the last period actual and indicates the consideration of only the most recent data and nothing else; the higher value of α indicates that the forecast is more sensitive to changes in recent demand; for most industrial applications, the value of α lies between 0.1 and 0.2.

Smust An *Air Pollution* term; a mixture of *smoke* and *dust*.

S–N Curve A *Strength of Materials* term; a curve associated with a fatigue test which is plotted between stress amplitude (S) and the number of stress cycles (N) of alternate tension and compression required to produce a material's fatigue failure.

Soaking A *Manufacturing* term related to *heat treatment*; a process of prolonged heating of a metal at a selected temperature.

Social Need A *Management* term related to *motivation*; one of the five levels of human needs proposed by *Abraham Harold Maslow* (1908–1970) in his '*Need-Hierarchy Theory*' of motivation; represents the *need to belong to a group* (relationships with co-workers and supervisors), acceptance (recognition) by one's peers as their equal, and exchange of affection.

Soda Glass A *Chemistry* term; see **Ordinary Glass.**

Sodium Tetraborate Decahydrate A *Chemistry* term; see **Borax.**

Soft Automation An *Automation* term; also called **Flexible Automation**; a process of using computer-driven automated machines which is easily reprogrammed to *produce a variety of products*; see also **Hard Automation.**

Soft Glass A *Chemistry* term; see **Ordinary Glass.**

Solar Battery A *Physics* term; *lead-acid* or *nickel–cadmium* battery used in solar electric systems to store the electricity generated during daylight hours for later use.

Solar Cell A *Physics* term; also called **Photocell** or **Photovoltaic Cell** or **PV Cell**; a semiconducting device which generates a voltage as it absorbs photons of light; directly converts sunlight into electricity without any intermediate mechanism; relatively *expensive*, as being made from processed *silicon*, just like *integrated circuits* and *computer microprocessors*; their higher costs have always been a *hindrance* in making them an alternative source of electricity generation; mostly confined to *space satellites*, *hand-held calculators*, *roadside emergency telephones*, *remote houses*, and other situations, where electricity is not otherwise available.

Solar Constant A *Physics* term; the amount of solar radiation which reaches the earth's atmosphere; equals to 1367 W/m^2.

Solar Energy The energy associated with the sun; a type of *thermal energy* as it produces *heat* and *temperature*; a *renewable* and a huge source of cheap energy which is *inexhaustible*; on a clear day when the sun is directly overhead, roughly one kilowatt of sunlight strikes each square metre of the ground; every minute the sun provides the earth with more energy than the entire world consumes in a year; environment-friendly, but suffers from non-uniform availability due to seasonal change and diurnal limitation.

Solder A *Materials Engineering* term; an alloy of *tin* (Sn) and *lead* (s) consisting of 67% (or 50%) Sn and 33% (or 50%) Pb; used in *soldering* to join light electrical components.

Soldering A *Joining* term; a process of joining metals and alloys without melting them; uses a fusible alloy, called *solder* (a *tin–lead alloy* which melts below 450 °C) which *acts as a filler metal* that reaches the gap between the two parts being joined by *capillary action*; the strength of a soldered joint is primarily due to formation of metallic bond, although adhesion and mechanical attachment also contribute; cleaning of the joining surfaces before soldering (using a flux) is essential (to make them free from impurities such as *oxide*, *dirt*, *oil* or *grease*); the various soldering methods include *iron*, *torch*, *dip*, *wave*, *induction*, *resistance*, *ultrasonic*, and *furnace* (oven) soldering; see also **Brazing**.

Solenoid A *Mechatronics* term; an electromechanical device which uses the principles of electromagnetism to control the motion of an actuator.

Solid Carbon Dioxide A *Chemistry* term, abbreviated as **Solid CO_2**; commonly called **Dry Ice** or **Cardice**; a soft, white, snow like substance with specific gravity 1.53 and melting point − 56.6 °C at 5.2 atmospheres of pressure; a *commercial refrigerant* sold under the trade name *Drikold*, which is used as a substitute of ice in cold drinks and in the preparation of ice cream; when mixed with ether or acetone, it provides freezing mixtures with exceedingly low temperatures (− 77 °C when mixed with acetone and − 108 °C when mixed with ether) which find use in making cold baths in the laboratory; finds extensive use as a *coolant* for *preserving perishable food-stuffs* during their transport due to its ability to provide cold as well as the inert atmosphere which helps in killing the undesirable bacteria, that otherwise may spoil the food; additionally used for curing local burns and in hospitals for surgical operations of *sores*; also used in *artificial rain-making* by spraying *small pellets of dry ice* from aeroplanes over clouds to produce *condensation*.

Solid CO_2 A *Chemistry* term; see **Solid Carbon Dioxide**.

Solid Film Lubrication A *Fluid Mechanics* term; a type of lubrication in which the load-carrying surfaces are separated by a *solid-film lubricant* like *graphite* or *molybdenum disulphide*; see also **Boundary Lubrication**.

Solid Friction A *Physics* term; see **Dry Friction**.

Solid Injection Cycle A *Thermodynamics* term related to *air standard cycle*; see **Dual Cycle**.

Solid Mechanics See **Strength of Materials**.

Solid Pattern A *Casting* term; also called **Single Piece Pattern**; the simplest of all the patterns; used to produce moulds for making simple castings, often made in one piece and in smaller numbers; usually made of wood.

Solid Solution A *Materials Engineering* term; the atomic-scale intermixing of more than one atomic species in the solid state.

Solid-State Welding A *Welding* term; a welding process performed in the solid state of materials at room temperature or elevated temperatures with the application of pressure; a few examples include *cold welding*, *friction welding*, *ultrasonic welding*, and *resistance welding*.

Solid A *Physics* term; a substance which can resist deformation when subjected to shear forces; has the tendency to regain its initial state of equilibrium on the removal of shear forces; a state of matter which has a *definite shape* and a *definite volume*; see also **Liquid** and **Gas**.

Solidus A *Materials Engineering* term; the line below which only one or more solid phases are present in a *phase diagram*; see also **Liquidus**.

Solubility A *Chemistry* term; the maximum amount of solute which can be dissolved in a solvent (liquid); expressed as:

$$\text{Solubility} = \frac{\text{Maximum mass of the solute}}{\text{Mass of the solvent}}$$
$$= \frac{\text{Maximum moles of the solute}}{\text{Moles of the solvent}}$$

SOM See **Strength of Materials**.

Sonic Flow A *Fluid Mechanics* term; a fluid flow in which the fluid velocity equals to the velocity of sound in that fluid; the *Mach number* (*M*) for a sonic flow is equal to 1.

Sonic Velocity A *Fluid Mechanics* term; also called **Acoustic Velocity**; the speed of sound.

Soot A *Power Plant* term; the *fine carbon particles* which contain small, varying amounts of complex hydrocarbons and mineral matter obtained from the combustion of fossil fuels.

S-OS Analysis An *Operations Management* term related to *inventory*; a *selective inventory control technique* which classifies the inventories based on their use in a specific season into two categories: *S* (*seasonal*) items and *OS* (*off-seasonal*) items; this analysis helps in determining suitable procurement strategies for seasonal items.

Sound Wave A *Physics* term; see **Audible Wave**.

Space Truss A *Strength of Materials* term related to a *truss*; a truss in which forces through the members and the external loads acting at the joints lie in different planes; a few examples include *shear legs, transmission line,* and *communication towers*; see also **Plane Truss**.

Span of Control A *Management* term; also called **Span of Management**; the number of people who are reporting to a manager in an organization; a manager with a small span of control can maintain close control over workers and stay in contact with daily operations, whereas a manager with a wide span of control loses control over subordinates and cannot hold them accountable for their actions.

Span of Management A *Management* term; see **Span of Control**.

Spark Plug An *IC Engine* term; a part of a petrol engine used to produce a spark to initiate combustion of fuel in the cylinder.

Spatter A *Welding* term; the *metal particles* expelled from the weld metal around the arc in different directions during a welding operation; found in *gas* and *arc welding*.

SPC A *Quality Control* term; see **Statistical Quality Control.**

Special Cause Variation A *Statistics* term; see **Assignable Variation.**

Speciality Goods An *Economics* term; a type of consumer goods of unique features or characteristics which are considered prized and branded products; involve prestige when possessed; often sold through a few prestigious outlets rather than being widely distributed; for example, a *Mercedes-Benz car* or a *Harley-Davidson motorcycle*; see also **Shopping Goods** and **Convenience Goods.**

Specific Enthalpy A *Thermodynamics* term; the enthalpy per unit mass of a working fluid; the SI unit is J/kg or kJ/kg.

Specific Entropy A *Thermodynamics* term; the entropy per unit mass of a working fluid; the SI unit is *J/kg-K* or *kJ/kg-K*.

Specific Fuel Consumption An *IC Engine* term, abbreviated as **sfc**; also called **Brake Specific Fuel Consumption**; a parameter used to measure the fuel consumption as a flow rate; expressed in fuel flow rate per unit power output; a measure of how efficiently an engine is using the fuel supply to produce work; lower values of sfc is desirable; the typical best values of sfc for SI engines are 75 μg/J (270 g/kW); for CI engines, the best sfc values are lower; for the larger engines, the sfc values can go below 55 μg/J (200 g/kW).

Specific Gravity A *Fluid Mechanics* term; a *dimensionless* quantity; a ratio of the specific weight (or the *mass density*) of a substance to the specific weight of some standard substance at a specified temperature; for liquids, water at 4 °C is considered as the standard substance, and for gases, the standard substance is air at 0 °C or hydrogen at 0 °C; the specific gravity of water is 1.00 and that of mercury is 13.6, which indicates that mercury is 13.6 times heavier than water.

Specific Heat at Constant Pressure A *Physics* term; the rate of change of enthalpy of a substance with respect to its temperature at constant pressure; denoted by C_p; given as:

$$C_p = \left(\frac{\partial h}{\partial T}\right)p$$

where

∂h = change in enthalpy
∂T = change in temperature
p = pressure.

Since the enthalpy of an ideal gas is a function of temperature only, hence the C_p of an ideal gas is also a function of *temperature* alone; the SI unit is J/kg-K; see also **Specific Heat at Constant Volume.**

Specific Heat at Constant Volume A *Physics* term; the rate of change of internal energy of a substance with respect to its temperature at constant volume, denoted by C_v; given as:

$$C_v = \left(\frac{\partial u}{\partial T}\right)v$$

where

∂u = change in internal energy
∂T = change in temperature
v = volume condition.

Since the internal energy of an ideal gas is a function of temperature alone, hence the C_v of an ideal gas is also a function of *temperature* alone; the SI unit is J/kg-K; see also **Specific Heat at Constant Pressure**.

Specific Heat A *Physics* term; the quantity of heat required to raise the temperature of unit mass of a body through unit rise in its temperature; the SI unit is J/kg-K; see also **Specific Heat at Constant Pressure** and **Specific Heat at Constant Volume**.

Specific Humidity A *Thermodynamics* term related to *psychrometry*; also called **Absolute Humidity** or **Humidity Ratio**; a *psychrometric* property; a ratio of the *mass of water vapour* to the *mass of dry air* in a given volume of the mixture; denoted by ω, given as:

$$\omega = \frac{m_v}{m_a} = 0.622\frac{p_v}{p - p_v} = 0.622\frac{p_v}{p_a}$$

where

m_v = mass of the water vapour
m_a = mass of the dry air
p_v = partial pressure of the water vapour
p_a = partial pressure of the dry air
p = total atmospheric pressure; see also **Relative Humidity**.

Specific Internal Energy A *Thermodynamics* term; the internal energy per unit mass of a body; the SI unit is J/kg.

Specific Property A *Thermodynamics* term; an *intensive* property which is independent of mass of a system; a few examples include *specific volume*, *specific enthalpy*, and *specific internal energy*; see also **Molar Specific Property**.

Specific Resistance An *Electrical* term; see **Resistivity**.

Specific Speed A *Fluid Mechanics* term; also called **Type Number** as its value represents a number used to describe the type of a *fluid machine*; a *nondimensional number* which represents the speed of a geometrically similar fluid machine operating under a unit head and handling a unit discharge or a unit power at the optimum operating conditions; classified into *discharged-based* and *power-based* specific speeds (nearly equal); discharged-based specific speed is used in case of a pump, and power-based specific speed is used in case of a turbine, and are expressed as:

$$N_s = \frac{N\sqrt{Q}}{(gH)3/4} \quad \text{(based on discharge, } Q\text{)}$$

$$N_s = \frac{N\sqrt{P/\rho}}{(gH)5/4} \quad \text{(based on power, } P\text{)}$$

where

H = head of the fluid across the machine (m)
ρ = density of the fluid (kg/m^3)
P = power for the machine (W)
Q = discharge through the machine (m^3/s)
N = speed of rotation (rad/s).

Specific Volume A *Fluid Mechanics* and *Thermodynamics* term; the volume of a body per unit its mass; the reciprocal of mass density; the SI unit is m^3/kg; has the dimension of [$M^{-1}L^3$]; see also **Molar Specific Volume**.

Specific Weight A *Fluid Mechanics* and a *Thermodynamics* term; the weight of a substance per unit its volume; equals to the product of mass density (ρ) and the acceleration due to gravity (*g*); denoted by ϒ (gamma); the SI unit is N/m^3.

Specification of a File A *Manufacturing* term; the descriptions indicating the *size*, *cut*, and *shape* of a file; the size of a file indicates its length between point and heel which may vary between 100 and 200 mm for *fine work*, and between 200 and 450 mm for *rough work*; the cut of a file implies single or double cut, which is further divided into *rough, bastard, smooth, dead smooth,* or *super smooth* category; the shape of a file indicates its cross section which may be *square, triangular* or *round*.

Specification of a Lathe A *Manufacturing* term; a lathe can be specified by any one of the parameters: (a) the *maximum distance between the live and dead centres*, which also represents the *maximum length of the work* which can be handled by the lathe (sometimes called length of the bed) (b) *swing*, which also represents the maximum diameter of the work which can be machined on the lathe.

Specification of a Milling Machine A *Manufacturing* term; a milling machine can be specified by any one of the parameters: (a) the length and breadth of the work table (b) the maximum length of the longitudinal, cross, and vertical travel of the work table.

Specification A *Design* term; a detailed description of a material, part or product including parameters like *viscosity, surface finish, pH rating*, and *physical dimensions.*

Speed of Sound A *Physics* term; the speed at which sound travels in a medium; the speed of sound in air at STP is about 340 m/s.

Speed A *Physics* term; a ratio of the total distance covered by a moving body or its total path length to the corresponding time taken by the body; a *scalar quantity*; can be zero or positive, but never negative; the SI unit is metre/sec (m/s); see also **Velocity**.

Spherometer A *Physics* term; an instrument used to measure either very small thickness or radius of curvature of a spherical surface; its least count is 0.001 cm.

Spin Ratio A *Fluid Mechanics* term; a parameter related to the spin of a smooth sphere in a fluid; a ratio of the surface speed to the free stream flow speed.

Spinning A *Manufacturing* term; a cold metal working process used to produce *axisymmetric parts* by pressing a flat sheet metal blank against a rotating shaped-mandrel using a spinning tool; resembles to the process of forming clay on a potter's wheel; a few typical spinned products include *lamp reflectors, cooking utensils, bowels*, and *bells of musical instruments*; see also **Shear Spinning**.

Spiral Gear A *Theory of Machines* term related to a *gear*; also called **Skew Gear**; a type of toothed gear used for transmitting motion or power between *two non-parallel* and *non-intersecting shafts.*

Spiral Spring A *Strength of Materials* term related to a *spring*; a spring made of a flat strip of rectangular section wound in the form of a spiral; used to produce torsion on an axle to which one of its ends (inner one) is attached.

Splined Shaft A *Design* term; a shaft over which *splines* are formed; see also **Splines**.

Splines A *Design* term; the stubby gear teeth formed by cutting equi-spaced grooves of uniform cross section on the surface of a shaft (called *a splined shaft*) which fit into the corresponding recesses in the hub of a load-transmitting component; *typically used to transfer high torques.*

Split Pattern A *Casting* term; a type of pattern used to produce moulds by splitting the pattern in two parts along the parting line of the moulds to facilitate the pattern's withdrawal from the moulds—one-half of the pattern is meant for the cope part of the mould, and the other half of the pattern is used for the drag part of the mould.

Spot Welding A *Welding* term; also called **Resistance Spot Welding**; the simplest and the most commonly employed resistance welding in which a joint is produced at a very small area (called *spot*) of the work parts held together under pressure; used as a replacement for *riveting.*

Spray Transfer A *Welding* term; a *mode of metal transfer* in *gas metal arc welding* (GMAW) in which the consumable electrode is propelled *axially* across the arc in *small droplets*; see also **Globular Transfer**.

Spreadsheet An *Operations Management* term; a table consisting of rows and columns in which numbers and text are input to organize information and perform mathematical analysis.

Spring Constant A *Design* term related to a *spring*; also called **Spring Stiffness**; the force required to produce unit deflection in a spring; the SI unit is N/m.

Spring Stiffness A *Design* term related to a *spring*; see **Spring Constant**.

Spring Work A *Physics* term; the work produced by changes in length (either expansion or compression) of a spring by the application of a force on the spring; equals to the product of force applied and the change in length of the spring.

Spring A *Strength of Materials* term; an elastic body or a device used to absorb or store (strain) energy and release it when required; has wide applications such as in *brakes* and *clutches* (to apply force and control motion), *weighing machines*, *mechanical clocks*, *shock absorbers*, *reciprocating mechanisms*, and other devices; also used to cushion *shocks*; classified into *helical*, *leaf*, *flat*, and *spiral springs*.

Sprocket Wheel A *Chain Drive* term; also called simply **Sprocket**; a wheel over which a chain runs and has projecting teeth which get fitted into the suitable recess in the links of the chain; used to transfer rotational speed and torque from an input shaft to an output shaft in a mechanical system.

Sprocket A *Chain Drive* term; see **Sprocket Wheel**.

Sprue A *Manufacturing* term related to *casting*; also called **Downsprue**; a part of the *gating system*; a *vertical tapered channel* connected to the pouring basin through which molten metal flows in the downward direction; see also **Runner**.

SPT Rule An *Operations Management* term; one of the priority rules for sequencing jobs which stands for *shortest processing time*; jobs with the shortest processing times are assigned first; the best technique for minimizing job flow and minimizing the average number of jobs in the system; on the demerit side, long-duration jobs may be continuously pushed back in priority in favour of short-duration jobs.

Spur Gear A *Theory of Machines* term related to a *gear*; the simplest type of toothed gear used to transmit motion or power between *two parallel shafts*; the spur gear has straight teeth with face parallel to the shaft on which the gear is mounted; the smaller (driving) gear in a spur gearset is called the *pinion*, and the other (driven) is simply called the *gear*; the engagement and disengagement of the gear teeth take place along the full width of each tooth suddenly that *produces more noise and vibration* than other types of gears.

SQC A *Quality Control* term; see **Statistical Quality Control**.

Square File A *Manufacturing* term; one of the types of a file used in the fitting shop which has a *square cross section*; consists of double cut on its both faces and edges; used in the filing of *square corners* and *slots*.

Square A *Mathematics* term; a parallelogram whose all the sides are equal and each angle is a right angle, that is, equal to 90°; see also **Parallelogram**.

Squeeze Casting A *Manufacturing* term related to *casting*; also called **Liquid–Metal Forging**; a casting process involving both *casting* and *forging* principles; the molten metal poured in the lower half of the die is pressed gradually through the upper half of the die before its complete solidification, thus reducing a less pressure as compared to a conventional forging process; the cast parts have improved mechanical properties because of less chance of gas or shrinkage porosity; used to produce intricate shapes from both *ferrous* and *nonferrous* alloys; the typical products made by this method include *automotive wheels* and *mortar bodies* (*a short-barrelled cannon*).

Stability The ability of an object to retain its equilibrium configuration.

Stable Equilibrium A *Physics* term; a type of *translational static equilibrium* which ensures a body to regain its equilibrium position after being slightly displaced and released; for example, a table or chair lying on the ground; see also **Unstable Equilibrium** and **Neutral Equilibrium**.

Stack A *Power Plant* term; a *vertical duct* used for the discharge of products of combustion to the atmosphere.

Staff Role A *Management* term; a manager who is in charge of a specific organizational function such as *sales* or *R&D*; see also **Line Role**.

Stagnation Point A *Fluid Mechanics* term; a point where the velocity of a fluid is reduced to zero and the pressure becomes maximum, called the stagnation pressure.

Stagnation Pressure A *Fluid Mechanics* term; the sum of the *static* and *dynamic* pressures of a fluid; expressed as:

$$P_{\text{stag}} = P + \rho v^2/2$$

P = static pressure of the fluid
ρ = density of the fluid
v = velocity of the fluid.

The pressure of a fluid at a point (called the *stagnation point*), where the fluid is brought to a complete rest *isentropically*; see also **Static Pressure**, **Dynamic Pressure,** and **Hydrostatic Pressure**.

Stainless Steel A *Materials Engineering* term; an important alloy steel commonly termed as *chromium-nickel steel*; contains chiefly *iron* and *carbon* along with *chromium* (usually above 15%); a widely used material which is tough and strong, highly resistant to *corrosion* (because of chromium), *high temperatures*, *oxidation*, and *scaling*; finds use in making *kitchen equipment*, *cutlery*, and *surgical instruments*.

Stamping A *Manufacturing* term related to *forming*; a group of punch-press operations usually performed in cold states such as *blanking*, *coining*, and *forming*.

Standard Atmospheric Condition A *Physics* term; a pressure of 101.3 kPa and a temperature of 15 °C.

Standard Deviation A *Statistics* term; the most frequently used parameter for characterizing both populations and samples; a value which identifies how close the values in a data set are to the mean value of that set; indicates the width of the normal distribution (also called Gaussian distribution); equals to the square root of the variance; denoted by σ (sigma); the smaller value of σ indicates the data are more precise; calculated by squaring the deviation from the mean (eliminating signs), adding them, finding their arithmetic mean, and then taking the square root of the result.

Standard Time A *Time Study* term; the sum of the normal time and the allowance time for a job; see also **Normal Time** and **Allowance Time**.

Standard Trusses A *Strength of Materials* term related to a *truss*; the most widely used trusses; include *Warren truss*, *Pratt truss*, *Howe truss*, *K-truss*, and *fink truss*.

Standard Wire Gauge A *Metrology* term, abbreviated as **SWG**; a round-shaped sheet metal tool used to measure the thickness of a metal sheet or the diameter of a wire using gauge number; the higher gauge number indicates lower thickness of the sheet.

Standard A *Metrology* term; a set of specifications for parts, materials, or processes used to achieve uniformity, efficiency, and a specified quality; limits the scope of variations which can arise from the arbitrary creation of a part, material, or process; see also **Code**.

Standardization A *Design* term; a design activity used to reduce variety among a group of products or parts.

State of a Matter A *Physics* term; a physical condition in which a matter exists; the three important states of a matter include *solid*, *liquid*, and *gas*; the change of state is often accompanied by either absorption or evolution of heat; for example, a solid can be converted into a liquid or a liquid into a gas by absorption of heat.

State A *Thermodynamics* term; a condition of a material defined in terms of its *pressure*, *temperature*, and *composition*.

Static Equilibrium A *Physics* term; also called **Translational Static Equilibrium**; the equilibrium connected to a rigid body at rest; classified into *stable equilibrium, unstable equilibrium*, and *neutral equilibrium*; see also **Dynamic Equilibrium**.

Static Friction A *Physics* term; the friction occurring between two surfaces of the bodies at rest due to impending motion; its maximum value is called the *limiting friction.*

Static Load A *Physics* term; a load which remains constant with time; also called **Gradually Applied Load** as it is applied slowly on a member so that it increases from zero to its maximum value, and thereafter it remains constant; causes no vibrational or dynamic effects in a member; see also **Dynamic Load**.

Static Pressure A *Fluid Mechanics* term; the pressure of a fluid which does not incorporate any dynamic effects; the actual thermodynamic pressure of a fluid used in property tables; see also **Dynamic Pressure, Hydrostatic Pressure** and **Stagnation Pressure**.

Statically Determinate Beam A *Strength of Materials* term related to a *beam*; a beam involving only three unknown reactions which can be determined using equations of statics; simple, cantilever, and overhanging beams are the examples of *statically determinate beams*; see also **Statically Indeterminate Beam**.

Statically Indeterminate Beam A *Strength of Materials* term related to a *beam*; a beam involving more than three unknown reactions that cannot be determined using equations of statics; calculation of unknown reactions of these beams include the consideration of their properties with regard to their resistance to bending; continuous, fixed, and propped cantilever beams are the examples of *statically indeterminate beams*; see also **Statically Determinate Beam**.

Statics A *Physics* term; a branch of mechanics which deals with the action of forces on a body at rest; see also **Dynamics**.

Stationary Wave A *Physics* term; a wave produced by the superposition of two waves of same frequency and same amplitude which travel with same velocity in opposite directions.

Statistical Process Control A *Quality Control* term; see **Statistical Quality Control**.

Statistical Quality Control A *Quality Control* term, abbreviated as **SQC**; also called **Statistical Process Control**, abbreviated as **SPC**; a statistical method of measuring and analysing deviations in manufactured products using various control charts and distribution curves; finds extensive applications in industries such as *metal casting, forging*, and other *metal working units.*

Statistics The collection, analysis, and interpretation of quantitative data; finds extensive applications in engineering.

Steady Flow A *Fluid Mechanics* term; a fluid flow is said to be steady, when its physical properties like *velocity*, *pressure* or *density* does not change with time at any point in the flow field; also called **Streamlined Flow** as the path of fluid particles is called *streamline* and the tangent at any point on the streamline gives the direction of fluid flow at that point; examples of steady flow include *constant discharge in a pipe or through an open channel*; see also **Unsteady Flow**.

Steady-State Condition A *Fluid Mechanics* and a *Thermodynamics* term; a physical condition in which any parameter of a body does not change with time; see also **Unsteady-State Condition**.

Steady-Flow Process A *Fluid Mechanics* and a *Thermodynamics* term; a process during which a fluid flows through a control volume steadily, that is, the fluid properties do not change with time.

Steam Coal A *Chemistry* term; a type of coal which contains 93% of *carbon*.

Steam Gauge A *Power Plant* term related to a *boiler*; see **Pressure Gauge**.

Steam Generator A *Power Plant* term; see **Boiler**.

Steam Point A *Thermodynamics* term; the boiling temperature of water; a state of equilibrium between pure water and water vapour (steam) at a pressure of one atmosphere (101 kPa); equals to 100 °C or 212 °F; see also **Ice Point**.

Steam Separator A *Power Plant* term related to a *boiler*; a *boiler accessory* used to separate water particles from the steam before it enters an engine.

Steam Stop Valve A *Power Plant* term related to a *boiler*; a *boiler mounting* used to regulate the steam flow from the boiler to an engine and stop its supply when not required; when it changes the direction of steam flow in a perpendicular direction, it is called a *junction valve*.

Steam Table A *Thermodynamics* term; a table which shows the *properties of steam* in its *saturated* and *supersaturated* states, such as *specific volume*, *internal energy*, *enthalpy*, and *entropy* at a given temperature and pressure.

Steam Turbine A *Thermodynamics* term; a work-producing device (*turbomachine*) which converts the heat energy of steam (the working fluid) into work (mechanical energy of a rotating shaft), which is further used in generating electricity in coal or nuclear power plants; see also **Gas Turbine** and **Hydraulic Turbine**.

Steam A *Thermodynamics* term; the gaseous phase of water which is used as a working substance for heat engines; generated in a boiler at *constant pressure*; available in *wet*, *dry*, *superheated* or *supersaturated* conditions.

Steel A *Materials Engineering* term; a ferrous alloy of *iron* (Fe) and *carbon* (C) with carbon content varying between 0.4 and 1.5%; obtained by removing the impurities from pig iron using *limestone* and then adding certain alloying elements (*chromium, nickel, manganese, silicon, molybdenum,* and *tungsten*) in predetermined amounts to arrive at the properties desired in the finished metal; higher carbon content increases its *hardness, strength,* and *wear resistance,* but at the same time reduces its *ductility*, weldability, and toughness; *offers good mechanical properties over cast iron,* for example, *steel is two to three times stiffer, three to six times stronger in tension* and *about four times fatigue resistant,* and *can withstand heavy impact*; the most commonly used engineering material; mostly made by *continuous casting*; also made form *cast ingots*; classified into *stainless steel, cobalt steel, nickel steel,* and others; available mainly in the form of *plates, sheet, bars, tubes, castings,* and *forgings.*

Stefan-Boltzmann Constant A *Heat Transfer* term; a constant of proportionality which appears in the expression for emissive power of a body; denoted by σ (sigma); equals $5.67 \times 10^{-8}\ \text{Wm}^{-2}\ \text{K}^{-4}$.

Stefan-Boltzmann Law A *Heat Transfer* term; a *radiation law* named after a *Slovenian* physicist *Josef Stefan* (1835–1893) and an *Austrian* physicist *Ludwig Boltzmann* (1844–1906) who proposed it in 1884; states that the rate of radiation emitted from the surface of a body varies directly proportional to the fourth power of its absolute temperature; given as:

$$Q = \sigma.A.T^4$$

where

Q = rate of radiation emitted
σ = Stefan-Boltzmann Constant
= $5.67 \times 10^{-8}\ \text{Wm}^{-2}\ \text{K}^{-4}$
A = surface area
T = absolute surface temperature.

Stellite A *Materials Engineering* term; also called **Cast Cobalt Alloy** as it is produced by *casting*; an alloy of *cobalt* (Co), *tungsten* (W), *chromium* (Cr) and *molybdenum* (Mo) consisting of 55% Co, 25% W, 15% Cr and 5% Mo; *cobalt* is the main constituent besides *chromium, tungsten,* and *carbon*; designed for *wear resistance*; finds use in making *surgical instruments* and *cutting tools,* which can be operated at high cutting speeds without affecting their hardness at elevated temperatures.

Step-Down Transformer An *Electrical* term; a type of transformer which decreases the input voltage; see also **Step-Up Transformer**.

Step-Growth Polymerization A *Chemistry* term; see **Condensation Polymerization**.

Stepped Shaft A *Design* term; a shaft whose cross section changes (either increases or decreases) abruptly at any point along its length; see also **Tapered Shaft**.

Step-Up Transformer An *Electrical* term; a type of transformer which increases the input voltage; see also **Step-Down Transformer**.

Stick Electrode Welding A *Welding* term; see **Shielded Metal Arc Welding (SMAW)**.

Stiffness Test A *Strength of Materials* term; a *destructive* test used to know the modulus of elasticity (E) of a solid material; higher value of E is indicative of higher stiffness of a material; generally a harder material has higher stiffness.

Stiffness A *Strength of Materials* term; a *mechanical property* of a material which enables it to absorb shock or strain energy produced by an external load without undergoing permanent deformation; cast iron has higher stiffness than diamond; in *spring* terms, the load required to produce unit deflection in a spring; the SI unit is N/m.

Stirling Cycle A *Thermodynamics* term related to *air standard cycle*; named after *Robert Stirling* (1790–1878), a *Scottish* engineer who developed it in 1816; a gas power thermodynamic cycle which uses gas (hot air) as the working fluid; a closed cycle which consists of *two reversible isotherms* (constant temperature process) and *two reversible isochors* (constant volume process); heat addition and heat rejection take place at *constant volumes*, but compression and expansion processes take place at constant temperatures; a regenerative Stirling cycle has the same thermal efficiency as the Carnot cycle; see also **Ericsson Cycle**.

Stockholder A *Management* term; the people and groups who have an interest, claim or stake in an organization in every aspect of its functioning; see also **Shareholder**.

Stockless Production An *Operations Management* term; see **JIT Manufacturing**.

Stockout Cost An *Operations Management* term; also called **Shortage Cost** or **Backorder Cost** or **Lost Sale** due to the loss of goodwill of customers; the cost associated with demand when the stocks have already been depleted.

Stoichiometric Air An *IC Engine* term; also called **Theoretical Air**; the minimum amount of air (O_2) required for the complete combustion of a fuel.

Stokes Law A *Fluid Mechanics* term named after the *British* mathematician and physicist *George Gabriel Stokes* (1819–1903) who proposed it; used to find the frictional force, also called the *drag force*, on spherical objects moving through a viscous fluid having negligible effects of Reynolds number; mathematically given as:

$F = 6\,\pi\, r\, \mu\, V$ (r = radius of the sphere, μ = dynamic viscosity of the fluid, V = terminal velocity of the object through the fluid or the flow velocity relative to the object).

Stop Watch A *Physics* term; a time measuring instrument used to measure duration of an event or time interval between two events.

Stored Energy A *Chemistry* term; the energy contained or stored in various masses such as *batteries*, *flywheels*, and *tides*; a few examples include *potential energy*, *kinetic energy*, and *internal energy*; see also **Energy in Transition**.

STP A *Physics* term; the short form of *standard temperature and pressure*; indicates the standard atmospheric temperature of 15 °C or 288.15 K, a pressure of 101.325 kPa or 10.33 m of water, density of 1.225 kg/m^3 at the sea level, and the sonic velocity of 340.29 m/s in air.

Straight Angle A *Mathematics* term; an angle made on a straight line; equals to 180° or two right angles.

Straight Cut System A *Manufacturing* term related to *numerical control*; one of the three basic types of motion control system used in *numerical control* (NC); capable of moving a cutting tool parallel to one of the major axes at a controlled rate as required for machining; suitable for performing *milling operations* to fabricate workpieces of rectangular configurations; an NC machine capable of straight cut movements is also capable of point-to-point movements.

Straight-Line Formula A *Strength of Materials* term related to a *column*; a straight-line-based *empirical formula* used to find the *crippling* (*critical*) *load* for *intermediate-length columns* with centric loading; expressed as:

$$P = A\left[\sigma_c - K\left(\frac{l_e}{k}\right)\right]$$

where

P = crippling load
A = area of cross section of the column
σ_c = direct crushing stress (yield stress)
l_e = equivalent/effective length of the column
k = radius of gyration
K = a constant for the column material.

Straight Pin A *Design* term; see **Cylindrical Pin**.

Straight Polarity An *Electrical* term; a polarity setup in which workpiece is connected to the positive terminal and electrode to the negative terminal of the DC electrical circuit; about two-third of the heat in arc welding is generated near the workpiece in straight polarity; usually used with *non-consumable* electrodes; see also **Reverse Polarity**.

Strain Energy Density A *Strength of Materials* term; the strain energy stored in a body per unit its volume.

Strain Energy due to Direct Load A *Strength of Materials* term; the strain energy stored in a bar when subjected to a direct load (gradually applied or suddenly applied load or impact load) is given as:

$$U = \frac{\sigma^2 V}{2E}$$

where

U = strain energy stored in the bar
σ = stress produced = $\frac{\text{Load applied on the bar}}{\text{Cross-sectional area of the bar}}$
E = modulus of elasticity of the bar material
V = volume of the bar = cross-sectional area x length.

Strain Energy due to Principal Stresses A *Strength of Materials* term; the strain energy stored in a body per unit its volume when subjected to three mutually perpendicular stresses; given as:

$$U = \frac{1}{2E}\left[\sigma_1^2 + \sigma_2^2 + \sigma_3^2 - 2v(\sigma_1\sigma_2 + \sigma_2\sigma_3 + \sigma_3\sigma_1)\right] \quad (1)$$

For a biaxial stress system ($\sigma_3 = 0$)

$$U = \frac{1}{2E}\left[\sigma_1^2 + \sigma_2^2 - 2v\sigma_1\sigma_2\right] \quad (2)$$

For a uniaxial stress system ($\sigma_2 = 0$, $\sigma_3 = 0$, and assuming $\sigma_1 = \sigma$)

$$U = \frac{1}{2E}\sigma^2 \quad (3)$$

where

$\sigma_1, \sigma_2, \sigma_3$ = principal stresses acting along three mutually perpendicular axes x, y and z respectively
E = modulus of elasticity of body material
υ = Poisson's ratio.

Strain Energy due to Pure Bending A *Strength of Materials* term; the strain energy stored in a body (*bar*, *shaft* or *beam*) due to its pure bending by an applied load; given as:

$$U = \int_0^L \frac{M^2}{2EI} dx$$

where

U = strain energy stored in the body
M = bending moment produced due to load applied on the body
E = modulus of elasticity of the body material
I = second moment of inertia of body's cross section
L = length of the body
dx = small incremental length of the body.

Strain Energy due to Shear A *Strength of Materials* term; equals to the work done by a shear force, given as:

$$U = \frac{\tau^2 V}{2G}$$

where

U = strain energy stored in the body
τ = shear stress induced by the shear force
V = volume of the body.

Strain Energy due to Torsion A *Strength of Materials* term; the strain energy produced and stored in a shaft (solid or hollow) under torsion; given as:

$$U = \frac{\tau^2 V}{4G} \quad \text{(for a solid shaft)}$$

$$U = \frac{T^2 L}{2JG} (\text{ in terms of } T) \quad \text{(for a solid shaft)}$$

$$U = \frac{\tau^2}{4G}\left(\frac{D^2 + d^2}{D^2}\right) x V \quad \text{(for a hollow shaft)}$$

where

U = strain energy stored in the shaft (solid or hollow)
τ = maximum shear stress induced in the shaft (solid or hollow)
V = volume of the shaft (solid or hollow)

$$= \pi R^2 L] \quad \text{(for a solid shaft)}$$

$$= \frac{\pi}{4}\left(D^2 - d^2\right) \times L \quad \text{(for a hollow shaft)}$$

L = length of the solid or hollow shaft
G = modulus of rigidity (shear modulus) of the shaft material
D = outside diameter of the hollow shaft
d = inside diameter of the hollow shaft

R = radius of the solid shaft
J = polar modulus of the solid shaft

$$= \frac{\pi}{2} R^4$$

Strain Energy due to Volumetric Strain A *Strength of Materials* term; the strain energy stored in a body due to volumetric strain caused by changes in its dimensions when subjected to equal stress intensity over its all dimensions; given as:

$$U = \frac{1}{2K} \sigma^2 V$$

where

U = strain energy stored in the body
σ = equal stress intensity acting on all the dimensions of the body
K = bulk modulus of elasticity of the body material
V = volume of the body.

Strain Energy A *Strength of Materials* term; the mechanical energy stored in an object, when it undergoes deformation (strain) under an applied load; represents the work done by the load; equals to the area below the load–displacement curve; classified into *elastic strain energy* and *inelastic strain energy.*

Strain A *Strength of Materials* term; a ratio of the change in configuration (shape and size) to the original configuration of a body, when a deforming force acts on the body; has no unit; classified into *longitudinal strain, shearing strain,* and *volumetric strain.*

Strand Casting A *Manufacturing* term related to *casting*; see **Continuous Casting**.

Strategic Alliance A *Management* term; a partnership of two or more *corporations* or *business units* to achieve strategically significant objectives which are mutually beneficial; needed to obtain *technology* and *manufacturing capabilities*, to obtain *access to specific markets*, to reduce *financial risk* or to achieve *competitive advantage.*

Strategic Business Unit A *Management* term, abbreviated as **SBU**; a distinct business unit of a large company setup to ensure that certain product or product lines are promoted and handled as though each were an independent business.

Strategic Management A *Management* term; a set of managerial *decisions* and *actions* which determines the *long-run performance* of an organization to give it a *competitive advantage*; emphasizes on the monitoring and evaluation of external *opportunities* and *threats* in the light of an organization's strength and weaknesses.

Strategic Planning A *Management* term; a *long-term planning* which decides the success or failure of a company; includes *policy formulation, capital investment*, and *physical facilities generation.*

Strategy A *Management* term; a specific pattern of decisions and actions which managers in an organization take to use their *core competencies* to achieve *competitive advantage* and *outperform competitors*; gives the direction in which human and material resources are effectively utilized for achieving the selected objectives; a typical business firm usually considers three types of strategy, namely *corporate*, *business*, and *functional* strategies.

Streakline A *Fluid Mechanics* term; the locus of a fluid particle which has passed sequentially through a prescribed point in the flow; the most common flow pattern generated in a physical experiment.

Stream Function A *Fluid Mechanics* term; a *scalar function* of *space* and *time* related to a *streamline flow*; the partial derivative of steam function with respect to any direction gives the velocity component of the fluid flow in the normal direction; defined only for two-dimensional flows and does not exist for three-dimensional flows; denoted by ψ; the velocity components of a two-dimensional *incompressible flow*, in terms of stream function, are expressed as:

$$V_x = \frac{d\psi}{dy}$$

$$V_y = -\frac{d\psi}{dx}$$

If the fluid flow is *compressible*, the velocity components are expressed as:

$$V_X = \frac{1}{\rho}\frac{d\psi}{dy}$$

$$V_y = -\frac{1}{\rho}\frac{d\psi}{dx};$$ see also **Velocity Potential**.

Streamline A *Fluid Mechanics* term; an imaginary line (curve) drawn through a flow field in such a way that tangent at its every point gives the direction of fluid flow that point; no two streamlines cut each other as well as there cannot be any flow across a streamline; the mass flow between two streamlines is confined; a fluid lamina or layer in a laminar flow is supposed to be bounded by two adjacent streamlines; see also **Pathline and Steamtube**.

Streamlined Flow A *Fluid Mechanics* term; see **Steady Flow**.

Streamtube A *Fluid Mechanics* term; an *imaginary cylindrical tube* having a finite cross section which consists of a *bundle of streamlines*, much like a communications cable that consists of a bundle of fibre-optic cables; no flow exists across a streamtube; see also **Streamline**.

Strength of Materials Also called **Mechanics of Solids,** abbreviated as **MOS** or **Solid Mechanics** or **Mechanics of Materials** or **Mechanics of Deformable Bodies or Strength of Materials,** abbreviated as **SOM**; a branch of applied mechanics and a core subject of mechanical as well as civil engineering which deals with the mechanical properties that describe the behaviour of solid materials when they are subjected to an external force; useful in the analysis of properties such as *hardness, tensile strength, compressive strength, ductility, malleability, fatigue*, and *creep*; which govern mechanical behaviour of solid materials and provide them strength; also used in finding stresses, strains, and deformations in structures and their components due to loads acting on them for their safe design.

Strength A *Strength of Materials* term; a general term which refers to a material's ability to resist applied loads; the ability of an object to resist permanent deformation or fracture.

Stress Element A *Strength of Materials* term; a small element of a material which shows the stresses acting on its all faces; useful in the stress analysis of the material.

Stress A *Physics* and a *Strength of Materials* term; the internal restoring force which acts on unit area of a deformed body; in case of elastic deformation of a body, the restoring force equals to the applied deforming force; classified into *normal stress* and *shear stress*; the SI unit is pascal (Pa) which is equivalent to N/m^2.

Stress–Strain Curve A *Strength of Materials* term; a graphical representation between longitudinal stress and longitudinal strain obtained from a tension or compression test; classified into *engineering* and *true* stress–strain curves based on original or actual cross section, respectively.

Stretch Forming A *Manufacturing* term; a cold metal working process used to produce the required curvature in a sheet metal blank by stretching it over a form block through stretcher jaws; useful in making prototype models of *aircrafts* and *automobile parts.*

Strike-off Bar A *Manufacturing* term; a *foundry tool* used to remove the excess sand from the top of the rammed sand mass in the moulding box to make its surface plane and smooth.

String Diagram A *Work Study* term; a model or a scale plan of a shop which shows every machine or equipment marked on it along with fixing a peg or pin representing that facility and uses a continuous coloured thread or string to trace the path covered by the materials or workers while performing a particular operation; used for conditions when movements of the materials and workers are repetitive and large in numbers, which makes the flow diagrams difficult to trace or understand.

Stroke Volume An *IC Engine* term; see **Swept Volume**.

Stroke An *IC Engine* term; the linear distance travelled by a piston in the cylinder of an engine from one dead centre position (say TDC) to another dead centre position (say BDC); equals to twice the crank radius.

Structural Design An engineering discipline related to the design of a structural member or system.

Structural Materials A *Materials Engineering* term; the materials mainly used for their mechanical properties which are suitable for structural applications like making of a beam or column; for example, *steel* and *cast iron.*

Structural Steel A *Materials Engineering* term; a low alloy steel with carbon content varying between 0.15 and 0.25%; possesses *high yield strength* ($\approx$ 350 N/ mm^2), *good ductility*, and *high fatigue resistance;* used in making structural members such as *beams, plates, channels*, and *angles.*

Structure A *Design* term; a stable assembly of connected components designed to withstand loads.

Stud Welding A *Welding* term; a welding process which uses heat of an arc as well as pressure to join studs (headless threaded bolts) or similar metal pieces to flat workpieces such as plates, usually made of *low carbon steels* and *low alloy steels*; the typical applications include attachment of handles to cookware, and conduits and piping to metal workpieces.

Stud A *Design* term; a headless threaded bolt.

Subcooled Liquid A *Thermodynamics* term; a liquid whose temperature is lower than its saturation temperature or whose pressure is higher than its saturation pressure, and is not about to vaporize.

Sublimation A *Thermodynamics* term; a process associated with the phase change of a substance from *solid* to *gas* without going through the liquid state as a result of absorbing heat; for example, *camphor* undergoes sublimation.

Submerged Arc Welding A *Welding* term, abbreviated as **SAW**; an *arc welding process* in which arc created between a consumable continuous bare electrode and the workpiece, remains submerged (hidden) under a blanket of granular, fusible flux that provides a shielding atmosphere around the joint; uses both *AC* and *DC* power supply with *reverse polarity*; there is no chance of weld spatter, harmful radiation or atmospheric contamination as the arc is completely hidden; involves a high deposition rate, as the process uses heavy currents; restricted to *flat horizontal welding position* in butt and fillet joints; not recommended for overhead welding because of danger of falling of molten metal in large quantity; used to join *thick plates* of low carbon and low alloy steels; not recommended for thin sheets because of burn-off possibility.

Subsidiaries A *Financial Management* term; the companies owned by another large company (holding company).

Subsonic Flow A *Fluid Mechanics* term; a fluid flow for which the *Mach number* (M) is less than 1.

Successor Activity A *Project Management* term; an activity which must occur after another activity in a project.

Suction Pressure A *Fluid Mechanics* term; see **Vacuum.**

Suddenly Applied Load A *Physics* term; a force (load) applied all of a sudden; the magnitude of the force does not change during its course of action; produces stress in a body equal to twice the stress produced by the same load, when applied gradually; work done (W) by a suddenly applied load is given by:

$$W = P \cdot \delta L$$

where

P = suddenly applied load
δL = linear deflection produced.

Sugar Charcoal A *Chemistry* term; the *purest form of amorphous carbon* obtained by the action of conc. H_2SO_4 on sucrose or cane sugar.

Sunk Key A *Design* term related to a *key*; a key which fits into the keyways provided in the hub of the mounting and in the shaft as well, where generally one-half of its thickness fits into the shaft keyway and another half in the hub keyway; *used for heavy-duty applications*; classified into *taper*, *parallel*, *feather*, and *woodruff* keys.

Superalloys A *Materials Engineering* term; also called **High-Temperature Materials**; the materials which are alloys of *nickel*, *iron-nickel*, and *cobalt* that contain large amounts of alloying elements; have excellent strength at elevated temperatures, offer resistance to creep at temperatures up to 1000 °C and are corrosion-resistant; used in making *vanes* and *blades* for *turbine and jet engines*, *cutting tools*, *heat exchangers*, *chemical reaction vessel components*, and *heat-treating equipment*.

Supercharger An *IC Engine* term; a device used to increase the air density by increasing its pressure before inducting it in the engine.

Supercharging An *IC Engine* term; a process used to increase the power output of an IC engine by increasing the air density through rise in its pressure, before inducting it in the engine.

Superconducting Magnet A *Physics* term; a magnet made from a superconducting material.

Superconductor An *Electrical* term; a material whose resistance to the flow of electric current is zero at temperatures close to absolute zero; a good electrical conductor at ordinary temperatures is not necessarily a superconductor; *lead* (Pb), *zinc* (Zn), *cadmium* (Cd), and *tin* (Sn) behave as superconductors at absolute zero temperatures, but offer high resistance to the flow of electricity at ordinary temperatures.

Supercritical Condition A *Thermodynamics* term; a thermodynamic phase which describes the state of a substance, where there is no clear distinction between liquid phase and gaseous phase; water reaches this state at a pressure of 220 bar (called supercritical pressure of water), and becomes a homogeneous mixture of *water* and *steam*.

Superheated Region A *Thermodynamics* term; the region to the right of the saturated vapour line.

Superheated Vapour A *Thermodynamics* term; the gaseous phase of a substance which cannot condense; see also **Saturated Vapour**.

Superheater A *Power Plant* term related to a *boiler*; an important *boiler accessory* used to extract (absorb) heat from the hot flue gases in the furnace and use it to superheat the saturated steam without increasing its pressure; increases the thermal energy of steam in terms of increased work output from the steam turbines; also reduces the possibility of rusting of turbine blades by the moisture which constitutes a basic limitation in the turbine design; fitted in the path of hot flue gases; see also **Economizer**.

Superposition Method A *Strength of Materials* term related to *deflection*; an analytical method used to find slope and deflection of a loaded beam at its various sections using principle of superposition; according to this method, the total deflection of a beam at any section is equal to the algebraic sum of the deflections produced separately at that section due to each load; considers only linearly elastic deformation of negligible amount; see also **Principle of Superposition**.

Superposition of Waves A *Physics* term; the phenomenon of intermixing of two or more waves to produce a new wave.

Supersaturated Steam A *Thermodynamics* term; the steam which exists in the wet region without containing any liquid.

Supersaturation A *Thermodynamics* term; a phenomenon in which the condensation of steam gets delayed but it still exists in the wet region without containing any liquid.

Supersonic Flow A *Fluid Mechanics* term; a fluid flow for which the *Mach number* (*M*) lies between 1 and 7; see also **Hypersonic Flow**.

Supersonic Wave A *Physics* term; see **Ultrasonic Wave**.

Supersonics A *Physics* term; see **Ultrasonic Wave**.

Supervisor A *Management* term; see **Low Level Manager**.

Supplementary Units A *Physics* term; the units of measurement of two physical quantities other than seven fundamental quantities which include *radian* (*rad*) for *plane angle* and *steradian* (*sr*) for *solid angle*; see also **Fundamental Units** and **Derived Units**.

Supplier Networks An *Operations Management* term; the cooperative association of suppliers and customers working over the long term for mutual benefit; an important requirement for JIT manufacturing and lean manufacturing.

Supplier A *Management* term; anyone whose output such as materials, information, and service becomes an input to another person or group in a process of work; may be external or internal to an organization; a general supplier usually supplies raw materials and component parts to a manufacturing organization to run its production activities smoothly.

Supply Chain Management An *Operations Management* term, abbreviated as **SCM**; a process of managing the *facilities*, *functions*, and *activities* of a supply chain involved in producing and delivering a product or service from *suppliers* to *customers*; the *fastest growing area* of business today which is at the core of success of most leading companies; companies such as *Apple Computer*, *Dell Computer*, *Toyota*, *Wal-Mart*, *P & G*, and *Amazon* have achieved world-class status because of strong focus on SCM.

Supply Chain Operations Reference An *Operations Management* term, abbreviated as **SCOR**; a benchmarking framework for supply chain management which provides guidelines to make a supply chain more responsive.

Supply Chain An *Operations Management* term; a *network of trading partners* to deliver *goods* and *services*; a manufacturing supply chain consists of *supplier*, *manufacturer*, *distributor* (*wholesaler*), *retailer*, and *customer*; a *service supply chain* consists of *fewer members*.

Surface Conductance A *Heat Transfer* term; see **Heat Transfer Coefficient**.

Surface Force A *Physics* term; an external force distributed over the *surface area* of a body by its surroundings through direct contact; usually expressed as force per unit area; for example, *pressure*; see also **Body Force**.

Surface Hardenability A *Manufacturing* term; the ability (property) of a material to get hardened easily on the surface; *mild steel* has good surface hardenability.

Surface Hardening A *Manufacturing* term related to *heat treatment*; see **Case Hardening**.

Surface Plate A *Measurement* term; an accurately machined casting or lapped granite block; used as an instrument to measure the *flatness* of a workpiece.

Surface Tension A *Fluid Mechanics* term; a *fluid property* which creates tension in a liquid surface thus behaving as a stretched membrane; results from the attractive forces between molecules; due to this property, the free surface of a liquid at rest tends to have the smallest possible surface area; calculated as the force per unit length of a liquid surface; the SI unit is newton/metre (N/m).

Surface A *Physics* term; the exterior planar boundary of a solid; for example, a ball has a curved surface.

Surge Tank A *Fluid Mechanics* term; an open vertical cylinder of large diameter that is connected to a pipeline at a convenient place near the hydroelectric turbine, and is used as a safety device, as it reduces the adverse effect of water hammer by storing excess water in case of sudden drop in load on the turbine, and can also be used to supply excess water to accelerate water in the pipeline leading to turbine, in case of sudden increase in load on the turbine.

Swage Block A *Manufacturing* term; a *forging tool* consisting of many curved surfaces on its faces, and holes and grooves on its sides; used for performing operations such as *bending, swaging, shaping,* and *finishing* while working on a workpiece.

Swaging A *Manufacturing* term; also called **Rotary Swaging** or **Radial Forging** as the force on the workpiece acts radially; a hot metal working process used to reduce the diameters or taper the end of *bars* and *tubes* by subjecting them to rotating dies; see also **Tube Swaging**.

Sweep Pattern A *Casting* term; a type of pattern used to produce large moulds for making castings which are axi-symmetrical or prismatic in nature such as *bell shaped* or *cylindrical*; made of wood.

Swell A *Casting* term; a type of *casting defect* which occurs in the form of some swollen part of a casting; occurs *due to enlargement of the mould cavity* because of hydrostatic pressure of the metal.

Swept Volume An *IC Engine* term; also called **Stroke Volume** or **Displacement Volume**; the volume a piston of an engine sweeps through in the cylinder in one stroke; equals to the product of surface area of the piston and its stroke length; also equals to the difference between the maximum volume and the clearance volume of a cylinder of an engine; see also **Clearance Volume**.

SWG A *Metrology* term; see **Standard Wire Gauge**.

Switch An *Electrical* term; a device for making, breaking or changing the connections in an electrical circuit.

Symmetrical Bending A *Strength of Materials* term; the bending of a beam which occurs about an axis (neutral axis) perpendicular to the plane of symmetry; see also **Unsymmetrical Bending**.

Synchronous Manufacturing An *Operations Management* term; a production system which uses the 'theory of constraints' for scheduling and operational control; stresses on synchronizing all parts of an organization to work together to achieve the organization's goal.

Synchronous Motor An *Electrical* term; see also **Induction Motor** and **BLDC Motor**.

Synthetic Diamond See **Artificial Diamond**.

System Dynamics A *Management* term; a computer-based *simulation methodology* used for developing and analysing models of systems and their behaviour.

System of Forces A *Physics* term; if a body is acted upon by a number of forces, then it constitutes a system of forces; represented by a resultant force whose effect is equivalent to the combined effect of all the forces of the system.

Systematic Error A *Metrology* term; also called **Fixed Error** or **Bias Error**; an error which occurs due to *faulty* or *improperly calibrated instruments*; may be linked to a specific cause; can be reduced or eliminated by careful choice and calibration of instruments; a few examples include *instrumental errors*, *errors due to wrong technique or procedure*, *least count errors*, and *personal errors*; see also **Random Error**.

T

Taconite A *Chemistry* term; a greenish low-grade iron ore which contains about 22–40% *iron* and a large amount of *silica.*

Taguchi Method A *Quality control* term; a method proposed by *Genichi Taguchi* (1924–2012), a well-known *Japanese quality expert*; the method emphasizes on the quality of a product through its robust design.

Taiichi Ohno (1912–1990) A *Japanese* industrial engineer who inspired *lean manufacturing* in the *USA*; better known as *'father of Toyota Production System (TPS)'*; also called *'father of JIT'*.

Tailstock A *Lathe* term; a principal part of a lathe which lies on the right side; used to hold either a workpiece at its end during its machining or a drill or similar tool to make a hole in the workpiece; see also **Headstock**.

Tall Organization A *Management* term; an organization in which the hierarchy has many levels relative to the size of the organization; suffers from the problems relating to *communication, motivation,* and *increased costs*; suitable for an organization which requires a high level of direct control or personal supervision over subordinates; for example, *nuclear power plants* have very tall managerial hierarchies so that managers can maintain effective supervision of operations, because any error can prove to be disastrous; see also **Flat Organization**.

Tangential Stress A *Physics* and a *Strength of Materials* term; see **Shearing Stress**.

Tap A *Manufacturing* term; a tool used to produce *internal threads* in a workpiece such as in case of a *nut*; see also **Die**.

Taper Pin A *Design* term; see **Conical Pin**.

Taper Sunk Key A *Design* term related to a *key*; a sunk key having a *rectangular* or *circular* cross section, uniform width, but tapered thickness; for easy assembly and removal of the joints, the bigger end of the key is sometimes provided with a gib.

D. K. Singh, *Dictionary of Mechanical Engineering*,
https://doi.org/10.1007/978-981-99-1722-8_20

Tapered Shaft A *Design* term; a shaft whose cross section changes (either increases or decreases) uniformly over its entire length; see also **Stepped Shaft**.

Tapping A *Manufacturing* term; a process of making *internal threads* in a drilled hole by means of a *tap* (e.g. in a *nut*); see also **Dieing**.

Task Force A *Management* term; a temporary committee set up to handle a specific problem; a useful tool for finding an early solution of problems.

Task of a Manager A *Management* term; see **Management Task**.

Taylor's Scientific Management A *Management* term; see **Scientific Management**.

T-Connector A *Fluid Mechanics* term; a fluid power component used to join three separate transmission lines in a pneumatic or hydraulic system.

TDC An *IC Engine* term; see **Top Dead Centre**.

T-dS Equations A *Thermodynamics* term; used to find the *entropy changes* for any process (reversible or irreversible) between two equilibrium states of a system from other more readily determined property data; given as:

$$\mathrm{TdS} = \mathrm{d}U + \mathrm{pdV}$$
$$\mathrm{TdS} = \mathrm{d}H - \mathrm{Vdp}$$

Team A *Management* term; a permanent task force or committee.

Teamwork A *Management* term; the working of people in a team to produce result which is sometimes difficult to achieve otherwise; enables workers to *avoid wasted effort*, *reduce errors*, and *react better to customers* thus resulting in more output for each unit of employee input; provides the flexibility to react quickly because of increasing global competition; major companies such as *Hewlett-Packard* (HP), *Xerox*, *Procter & Gamble*, and *General Motors* have been using teams as a primary means of accomplishing tasks for many years; *quality circles* (also called *quality control circles*) are the best examples of self-directed teams.

Technical Skill A *Management* term; a *management skill* proposed by *Robert L. Katz* (1933-2010); relates to a manager's ability to use the *tools*, *procedures*, and *techniques* of a *specialized field*; concerned with the knowledge of a particular subject and is of greatest importance at the supervisory level.

Technology Transfer A *Management* term; the transfer of *knowledge* from one country to another for the development of *new products* or for the *improvements* in a *production process*.

Technology A *Management* term; any *equipment*, *tools* or *operating methods* designed to make work more efficient; a tool as well as an important determinant of economic growth used by an organization to produce goods and services; new *management techniques*, *scientific discoveries*, and other *innovations* improve technology; technological advances allow the production of more output from a given amount of resources.

Teflon A *Chemistry* term; a *thermoplastic material* whose chemical name is *polytetrafluoroethylene* (PTFE); used in making *non-lubricating* or *self-lubricating bearings* because of its very low coefficient of friction.

TEL A *Fluid Mechanics* term; see **Energy Gradient Line**.

Temperature Scale A *Thermodynamics* term; a measuring scale which forms a common basis for temperature measurement; all the temperature scales are based on some easily reproducible states such as the freezing and boiling points of water which are respectively called the ice point and the steam point; some temperature scales include *Celsius*, *Kelvin*, *Fahrenheit*, and *Rankine* scales.

Temperature Strain A *Strength of Materials* term; also called **Thermal Strain**; the strain produced by a temperature change in a metallic body; can be tensile or compressive in nature.

Temperature Stress A *Strength of Materials* term; also called **Thermal Stress**; the stress produced in a material due to temperature difference; can be tensile or compressive in nature; given as:

$$\sigma_t = E.\alpha \cdot \Delta T$$

where

σ_t = thermal stress
E = elastic modulus of body's material
α = coefficient of linear expansion of the body
ΔT = change in temperature.

When the temperature of a gas enclosed in a vessel of any rigid material is changed, then the thermal stress produced is equal to the change in pressure (ΔP) of the gas, given as:

$$\sigma_t = \Delta P = K \cdot \gamma \cdot \Delta T$$

where

K = bulk modulus of elasticity of the gas
γ = coefficient of cubical expansion of the gas.

Temperature A *Thermodynamics* term; a measure of the kinetic energies of the molecules of a body; measured in *degree Celsius* (°C), *degree Fahrenheit* (°F), *Rankine* (R), and *kelvin* (K).

Temperature-Entropy Diagram A *Thermodynamics* term, abbreviated as **T-S Diagram**; a plot between absolute temperature on the vertical axis and entropy on the horizontal axis; the area under the T-S diagram gives the net heat transfer during a process; a very useful means of keeping track of the states of an isentropic flow of an ideal gas.

Tempering A *Heat Treatment* term; a process of heating steel to some predetermined temperature (between room temperature and the critical temperature), holding it at that temperature for a certain length of time, and then allowed to cool in *air* or *water*; used to reduce *hardness*, *brittleness*, and *residual stresses*; also reduces strains caused by *quenching* thereby improving *ductility* and *toughness* of hardened steel.

Temporary Fastener A *Design* term; a fastening device used to join two parts which can be easily separated from each other when required without damaging the parts or the device itself; widely used in joining machine parts and other engineering components where frequent dismantling is required; a few examples include *screw, key, cotter, pin, nut, and bolt*; see also **Permanent Fastener**.

Tensile Strain A *Strength of Materials* term; a type of longitudinal strain produced on account of increase in length of a body by a tension force; calculated as a ratio of increase in length of a body to its original length; see also **Compressive Strain**.

Tensile Stress A *Strength of Materials* term; a *positive normal stress* caused by a tension force; results in increase in length (called extension) of a body in the direction of force applied; see also **Compressive Stress**.

Tension Test A *Strength of Materials* term; a *destructive test* conducted on a *universal testing machine* (UTM) on a 10- or 12-mm diameter mild steel test specimen to know its response under uniaxial tension loading; helpful in determining important mechanical properties such as *ductility*, *elasticity*, *plasticity*, *yield point*, *ultimate strength*, and the *modulus of elasticity* of *ductile materials*, by means of a *stress–strain curve*.

Tension A *Physics* term; the act of two forces pulling on an object in opposite directions, causing it to stretch; see also **Compression**.

Terminal Speed A *Fluid Mechanics* term; the steady speed attained by a falling body.

Tetrahedron A *Mathematics* term; a pyramid whose base is a triangle, and has six edges and four faces; see also **Pyramid** and **Regular Tetrahedron**.

TFP A *Management* term; see **Total Factor Productivity**.

Theoretical Air An *IC Engine* term; see **Stoichiometric Air**.

Theories of Failure A *Strength of Materials* term; see **Failure Criteria**.

Theory of Constraints An *Operations Management* term, abbreviated as **TOC**: a philosophy and production control approach of managing bottlenecks or capacity constraints for complex job-shop environments; originated by an Israeli business management guru *Eliyahu Goldratt* (1947–2011).

Theory of Elasticity A *Strength of Materials* term; see **Hooke's Law**.

Theory of Machines See **KDM**.

Theory X A *Management* term; a *motivation theory* proposed by *Douglas McGregor* (1906–1964) in 1960; based on the *negative approach* of human mind; states that the average person dislikes work, avoids responsibility and has little ambition, wishes for job security and works only for monetary gain, and hence he must be coerced to perform and closely supervised; see also **Theory Y**.

Theory Y A *Management* term; a *motivation theory* proposed by *Douglas McGregor* (1906–1964) in 1960; based on the *positive approach* of human mind; states that the average person is creative, enjoys work, seeks responsibility and is motivated by achievements, and wants to have the freedom to act according to his own judgement, but resents close supervision; see also **Theory X**.

Theory Z A *Management* term; an approach proposed by an *American* professor *William Ouchi* (1943-) in the early 1980s; stresses on the contribution of every employee in an organization in solving problems through group consensus; combines the advantages of traditional American and current Japanese methods.

Therbligs An *Operations Management* term related to *work study*; the fundamental hand motions of the work cycle in a job proposed by *Frank Gilbreth* (1868–1934); named when *Gilbreth* is spelled backwards with *t* and *h* reversed; include activities such as *select*, *grasp*, *position*, *assemble*, *reach*, *hold*, *rest*, and *inspect*; each therblig is represented by a symbol and a definite colour code.

Thermal Conductivity A *Physics* term; a physical property of the material of a body; a proportionality constant in the expression of heat transfer by conduction; represents the amount of heat flowing per unit time, per unit area normal to the direction of flow of heat, through a unit thickness of the material maintained at unit temperature difference; its higher value is indicative of a material's ability to allow heat flow quickly; metals generally have higher thermal conductivity and insulators have low values; denoted by k; the SI unit is W/(m–K); the thermal conductivity of certain materials are listed below:

Material	Thermal Conductivity (W/(m–K)
Steel	45
Copper	390
Aluminium	200
Glass	0.85
Wood	0.30

Thermal Conductor A *Physics* term; see **Heat Conductor**.

Thermal Efficiency A *Thermodynamics* term; a ratio of the *network output* of a heat engine to the *total heat supplied* to the engine; a measure of how efficiently a heat engine converts the heat which it receives to work; always less than unity; given as:

$$\eta_{\text{th}} = \frac{W_{\text{net}}}{Q_s} = \frac{Q_{s-Q_r}}{Q_s} = 1 - \frac{Q_r}{Q_s}$$

where

W_{net} = net work output of the engine
Q_s = heat supplied to the engine
Q_r = heat rejected by the engine; see also **Mechanical Efficiency**.

Thermal Equilibrium A *Thermodynamics* term; a state of equilibrium which occurs when there is a uniformity of temperature or absence of temperature gradient in the system; ensures that the system is incapable of making any spontaneous change on being isolated from its surroundings; see also **Mechanical Equilibrium** and **Chemical Equilibrium**.

Thermal Radiation A *Heat Transfer* term; a form of radiation emitted by a body because of its temperature; a volumetric phenomenon where all solids, liquids and gases emit, absorb or transmit thermal radiation of varying degree at a temperature above absolute zero; differs from other forms of *electromagnetic radiation* such as *X-rays*, *gamma rays*, *microwaves*, *radio waves*, and *television waves* which are not related to temperature; see also **Electromagnetic Radiation**.

Thermal Shock A *Physics* term; the fracture (partial or complete) of a material as a result of sudden change in temperature (usually a *sudden cooling*).

Thermal Strain A *Strength of Materials* term; see **Temperature Strain**.

Thermal Stress A *Strength of Materials* term; see **Temperature Stress**.

Thermal Vibration A *Physics* term; the periodic oscillation of atoms in a solid at a temperature above absolute zero.

Thermistor An *Electrical* term; a thermally sensitive variable resistor made of ceramic-like semiconducting materials consisting of oxides of *manganese*, *nickel*, and *cobalt*; has resistivity varying between 100 and 450,000 Ω-cm; used as a *temperature-detecting element* (for the purpose of measurement or control) and as an *electric-power sensing device* (using resistance–power relationship).

Thermit Welding A *Welding* term; a *fusion welding* process; uses a *thermit mixture* consisting of *aluminium powder* and *iron oxide* in the ratio of 1:3; the thermit mixture is ignited to about 1300 °C to produce superheated molten iron, which can fill up the gap between the parts being joined; used for connecting *railroad rails* and for *repair works in casting and forging*.

Thermocouple A *Physics* term; the most versatile and the most widely used *temperature measurement device* consisting of two wires, which operates on *Seebeck effect* in which temperature can be measured by simply measuring voltages.

Thermodynamic Cycle A *Thermodynamics* term; a closed thermal cycle consisting of only reversible processes, where both initial state and final state conditions coincide; a few examples include *Otto cycle*, *Diesel cycle*, and *Brayton cycle*.

Thermodynamic Equilibrium A *Thermodynamics* term; a state of *complete equilibrium* of a system which requires a system to be in *mechanical*, *chemical*, and *thermal* equilibrium separately.

Thermodynamic System A *Thermodynamics* term; a certain quantity of matter or a definite volume in space which can be analysed in the solution of a problem; a system is separated from the surroundings by its boundary.

Thermodynamic Temperature Scale A *Thermodynamics* term; a *temperature scale* which is independent of the properties of a substance; developed in conjunction with the second law of thermodynamics; examples include *Kelvin scale* and *Rankine scale*.

Thermodynamics A core subject of mechanical engineering which deals with heat and temperature, and includes the changes in temperature, internal energy and entropy of a system through external work along with modes of heat transfer and efficiency of heat engines; finds extensive applications in power producing devices and other components such as *steam* and *gas turbines*, *internal combustion engines*, *refrigerators*, *air-conditioning units*, and others.

Thermoelectric Circuit A *Physics* term; a circuit which involves both *thermal* and *electrical* effects.

Thermoelectric Device A *Physics* term; a device operating on a *thermoelectric circuit*, which incorporates both *thermal* and *electrical* effects.

Thermometric Properties A *Thermodynamics* term; the properties which change due to change in temperature; a few examples include *entropy*, *enthalpy*, and *internal energy*.

Thermometry A *Thermodynamics* term; the science of measurement of temperature (including the lowest possible temperature of 0 K and the highest temperature of 100,000 K).

Thermoplastic Polymer A *Chemistry* term; see **Thermoplastic.**

Thermoplastic A *Chemistry* term; also called **Thermoplastic Polymer**; a *highly polymerized substance* which can be repeatedly softened on heating and hardened on cooling; can be easily moulded and remoulded into any shape just like glass; composed of chain molecules using *addition polymerization* process in which interlocking of their molecules gives them hardness and insolubility; its transparent variety has suitably replaced glass in the recent years for many applications; used in making *sheets, pipes, bottles, buckets, helmets, synthetic fibres, wire insulation*, etc.; a few examples include *polyethylene (polythene), polystyrene, polyvinyl, Teflon*, and *nylon*; see also **Thermosetting Plastic.**

Thermosetting Plastic A *Chemistry* term; also called **Thermosetting Polymer** or simply **Thermosetting**; a *polymerized substance* which becomes *hard and rigid upon heating*, and *sets during moulding process and cannot be softened again*; composed of *cross-linked molecules* using *condensation polymerization* process; has *good strength, stiffness*, and *hardness*, but has *poor ductility* and is *brittle*; cannot be mechanically deformed or softened at high temperatures, as their plasticity does not change with temperature, hence cannot be remoulded into any new shape; used in making *insulation for electrical equipment, plastic crockery, automobile parts, varnishes*, etc.; a few examples include *bakelite, polyester, phenolics, urea formaldehyde*, and *epoxies*; see also **Thermoplastic.**

Thermosetting Polymer A *Chemistry* term; see **Thermosetting Plastic.**

Thermosetting A *Chemistry* term; see **Thermosetting Plastic.**

Third Angle Projection An *Engineering Drawing* and a *Machine Drawing* term; a projection method in which an *object is imagined to be placed in the third quadrant* for the purpose of its projection; as the observer normally looks from the right side of the quadrant to obtain the front view of the object, here the *vertical projection plane has to be transparent* which lies between the observer and the object, and the intersection of this plane with the projectors (the projection lines) from all the points of the object on being joined together in order forms its image on the transparent plane; see also **First Angle Projection.**

Third Law of Thermodynamics A *Thermodynamics* term; states that the entropy of a pure crystalline substance is zero at the absolute zero temperature (0 K); a law which provides a common base for the entropy of all substances.

Third-Class Lever A *Physics* term; a lever in which the effort is positioned between the fulcrum and the load; the load force and the effort force act in opposite directions and its mechanical advantage value lies between 0 and 1; a few examples of third-class lever include *barbecue tongs* (used in kitchens), *shovels, backhoes* (excavating equipment or digger), and *tweezers.*

Third-Party Logistics An *Operations Management* term, abbreviated as **3PL**; an *outsourcing activity* in which an outside company performs all or part of a firm's materials management and product distribution functions; the 3PL providers allow a firm to focus on its core competencies, leaving the logistics expertise to the logistics companies because of their improved technological capabilities; may also own assets such as trucks and warehouses for better materials management and efficient delivery; the use of 3PL is most prevalent among large companies; for example, firms such as *3M*, *Eastman Kodak*, *Dow Chemical*, *Time Warner*, *General Motors*, etc. have outsourced their larger logistics operations to 3PL providers; *Ryder Dedicated Logistics* currently has a multiyear agreement to design, manage, and operate all of Whirlpool Corporation's inbound logistics; *Safe Express* also provides 3PL services to many firms in India; the most obvious disadvantage of the use of 3PL providers is the loss of control inherent in outsourcing a particular function; see also **Fourth-Party Logistics**.

Thixotropic Fluid A *Fluid Mechanics* term; a fluid whose apparent viscosity (that depends on shear rate) decreases with time under a constant applied shear stress; for example, *many paints*; see also **Rheopectic Fluid**.

Thread Rolling A *Manufacturing* term; a cold working process used to produce threads on a metal blank by rolling it through hardened dies, which cause the metal to flow radially into the desired shape.

Threading A *Manufacturing* term; see **Dieing**.

Three-Dimensional Motion A *Physics* term; the motion of a body along three directions (in a *space*); a few examples include the *flying of an aeroplane* or *a bird*, the *random motion of a gas molecule*, and the *flying of a kite in the sky*; see also **One-Dimensional Motion** and **Two-Dimensional Motion**.

Throttling A *Thermodynamics* term; an *irreversible process* in which a fluid flows across a restriction in such a manner that there is a drop in its pressure without any change in its *enthalpy*, *kinetic energy* or *potential energy*; does not involve work or heat transfer; a few examples include *flow through a porous plug*, *flow through a partially closed valve* or *flow through a small orifice*; see also **Free Expansion**.

Thrust Bearing A *Design* term; a bearing used to support a shaft subjected to an axial load; classified into *pivot bearing* (*foot-step bearing*) and *collar bearing*.

Thyristor An *Electronics* term; a *semiconductor device* having three or more junctions; acts as a switch without any bias; can be fabricated to have voltage ratings of several hundred volts and current ratings from a few amperes to almost thousand amperes.

TIG Welding A *Welding* term; see **Tungsten Inert Gas Welding**.

Time of Ascent A *Physics* term related to a *projectile*; the time (t) taken by a projectile to go from its point of projection to the highest point where it attains the maximum height; equals to the time of descent or the one-half of the time of flight (T); given as:

$$t = \frac{u \sin \theta}{g}$$

where

u = velocity of projection
θ = angle of projection
g = acceleration due to gravity; see also **Time of Descent**.

Time of Descent A *Physics* term related to a *projectile*; the time (t) taken by a projectile to return from the highest point where it attains the maximum height to the point of hitting the ground on the same level of projection; equals to the time of ascent or the one-half of the time of flight (T); given as:

$$t = \frac{u \sin \theta}{g}$$

where

u = velocity of projection
θ = angle of projection
g = acceleration due to gravity; see also **Time of Ascent**.

Time of Flight A *Physics* term related to a *projectile*; the total time for which a projectile is in flight (remains in air) between the point of projection and the point of hitting the ground; equals to the *sum of time of ascent and time of descent*; *independent of mass of the projectile*; denoted by T; given as:

$$T = \frac{2u \sin \theta}{g}$$

where

u = velocity of projection
θ = angle of projection
g = acceleration due to gravity; see also **Time of Ascent** and **Time of Descent**.

Time Period A *Physics* term; the time taken to complete one oscillation (or one cycle); represents the least interval of time after which the periodic motion of a body repeats itself; denoted by T; the SI unit is second (s).

Time Study An *Operations Management* term related to *work study*; the method of establishing *time standards* by using a stopwatch to time the operations performed by workers at a defined level of performance; pioneered by *F. W. Taylor* (1856–1915) in 1881; see also **Motion Study** and **Method Study**.

Tin Alloys A *Chemisrty* term; the alloys of *tin* (Sn) which contains tin as the main constituent, besides other elements present in various amounts; important tin alloys include *solder*, *pewter*, *babbitt metal*, *britannia metal*, *white metal*, *bell metal*, and *rose metal*.

Tin Plate A *Materials Engineering* term; a mild steel plate coated with *tin*.

Tin A *Chemistry* term; a metallic element having symbol Sn with atomic number 50; melts at a relatively low temperature of 232 °C but boils at a relatively high temperature of 2270 °C; exists in three allotropic forms, namely *grey*, *white*, and *rhombic*; when bent, the metal produces a cracking noise, called '*tin cry*' which occurs due to rubbing of metal crystals over one another; used as an important alloying element, and its *addition to steel increases its corrosion resistance* by forming a silvery-white protective coating over iron (the process is called tin plating) which does not tarnish on being exposed to air; also used in the manufacture of a number of alloys such as *solder*, *bell metal*, *rose metal*, *babbitt metal*, *britannia metal*, *white metal*, and *pewter*; thin tin plates are used in the manufacture of tin cans, kerosene oil containers and other similar articles; also extensively used in tinning (thin coating of tin layer) of brass utensils; also finds use in the preparation of *collapsible tubes* for tooth pastes and various ointments; tin amalgam is used in making mirrors and tin foil is used for wrapping cigarettes and other articles.

Tincal A *Chemistry* term; a *naturally occurring borax* obtained from dried up lakes; contains about 50% *borax* along with some other insoluble impurities such as clay, sand, etc.

Title Block An *Engineering Drawing* and a *Machine Drawing* term; an enclosed space made at the bottom right hand corner in a drawing sheet; contains the identification details of a drawing such as the *title of the drawing*, the *sheet number*, the *scale used*, the *symbol* (which denotes the method of projection), the *name of the firm*, and the *initials of staff* who designed, drawn, checked, and approved the drawing.

TOC An *Operations Management* term; see **Theory of Constraints**.

Tolerance A *Metrology* term; the difference between the maximum and minimum size limits of a part giving variation in its size to be tolerated to cover reasonable imperfections in workmanship; varies with different grades of work; may be unilateral or bilateral, where unilateral tolerances are represented as $35^{+0.00}_{-0.10}$, and bilateral tolerances as 35 ± 0.06 or $35^{+0.06}_{-0.05}$, that is, with equal or unequal variations.

Ton Refrigeration A *Refrigeration* term; see **Ton**.

Ton A *Refrigeration* term; also called **Ton Refrigeration**, abbreviated as **TR**; the *standard unit of refrigeration*; equivalent to producing cold at the rate of removing heat from 1 ton (1000 kg) of water at 0 °C to convert it into ice at 0 °C in one day or 24 h; 1 TR = 50 kcal/min = 12,000 Btu/hr = 3.5 kW.

Tool Life A *Manufacturing* term related to a *cutting tool*; the total time during which a cutting tool can be successfully used for cutting operations; usually expressed in minutes; finds mention in Taylor's tool life equation, VT^n = constant, which is used for various metal cutting operations (V = velocity of cutting tool, T = time of cutting and n = an exponent which depends on the material being cut).

Tool Post A *Lathe* term; a principal part of a lathe mounted on the compound slide; used to hold the cutting tool firmly in a particular position.

Tool Signature A *Manufacturing* term related to *machining*; a parameter used to define a cutting tool completely; consists of seven elements placed in order such as *back rake angle, side rake angle, end relief angle, side relief angle, end-cutting edge angle, side-cutting edge angle*, and *nose radius* of which first six elements are angles in *degrees* and the seventh element is expressed in *inch* or *mm*.

Tool Steel A *Materials Engineering* term; a *high carbon steel* whose hardness is obtained by *quenching* and *tempering*; used in making *cutting tools* for various machining operations, *dies for die casting, forming dies*, and other uses in which a combination of high strength, hardness, toughness, and temperature resistance is required.

Toothed Gearing A *Theory of Machines* term related to a *gear*; a means to transmit motion or power between two shafts without any loss; a few examples include *spur gearing, bevel gearing, spiral gearing*, and *helical gearing*.

Top Dead Centre An *IC Engine* term, abbreviated as **TDC**; a point where the piston is at the top of the cylinder of an engine; the position of the piston when it forms the smallest volume in the cylinder; occurs when the connecting rod and crank are in-line with one another; see also **Bottom Dead Centre**.

Top Manager A *Management* term; see **High Level Manager**.

Top Riser A *Manufacturing* term related to *casting*; see **Open Riser**.

Top View An *Engineering Drawing* and a *Machine Drawing* term; the view of an object obtained as a projection on a *horizontal plane* of projection by looking at the object normal to its surface from the *top*; see also **Front View**.

Torque Meter A *Theory of Machines* term; see **Transmission Dynamometer**.

Torque Ratio A *Gear* term; a ratio between the torque values on the input and output gears.

Torque A *Physics* term; see **Moment**.

Torr A *Physics* term; a unit of *pressure*, named after the *Italian* physicist *Evangelista Torricelli* (1608–1647) who invented the *barometer*; 1 torr = 1 mm of Hg = 0.133 Pa; finds use in *medicine* and *physiology*.

Torsion Equation A *Strength of Materials* term; also called **Torsion Formula**; an equation used to find the maximum shear stress produced in a rotating circular shaft when subjected to a torque: can be applied to solid and hollow circular shafts; given as:

$$\frac{T}{J} = \frac{\tau}{R} = \frac{G\theta}{L}$$

where

T = torque applied on the shaft
J = polar moment of inertia of shaft's cross-section
τ = maximum shear stress on the shaft
R = radius of the shaft
G = modulus of rigidity of the shaft material
L = length of the shaft
θ = angle of twist produced in the shaft.

Torsion Formula A *Strength of Materials* term; see **Torsion Equation**.

Torsion Meter A *Metrology* term; a device used to measure *torque* by measuring the angle of twist over a fixed length of a shaft; classified into mechanical, optical, and electrical torsion meters.

Torsion Spring A *Strength of Materials* term related to a *spring*; see **Helical Spring**.

Torsion Test A *Strength of Materials* term; a *destructive* test used to measure the modulus of rigidity (G) of solid materials; especially useful for studying a material's flow at large plastic (shear) strains as produced in many operations such as during punching of holes in a sheet metal and metal cutting.

Torsion A *Strength of Materials* term; a twisting action produced on a rotating element (shaft) about its longitudinal axis; the torsional effect on circular members was first noticed by the French physicist *Charles a. Coulomb* (1736–1806) in 1784.

Torsional Flexibility A *Strength of Materials* term; the angle of twist per unit length of a bar or shaft; see also **Torsional Stiffness** and **Torsional Rigidity**.

Torsional Rigidity A *Strength of Materials* term; the product of the *polar moment of inertia* (J) of a shaft's cross-section and the *modulus of rigidity* (G) of the shaft material; *a measure of strength of a shaft against torsion*; equals to the torque acting on a shaft of unit length and producing unit angle of twist; see also **Torsional Flexibility** and **Torsional Stiffness**.

Torsional Stiffness A *Strength of Materials* term; the reciprocal of the *torsional flexibility*; see also **Torsional Flexibility** and **Torsional Rigidity**.

Torsional Strength A *Strength of Materials* term; the ability of a material to withstand twisting moment; measured in terms of torsional shear strength (τ) found by conducting a torsion test; given as:

$$\tau = \frac{16T}{\pi d^3} \quad \text{(for a solid circular specimen)}$$
$$= \frac{16T d_o}{\pi\left(d_o^4 - d_i^4\right)} \quad \text{(for a hollow circular specimen)}$$

where

T = twisting moment
d = diameter of the solid specimen
d_o = outside diameter of the hollow specimen
d_i = inside diameter of the hollow specimen.

Total Cost An *Operations Management* term related to *cost analysis*; the sum of *factory cost*, *selling overhead*, *distribution overhead*, and *administration overhead*.

Total Derivative Operator A *Mathematics* term; denoted by *d*; used to find the rate of change of one parameter with respect to another; for example, acceleration ($\boldsymbol{a}$) of an object can be expressed as the time rate of change of velocity ($\mathbf{v}$), and is shown as: $\boldsymbol{a} = \frac{dV}{dt}$, where both velocity and acceleration are shown in bold to show their vector nature; see also **Partial Derivative Operator**.

Total Energy Line A *Fluid Mechanics* term; see **Energy Gradient Line**.

Total Factor Productivity A *Management* term, abbreviated as **TFP**; also called **Multi-factor Productivity**; a ratio of the economy's outputs to its total inputs of *labour*, *capital*, *materials*, and *energy*; describes the overall productivity of an economy; see also **Partial Factor Productivity**.

Total Head A *Fluid Mechanics* term; the sum of the *pressure head* ($P/\rho g$), the *velocity head* ($v^2/2\,g$), and the *elevation head* (z) in the *Bernoulli equation* (P = pressure of the fluid, g = acceleration due to gravity, ρ = density of the fluid, v = velocity of the fluid, and z = elevation of the fluid column); the total head for an ideal streamline flow is constant.

Total Product Function An *Operations Management* term; see **Production Function**.

Total Productive Maintenance An *Operations Management* term, abbreviated as **TPM**; a system which combines the practice of preventive maintenance with the concepts of total quality in order to improve a production system.

Total Quality Control A *Management* term, abbreviated as **TQC**; a process of improving the quality of a product right from the start of its manufacturing, that is, from the raw material stage, called 'quality at the source'; a *Japanese* approach to quality control evolved by an American quality expert *Armand V. Feigenbaum* (1920–2014) in 1983; stresses on continuous improvement through attention to manufacturing details rather than attainment of a fixed quantitative quality standard; see also **Total Quality Management**.

Total Quality Management A *Management* term, abbreviated as **TQM**; also called **Continuous Quality Improvement**; an *integrative management philosophy* aimed at continuously improving the quality of products, processes or services through the use of various means such as statistical tools, competitive benchmarking, participative management, teamwork, and continuous training to achieve ultimate customer satisfaction by making quality everybody's concern in the organization; requires quality improvement at all levels of management and across all areas; the concept evolved in manufacturing companies like *Toyota*, *IBM*, and *Motorola*; the principles of TQM are equally applicable for manufacturing and service operations organizations; see also **Total Quality Control**.

Total Slack A *Project Management* term; also called **Slack**; the spare (slack) time which is available for carrying out an activity without delaying the project completion; exists for non-critical activities only; for critical activities the total slack is zero; calculated as:

$$\begin{aligned}\text{Slack (S)} &= \text{LST} - \text{EST}\\ &= \text{LFT} - \text{EFT}\end{aligned}$$

where

LST = latest start time
EST = earliest start time
LFT = latest finish time
EFT = earliest finish time; see also **Free Slack**.

Toughness A *Strength of Materials* term; a *mechanical property* of materials which defines its capacity to absorb energy without fracture; see also **Modulus of Toughness**.

Toyota Production System An *Operations Management* term, abbreviated as **TPS**; also called **Toyota System**; the most admired production system in the world, evolved by a Japanese company *Toyota Motor Company*; an example of excellence in automobile manufacturing; one of the most admired lean manufacturing systems in existence which requires constant improvements to increase efficiency and reduce waste; consists of three important Japanese-invented techniques, namely *jidoka*, *andon*, and *poka-yoke*.

Toyota System An *Operations Management* term: see **Toyota Production System**.

Toyota A *Japanese* company which is exemplary of excellence in *automobile manufacturing*; well known today for its *Toyota production system* (TPS), one of the most admired lean manufacturing systems in existence which has created a system which stimulates employees to experiment with their environment by seeking better ways (using scientific methods) whenever things go wrong.

TPM An *Operations Management* term; see **Total Productive Maintenance**.

TPS An *Operations Management* term; see **Toyota Production System**.

TQC A *Management* term; see **Total Quality Control**.

TQM Tools A *Quality Management* term; the tools used in the quality management of products and services; include *Pareto charts, check sheets, fishbone diagrams,* and *process flowcharts.*

TQM A *Quality Management* term; see **Total Quality Management**.

TR A *Refrigeration* term; see **Ton**.

Tracking Signal A *Management* term related to *forecasting*; a measurement method which shows whether a forecast has had any built-in biases over a period of time.

Trade Deficit An *Economics* term; a condition of an economy in which imports exceed exports, that is, a country imports more than it exports.

Train Value A *Theory of Machines* term related to *gear*; also called **Gear Train Value**; a ratio of the product of number of driving tooth to the product of number of driven tooth; a ratio of the speed of driven gear (follower) to the speed of driver; the reciprocal of velocity ratio.

Training A *Management* term; a process of improving performance in the current job.

Trajectory A *Physics* term; the *path described by a projectile* during its motion; has *parabolic* shape; at the highest point of the projectile path, the velocity and acceleration are perpendicular to each other, and the projectile possesses velocity only along the horizontal direction.

Transducer An *Electronics* term; a device used to convert one form of energy to another form.

Transformer An *Electrical* term; an electrical device which changes the input voltage; classified into *step-up transformer* and *step-down transformer*.

Transistor An *Electronics* term; a *semiconductor device* and a fundamental component of almost all the electronic devices used to control large electrical output signals using a small input signal; a solid-state amplifier.

Transitional Energy A *Power Plant* term; the energy obtained from objects in motion; for example, *wind energy* and *hydel energy.*

Translational Dynamic Equilibrium A *Physics* term; see **Dynamic Equilibrium.**

Translational Equilibrium A *Physics* term; the equilibrium which ensures that a rigid body remains at rest or moves with a constant velocity in a particular direction; requires the net external force or the vector sum of all the external forces acting on the body is zero; classified into *static equilibrium* and *dynamic equilibrium*; mathematically expressed as:

$\Sigma F_x = 0, \Sigma F_y = 0,$ and $\Sigma F_z = 0;$ see also **Rotational Equilibrium.**

Translational Static Equilibrium A *Physics* term; see **Static Equilibrium.**

Transmissibility of Forces A *Physics* term; see **Principle of Transmissibility.**

Transmission Dynamometer A *Theory of Machines* term; also called **Torque Meter**; a passive type of dynamometer in which the work done by a prime mover under test is not absorbed, but is transmitted for use during measurement; just used for the purpose of sensing a torque at a well-defined location of the prime mover, and neither adds to nor subtracts from the transmitted energy or power; examples include *belt transmission dynamometer* and *torsion dynamometer*; see also **Absorption Dynamometer.**

Transmission Line A *Fluid Mechanics* and an *Electrical* term; in *fluid mechanics* terms, a pipe or tube which serves as a connection between two fluid power components; in *electrical* terms, an *overhead high voltage electrical wire* used for the transmission of electricity between two or more places from a power plant .

Transonic Flow A *Fluid Mechanics* term; a fluid flow involving both *subsonic* and *supersonic* regions; the *Mach number* (*M*) for a *transonic flow* lies between 0.9 and 1.1.

Transportation Method An *Operations Research* term; a *linear programming solution method* used to find the minimal cost of shipping products from several sources to several destinations.

Transportation Problem An *Operations Research* term; also called **Integer Linear Programming Problem** as its variables take only integer values; a special class of linear programming problem in which the objective is to transport a single commodity from various sources to different destinations at a minimum total cost; assumes that total capacity equals to total requirement.

Trapezium A *Mathematics* term; a quadrilateral which has exactly one pair of parallel sides.

Trapezoidal Weir A *Fluid Mechanics* term; the combination of a *rectangular weir* and a *triangular weir*; used to measure and control flow (discharge) in an open channel.

Tresca's Yield Criterion A *Strength of Materials* term related to *theory of material's failure*; see **Maximum Shear Stress Theory**.

Triangle Law of Forces A *Physics* term; see **Law of Triangle of Forces**.

Triangle A *Mathematics* term; a closed geometric plane figure consisting of three non-parallel sides; the simplest polygon; classified into *acute-*, *obtuse-*, and *right-*angled triangles on the basis of measures of angle; also classified into *scalene*, *isosceles*, and *equilateral* triangles on the basis of measures of length of sides.

Triangular File A *Manufacturing* term; one of the types of a file used in the fitting shop which has a *triangular cross section* and double cut on all faces; used in the filing of *square corners* and *sharpening of woodworking saws.*

Triangular Load A *Strength of Materials* term; a distributed load in the shape of a triangle whose intensity varies linearly from zero to a maximum intensity (w); liquid pressure can be represented by a triangular load; see also **Uniform Load** and **Distributed Load**.

Triangular Pyramid A *Mathematics* term; a pyramid whose base is a triangle; see also **Pyramid**.

Triangular Weir A *Fluid Mechanics* term; also called **V-Notch Weir**; a weir with a *triangular opening*; used to measure the liquid flow rate in open channels; maintains a high weir head even with small flow rates, and hence it can be used to measure a wide range of flow rates accurately.

Tridymite A *Chemistry* term; one of the three *naturally occurring crystalline forms of silica* (SiO_2), other two being *quartz* and *cristobalite*; quartz when heated above 870 °C changes into tridymite and remains stable in this form till 1470 °C.

Trigonometric Ratios A *Mathematics* term; the ratios with respect to a right triangle used in trigonometry; include *sine*, *cosine*, *tangent*, *cosecant*, *secant*, and *cotangent*.

Triple Point A *Thermodynamics* term; the temperature at which all the three phases of a substance coexist in equilibrium; for example, the triple point of water is 273.16 K or 0.01 °C (pressure of 4.58 mm of Hg or 610.65 Pa), where the three phases of water such as *ice* (solid form), *liquid water*, and *water vapour* (gaseous form) all coexist.

Trowel A *Manufacturing* term; a *foundry tool* used to finish or repair a mould.

True Centrifugal Casting A *Manufacturing* term related to *casting*; *a type of centrifugal casting* used to produce pipes, cylinder liners and hollow symmetrical objects (without using cores) using a *rotating mould* (about its horizontal or vertical axis); see also **Semi-Centrifugal Casting** and **Centrifuging**.

True Strain A *Strength of Materials* term; also called **Natural Strain** or **Logarithmic Strain**; the strain related to the *actual length* of a test specimen, when there is reduction in its cross sectional area with increase in applied load; a ratio of the sum of incremental elongations to the instantaneous gauge length; given as:

$$\epsilon = \int_{l0}^{l} \frac{dl}{l} = \ln \frac{l}{l_o}$$

where

l = instantaneous length
l_o = original length; see also **Nominal Strain**.

True Stress A *Strength of Materials* term; a *direct* or *normal* stress obtained by dividing load by the actual cross-sectional area of a metal test specimen at a particular point of time during tension or compression test; see also **Engineering Stress**.

Truss A *Strength of Materials* term; a *stationary civil structure* which consists of straight steel members such as *bars*, *angles*, and *channels* (the angle section is the most preferred section) joined together by welding or riveting; used to support a long span in buildings and bridges; the load in a truss is applied at its joints only, whereas its members are subjected to either a tension or compression force; classified into *plane truss* and *space truss*.

Try Square A *Manufacturing* term; a *fitting tool* used to check an edge or surface exactly at right angles to an already finished surface or for laying out work.

T-S Diagram A *Thermodynamics* term; see **Temperature-Entropy Diagram**.

T-Series A *Materials Engineering* term; the second series of *high speed steel* (HSS) that contains higher content of *tungsten*; the 18-4-1 HSS is the most widely used T-series, which consists of 18% *tungsten*, 4% *chromium*, and 1% *vanadium*; see also **M-Series**.

Tsunami A *Fluid Mechanics* term; a strong water wave which has very long wavelengths (of the order of 100 km) and travels with extremely large wave speeds; produced by a disturbance like an *earthquake*, *volcanic eruption* or *meteorite impact*, which vertically displaces the water column; has extremely damaging effects.

Tube Drawing A *Manufacturing* term; a metal working process used to reduce the diameter of a tube by pulling it through a die using or without a mandrel.

Tube Extrusion A *Manufacturing* term; a hot metal working process used to produce tubular products from a heated workpiece by forcing it to go through a shaped-die opening using a mandrel; see also **Tube Swaging**.

Tube Spinning A *Manufacturing* term; a cold metal working process used to reduce the thickness of walls of cylindrical parts by pressing a flat sheet metal blank against a rotating cylindrical mandrel using rollers; typical tube spinned products include *pressure vessels*, *high-pressure gas cylinders*, and *kitchenwares* such as *pots* and *pans*; see also **Shear Spinning**.

Tube Swaging A *Manufacturing* term; a hot metal working process and a variation of swaging used to reduce the diameter of tubes by subjecting them to rotating dies using or without a mandrel; see also **Tube Extrusion**.

Tube Turbine A *Fluid Mechanics* term; *an inclined-axis propeller turbine* that is particularly suitable for low head installations.

Tungsten Electrode A *Welding* term; a tungsten wire electrode used as a *nonconsumable electrode* in *arc welding* because of its high melting point (3410 °C).

Tungsten Inert Gas Welding A *Welding* term, abbreviated as **TIG Welding**; also called **Gas Tungsten Arc Welding**, abbreviated as **GTAW**; an arc welding process which uses a shielding atmosphere of inert gases such as *argon* and *helium* (either alone or in combination) to protect the arc produced between a nonconsumable tungsten electrode and the workpiece; uses both *ac* and *dc* power supply with *straight polarity*; previously used for the joining of *aluminium* and *magnesium alloys*, but currently also used for *copper*, *stainless steel*, *silver*, *tungsten*, *molybdenum*, *titanium*, and *high-temperature materials*; finds use in *aircraft industry*, and *chemical and nuclear plants*; see also **Metal Inert Gas Welding**.

Tungsten A *Chemistry* term; a *metallic element* having symbol *W* with atomic number 74; a hard metal with silvery white lustre which has the *highest melting point temperature* in the periodic table, close to 3410 °C; has high specific gravity (19.3), close to that of gold; resists the action of air, oxygen or chlorine unless heated to bright redness; not attacked by acids; due to its high melting point, it cannot be cast into rods and wires like most of the other metals; mostly used in the *powdered form* for shaping products; because of its extremely high melting point and low volatility, it finds use in making *filaments for electric lamps*; also used as a major *alloying element* for *steel*; increases hot strength and hot hardness of steel, if added in significant amount.

Tuning Fork A *Physics* term; an instrument used as a source of *standard frequency*; finds use in conducting experiments on sound using a *sonometer* or *resonance tube.*

Turbine Efficiency A *Thermodynamics* term; a ratio of the *work output of a turbine during actual adiabatic expansion* to the *work output during reversible adiabatic (isentropic) expansion* under similar conditions of temperature; a ratio of the *mechanical energy output* to the *decrease in the mechanical energy* of a fluid; see also **Pump Efficiency**.

Turbine A *Fluid Mechanics* term; a *fluid machine* (*turbomachine*) and a *work-producing device* which extracts mechanical energy from a fluid by dropping its pressure, and transfers most of this energy to some form of mechanical energy output, typically in the form of a rotating shaft; classified into *hydraulic*, *steam*, and *gas* turbines; see also **Pump**.

Turbofan Engine A *Fluid Mechanics* term; a multistage axial-flow *turbomachine* which utilizes both multistage axial-flow compressors and multistage axial-flow turbines; used to power modern commercial aeroplanes.

Turbomachine A *Fluid Mechanics* term; a *dynamic fluid machine* which either extracts energy from a fluid or adds energy to a fluid as a result of dynamic interactions between the device and the fluid using rotating propellers or vanes; offers high power density; has relatively few moving parts and reasonable efficiency; widely used in industry for power generation; a few examples include *hydraulic turbine*, *steam turbine*, *pump*, *compressor*, *fan*, and *blower*.

Turbomachinery A *Fluid Mechanics* term; a branch of fluid mechanics which deals with transfer of energy either to or from a fluid using a rotating shaft.

Turbulent Flow A *Fluid Mechanics* term; an irregular *three-dimensional flow* in which fluid particles have disorderly motion; involves a violent transverse exchange of momentum; most of the energy associated with this flow is used in setting up *eddy currents* in the fluid; a fluid flow becomes turbulent, when its Reynolds number exceeds 4000; see also **Laminar Flow**.

Turn Over A *Management* term; a ratio of the sales to the assets of a company.

Turret Lathe A *Manufacturing* term; a *type of lathe* which contains a *turret* to hold the cutting tools during machining; has no tailstock.

Twin Electrode Carbon Arc Welding A *Welding* term; a variation of the *carbon arc welding* in which arc is produced between two graphite (carbon) electrodes; see also **Carbon Arc Welding**.

Twin Primes A *Mathematics* term; the prime numbers (primes) which occur in pairs with a difference of two; a few examples include 3, 5 and 11, 13; see also **Prime Number**.

Twist Drill A *Manufacturing* term; the most common cutting tool used in *drilling* for which the point angle is 118°.

Two Stroke Engine A *Thermodynamics* term related to an *IC engine*; an internal combustion engine which works on a *two-stroke cycle* consisting of *compression* and *expansion* strokes; has no suction and exhaust strokes; the cyclic operation is completed in every one revolution of the crankshaft; widely employed for small power requirements, for examples, *scooters*, *motorcycles*, and *mopeds*; theoretically develops *twice* the power of a four-stroke engine of the same size and speed (actually develops 1.7–1.8 times the power as some power is used for compressing the charge and scavenging); see also **Four Stroke Engine**.

Two-Dimensional Collision A *Physics* term; also called **Oblique Collision**; a collision in which two bodies moving initially along the same straight line, when collide (without loss of kinetic energy), move along different directions in a pane after collision; see also **One-Dimensional Collision**.

Two-Dimensional Motion A *Physics* term; the motion of a body along two directions (in a *plane*); a few examples include the *crawling of an insect over the floor* and the *revolving of the earth around the sun*; see also **One-Dimensional Motion** and **Three-Dimensional Motion**.

Two-Factor Theory A *Management* term related to *motivation*; see **Hygiene Theory**.

Two-Force Member A *Strength of Materials* term related to a *truss*; a truss member which can be subjected to two forces only, either tensile or compressive.

Two-Stroke Cycle A *Thermodynamics* term related to an *IC engine*; a thermodynamic cycle which is completed in *two strokes* of a piston or in one revolution of the crankshaft of an internal combustion (*IC*) engine; consists of *compression* and *expansion* strokes only; has no suction and exhaust strokes; mostly used in petrol engines for small power requirements; see also **Four-Stroke Cycle**.

Type I Error A *Quality Management* term; also called **Alpha Error**; an error related to *rejection* of something which is acceptable; see also **Type II Error**.

Type II Error A *Quality Management* term; also called **Beta Error**; an error related to *acceptance* of something that should have been rejected; see also **Type I Error**.

Type Metal A *Chemistry* term; a *lead alloy* which contains 82% *lead* (Pb), 15% *antimony* (Sb), and 3% *tin* (Sn); used for making printing type.

Type Number A *Fluid Mechanics* term; see **Specific Speed**.

U

UCON A *Materials Engineering* term; a costly *cutting tool material* consisting of 50% *columbium*, 30% *titanium*, and 20% *tungsten*; mainly used in the machining of steel with higher speed, large depth of cut, and low feed; can be cast and rolled into sheets.

udl A *Strength of Materials* term; see **Uniform Load**.

U-Joint A *Design* term; see **Hooke's Joint**.

Ultimate Strength A *Strength of Materials* term; also called **Ultimate Stress** or **Ultimate Tensile Strength** or **Ultimate Tensile Stress**; a ratio of the maximum load which a material can take up before its failure to its original cross-sectional area; the SI unit is *pascal* (Pa) which is equivalent to N/m^2.

Ultimate Stress A *Strength of Materials* term; see **Ultimate Strength**.

Ultimate Tensile Strength A Strength of Materials term; see **Ultimate Strength**.

Ultimate Tensile Stress A *Strength of Materials* term; see **Ultimate Strength**.

Ultrasonic Inspection A *Materials Engineering* term; see **Ultrasonic Testing**.

Ultrasonic Machining A *Nonconventional Machining* term, abbreviated as **USM**; a nonconventional machining process which uses high-frequency motion of a shaped tool to remove the materials from *hard* and *brittle* workpieces; capable of producing *intricate shapes* and *complex profiles* which are difficult to produce by traditional machining processes; used to machine all materials whether they are electrically conductive or insulator.

Ultrasonic Testing A *Materials Engineering* term; also called **Ultrasonic Inspection**; a *nondestructive testing* in which defects are detected using high-frequency *acoustic waves*.

D. K. Singh, *Dictionary of Mechanical Engineering*,
https://doi.org/10.1007/978-981-99-1722-8_21

Ultrasonic Wave A *Physics* term; also called **Supersonic Wave** or **Supersonics** or **Ultrasonics**; the *longitudinal wave* having frequency more than 20,000 Hz (the upper limit of audibility), but has shorter wavelength (of the order of 1 cm); the human ear cannot detect this wave similar to infrasonic wave, but some animals like dogs and bats can do so; produced by the high-frequency vibrations of a *quartz crystal* under an alternating electric field (piezo-electric effect); because of its high frequency and small wavelength, it can transmit over long distances without appreciable loss of energy and diffraction; extensively used in detecting any *flaw in a metal*, *finding depth of sea, lakes*, etc. (using echo depth sounding and producing directional signalling), detecting *submarines*, *icebergs*, and other objects in the ocean; in medical fields, it can be used to lame or kill smaller animals like *rats*, *fish*, and *frogs* or *bacteria* and can also *destroy red blood corpuscles* (RBC); in industrial fields, it finds use in *cleaning clothes*, *watches*, etc.; see also **Infrasonic Wave**.

Ultrasonics A *Physics* term; see **Ultrasonic Wave.**

Ultrasound A *Physics* term; a sound wave of high frequency; used extensively in medical applications.

Unavailable Energy A *Thermodynamics* term; the energy which cannot be converted into work by any means even by reversible engines; for example, *energy of the atmosphere*; see also **Available Energy**.

Unbreakable Glass A *Chemistry* term; see **Safety Glass**.

Uncertainty Principle A *Chemistry* term; also called **Heisenberg's Uncertainty Principle**; one of the most celebrated results of quantum mechanics proposed by a German theoretical physicist Werner Heisenberg (1901–1976) in 1927; states that it is impossible to specify simultaneously, with arbitrary precision, both the momentum and the position of a particle.

Uncoated Electrode A *Welding* term; see **Bare Electrode**.

Undercut A *Welding* term; a *groove* melted into the base metal adjacent to the toe of the weld and left unfilled.

Underflow Gate A *Fluid Mechanics* term; an obstruction in the form of a vertical thin plate with an adjustable opening at the bottom, which allows a liquid to flow underneath it; used to measure the liquid flow rate in *open channels*; classified into *sluice gate* and *drum gate*; see also **Weir**.

Under-rigid Truss A *Strength of Materials* term related to a *truss*; a *non-rigid truss* which consists of less number of members than required to make it a rigid truss; for an under-rigid truss: $m < 2j - 3$, where m = number of members and j = number of joints in the truss; see also **Over-rigid Truss**.

Uniaxial Loading A *Strength of Materials* term; a condition of loading in which a member is subjected to loads in one direction; see also **Biaxial Loading**.

Uniaxial Stress A *Strength of Materials* term; a state of stress in which an element of a beam's cross section is subjected to only normal stress along one direction, and there is no shear stress; *tension test* is an example of uniaxial stress condition; see also **Biaxial Stress.**

Uniform Acceleration A *Physics* term; occurs if the velocity of a body changes by equal amounts in equal intervals of time; see also **Variable Acceleration.**

Uniform Circular Motion A *Physics* term; the motion of a particle (or a body) in a circular path with a *constant speed*; the velocity of the particle changes continuously in direction, but its magnitude always remains constant; the particle experiences an acceleration (a) whose magnitude is given as:

$$a = \frac{v^2}{r}$$

where

v = velocity of the particle
r = radius of the circle.

Uniform Flow A *Fluid Mechanics* term; a fluid flow is said to be uniform, when at every point of the flow field, the velocity is identical in magnitude and direction; practically difficult to obtain, but *flow through a prismatic channel or conduit can be regarded as a uniform flow*, as the average velocity of the flow at each cross section is same; see also **Non-uniform Flow.**

Uniform Load A *Strength of Materials* term; also called **Uniformly Distributed Load;** abbreviated as *udl*; a distributed load with a constant intensity (w); represented by a loading diagram in the shape of a rectangle; the total load equivalent to a uniform load is equal to the area of the loading diagram; the weight of a beam of uniform cross section is an example of uniform load; see also **Distributed Load.**

Uniform Speed A *Physics* term; the speed with which an object covers equal distances in equal intervals of time, howsoever small these intervals may be; see also **Variable Speed.**

Uniform Velocity A *Physics* term; the velocity with which an object undergoes equal displacements in equal intervals of time, howsoever small these intervals may be; object moving with a uniform velocity is said to be in uniform motion; see also **Uniform Speed.**

Uniformly Distributed Load *Strength of Materials* term; see **Uniform Load.**

Uniformly Varying Load A *Strength of Materials* term; a vertical load which increases or decreases at a constant rate along the length of a structural member (beam); has triangular or trapezoidal distribution or has also parabolic, cubic, or higher-order curves; the total load equivalent to this load is equal to the area of the distribution of uniformly varying load; see also **Uniform Load.**

Unit Cell A *Materials Engineering* term; *the smallest group of an atom* showing the characteristic lattice structure of a particular metal; the building block of a crystal forming the crystalline structure of a material when repeated by translation; a single crystal can have many unit cells.

Unit Vector A *Physics* term; a vector of *unit magnitude.*

Unity of Command A *Management* term; one of the 14 principles of management proposed by *Henri Fayol* (1841–1925); a component of *chain of command* which suggests that each person within an organization must have a clear reporting relationship to one and only one boss; see also **Scalar Principle.**

Unity of Direction A *Management* term; one of the 14 principles of management proposed by *Henri Fayol* (1841–1925); the system of having a single objective of every group of activities which should be guided by one plan and one head.

Universal Gas Constant A *Chemistry* term; a gas constant (R_u) appearing in the equation of state for an *ideal gas* ($PV = nR_uT$, where P = pressure, V = volume, T = absolute temperature, and n = number of moles of the gas); equals to 8.314 kJ/kmol K; the product of *characteristic gas constant* and the *molar mass* (also called *molecular weight*) of the gas; see also **Characteristic Gas Constant.**

Universal Gravitational Constant A *Physics* term; a proportionality constant which appears in the *Newton's law of gravitation*; equals to the force of attraction acting between two bodies each of unit mass and placed at unit distance apart; it was first determined experimentally in 1798 by a *British* scientist *Henry Cavendish* (1731–1810) who discovered *hydrogen*; its accepted value in SI unit is 6.67×10^{-11} $N.m^2.kg^{-2}$; a *scalar quantity* whose value remains same throughout the universe; does not depend on the nature and size of the bodies, as well as the nature of the intervening medium between the bodies.

Universal Joint A *Design* term; see **Hooke's Joint.**

Universal Testing Machine A *Strength of Materials* term; a multipurpose machine capable of performing several different types of destructive stress tests such as tensile tests, compression tests, direct shear tests, and flexure tests.

Unlike Fractions A *Mathematics* term; the fractions having different denominators; a few examples include $\frac{2}{11}$, $\frac{4}{13}$, $\frac{7}{17}$, etc.; see also **Like Fractions.**

Unlike Vectors A *Physics* term; vectors which are parallel but have opposite directions; see also **Like Vectors.**

Unlubricated Friction A *Physics* term; see **Dry Friction.**

Unsaturated Air A *Thermodynamics* term related to *psychrometry*; the moist air containing water vapour in the *superheated state*; see also **Saturated Air.**

Unstable Equilibrium A *Physics* term; a type of *translational static equilibrium* which ensures a body to get disturbed further after being slightly displaced and released; for example, a book standing on an edge; see also **Stable Equilibrium** and **Neutral Equilibrium**.

Unsteady Flow A *Fluid Mechanics* term; a fluid flow is said to be unsteady, when the fluid velocity changes with time at a point in the flow field; an unsteady flow can be created by varying the discharge in a closed conduit or in an open channel by manipulating a valve or a lock gate; *tidal bore* in a river is another example of unsteady flow that occurs in nature; see also **Steady Flow**.

Unsteady-State Condition A *Fluid Mechanics* and a *Thermodynamics* term; a physical condition in which any parameter of a body changes with time; see also **Steady-State Condition**.

Unsymmetrical Bending A *Strength of Materials* term; also called **Asymmetric Bending** or **Skew Bending**; the bending of a beam which occurs when its cross sections are not symmetric or when the loads do not act in a plane of symmetry; see also **Symmetrical Bending**.

Upper Critical Temperature A *Materials Engineering* term; the temperature at which the transformation of an alloy from one solid structure to another begins, as it is cooling; see also **Lower Critical Temperature**.

Upset Forging A *Manufacturing* term; a hot metal working process used to shape a limited portion of a heated workpiece in the form of wire, rod, or bar using a closed or open die by pressing only that portion; typically used to form *heads on fasteners* such as *bolts*, *screws*, *rivets*, and *nails* and to shape *valves* and *couplings*.

Upset Welding A *Welding* term; a type of *resistance welding*; also called **Resistance Butt Welding**; the joint is produced due to upsetting of metals when two abutting work surfaces are pressed after getting heated to their plastic states by electrical resistance heating; does not involve any flashing; see also **Flash Welding**.

Upward Thrust A *Fluid Mechanics* term; see **Buoyant Force**.

Urea Formaldehyde A *Chemistry* term; a *thermosetting polymer* produced by the chemical combination of *urea* and *formaldehyde*; has good bonding quality, mechanical properties, and dielectric properties; widely used in making *instrument dials*, *electric mixer housings*, *cosmetic boxes*, and *distributor heads*.

USM A *Nonconventional Machining* term; see **Ultrasonic Machining**.

U-Tube Manometer A *Fluid Mechanics* term; a device used to measure the *pressure of a fluid* (both above and below the atmosphere); consists of two tubes (usually called *limbs*) joined at one end to form a U-shaped tube having a uniform bore diameter which contains a manometric liquid (mercury or water); the difference between the levels of the manometric liquid (called *liquid column*) in the two limbs of the U-tube is used to measure the pressure.

V

Vacancy A *Materials Engineering* term; an unoccupied atom site which results when an atom or ion is missing from its normal site in a crystal structure; a type of *material defect* caused during solidification at high temperatures or as a consequence of radiation damage.

Vacuum Casting A *Manufacturing* term related to *casting*; also called **Vacuum Permanent Mould Casting** as it uses *vacuum* to draw the molten metal in the mould cavity of a metal mould (die); the use of vacuum in the process imparts uniform and improved mechanical properties in the cast parts; particularly suitable for producing castings of thin-walled complex shapes; an alternative to *investment*, *shell-mould*, and *green-sand* casting; used to produce *gas turbines parts*.

Vacuum Permanent Mould Casting A *Manufacturing* term related to *casting*; see **Vacuum Casting**.

Vacuum Pressure A *Physics* term; see **Vacuum**.

Vacuum A *Physics* term; also called **Vacuum Pressure** or **Negative Gauge Pressure** or **Suction Pressure**; the pressure below the atmospheric pressure; see also **Gauge Pressure** and **Absolute Pressure**.

Valence Electron A *Chemistry* term; the number of electrons in the valence shell (the outermost orbit) of an atom; for example, sodium (Na-11) has the electronic configuration $1s^2 2s^2 2p^6 3s^1$; hence, its valence shell is its *third orbit*, and its valence electron is 1 which also shows its *valency*.

Valence Shell A *Chemistry* term; the *outermost orbit* of an electron in an atom; for example, *sodium* (Na-11) has the electronic configuration of $1s^2 2s^2 2p^6 3s^1$; hence, its valence shell is its *third orbit*; see also **Valency**.

Valency A *Chemistry* term; the number of electrons in an atom which participates in a chemical bonding; see also **Valence Shell**.

D. K. Singh, *Dictionary of Mechanical Engineering*,
https://doi.org/10.1007/978-981-99-1722-8_22

Value Chain An *Operations Management* term related to a *supply chain*; also called **Value Network**; an improved supply chain; the activities that do not add value to the supply chain are bypassed or eliminated in the value chain; the concept of value chain was introduced by *Michael Porter*, a Harvard Business School Professor, in the 1980s.

Value Engineering A *Manufacturing* term; the process of analysing the operations of products or services with an objective to improve them so that the products or services can be delivered at reduced costs.

Value Network An *Operations Management* term related to a *supply chain*; see **Value Chain.**

Valve A *Fluid Mechanics* term; a device used to control the flow of fluid.

Van der Waals Equation A *Chemistry* term; an *equation of state* named after *Dutch* physicist *Johannes Diderik Van der Waals* (1837–1923) who proposed it in 1873; improves the ideal-gas model by incorporating the *intermolecular attraction forces* and the *volume occupied by the gas molecules* themselves; used to describe the behaviour of *real gases*; expressed as:

$$\left(p + \frac{a}{v^2}\right)(v - b) = RT$$

where

p = pressure of the gas
v = volume occupied by the gas
R = characteristic gas constant
T = absolute temperature of the gas
a, b = constants

The term $\frac{a}{v^2}$ accounts for the *intermolecular forces*, and b accounts for the *volume occupied by the gas molecules.*

Van der Waals Force A *Chemistry* term; named after the *Dutch* physicist *Johannes Diderik Van der Waals* (1837–1923); a *force of attraction* which exists between all types of molecules, whether *polar*, *nonpolar* or *nonpolar mono-atomic molecules* such as *helium (He)*, *neon (Ne)* or *argon (Ar)*; holds the molecules in position in a molecular crystal; has the maximum value in solids and decreases on passing from solid to liquid, and from liquid to gaseous states.

Vanadium A *Chemistry* term; a *chemical element* having symbol *V* with atomic number 23 and atomic weight 50.942; a silvery grey *nonferrous metal* (in its pure form) which is *hard* and *corrosion-resistant*; melts at 1900 °C; an important alloying element added to steel; has wide industrial applications, chiefly in the manufacture of *alloy steels* (in the form of *ferro-vanadium alloys*); acts as a *scavenger* for oxygen and nitrogen, and makes the steel stronger, tougher, shock-resistant and more ductile, and also raises its tensile strength; steel containing chromium and vanadium is used in making parts for *motor vehicles*; the alloys of vanadium with titanium, copper and aluminium are used in industry for high-temperature applications.

Vanes A *Fluid Mechanics* term; the curved blades attached to the impeller of a centrifugal pump around the shaft in a regular pattern.

Vaporization A *Thermodynamics* term; see **Evaporation**.

Vapour Power Cycle A *Thermodynamics* term related to a *power plant*; a thermodynamic cycle which uses *vapour* or *steam* as the working fluid; used in steam power plants; the *Rankine cycle* is the most important vapour power cycle; other vapour power cycles include *reheat* and *regenerative* cycles.

Vapour Pressure A *Thermodynamics* germ; the (partial) pressure exerted by the vapour or gas of a liquid in its phase equilibrium at a given temperature; a property of a pure substance; the atmospheric pressure is the sum of the partial pressure of dry air and the partial pressure of water vapour; when pressure in a liquid is reduced to less than the vapour pressure, the liquid suddenly changes its phase to vapour; increases with temperature due to increased molecular activity.

Vapour A *Thermodynamics* term; an intermediate phase between liquid and gas; for example, wet steam.

Vapour-Compression Refrigeration System A *Refrigeration* term; the most widely used refrigeration system; works on a closed cycle in which the refrigerant (in a vapour phase) is compressed by a compressor, which then flows into a condenser (where heat is removed to condense it to a liquid) which further flows through an expansion valve into an evaporator (where the temperature and pressure are low), and the evaporating refrigerant absorbs heat from the low-temperature region in the surroundings thus producing cooling effect.

Variable Acceleration A *Physics* term; occurs if the velocity of a body changes by unequal amounts in equal intervals of time; see also **Uniform Acceleration**.

Variable Characteristics A *Statistics* term; also called **Variable Measure**; a product characteristic which can be measured with quantification such as *weight*, *length*, or *volume*; see **Attribute Characteristics**.

Variable Costs An *Operations Management* term related to *cost analysis*; vary with the *volume of output* produced; the costs of material and labour used in a product or service are main variable costs; the total variable costs are expressed as unit cost of material or labour multiplied by their total units; see also **Fixed Costs**.

Variable Measure A *Statistics* term; see also **Variable Characteristics**.

Variable Speed A *Physics* term; the speed with which an object covers either equal distances in unequal intervals of time or covers unequal distances in equal intervals of time, howsoever small these intervals may be; see also **Uniform Speed**.

Variable A *Mathematics* term; a quantity which can take up different values during mathematical operations; classified into dependent and independent variables; see also **Constant**.

Variance A *Statistics* term; a parameter used as a *measure of deviation* from the *mean* in a sample or distribution; equals to the square of standard deviation (σ^2); has wide statistical applications; used in *program evaluation and review technique* (PERT) where it is given as:

$$\sigma^2 = \left(\frac{b-a}{6}\right)^2$$

where

a = optimistic time of an activity
b = pessimistic time of an activity.

Varignon's Theorem A *Physics* term related to *mechanics*; proposed by the *French* mathematician *Pierre Varignon* (1654–1722); states that the moment of a force about any point is equal to the algebraic sum of the moments of the components of the force about the same point.

Vector Product of Vectors A *Physics* term; also called **Cross Product of Vectors**; the product of the magnitudes of the two vectors and the sine of the angle between them; a *vector quantity* which has the direction perpendicular to the plane containing the two vectors; given as:

$$\boldsymbol{A} \times \boldsymbol{B} = AB \sin\theta \, \boldsymbol{n}$$

where

A = magnitude of vector $\boldsymbol{A}$
B = magnitude of vector $\boldsymbol{B}$
θ = angle between the two vectors $\boldsymbol{A}$ and $\boldsymbol{B}$
$\boldsymbol{n}$ = unit vector perpendicular to the plane of vectors $\boldsymbol{A}$ and $\boldsymbol{B}$

The physical vector quantities like *torque, angular momentum,* or *force* on a moving charge in a magnetic field can be described as the vector product of two vectors; follows the *distributive law*, but *does not follow the commutative law of algebra*; if either of the two vectors is multiplied by a scalar, the vector product also gets multiplied by the scalar; the vector product of two parallel vectors (like or unlike vectors) is a null vector (as $\sin 0° = \sin 180° = 0$); the vector product of a vector by itself is also a null vector (as sin 0° = 0); the vector product of unit orthogonal vectors are defined as:

$$\boldsymbol{i}\times\boldsymbol{i}=\boldsymbol{j}\times\boldsymbol{j}=\boldsymbol{k}\times\boldsymbol{k}=0$$
$$\boldsymbol{i}\times\boldsymbol{j}=-\boldsymbol{j}\times\boldsymbol{i}=\boldsymbol{k}$$
$$\boldsymbol{j}\times\boldsymbol{k}=-\boldsymbol{k}\times\boldsymbol{j}=\boldsymbol{i}$$
$$\boldsymbol{k}\times\boldsymbol{i}=-\boldsymbol{i}\times\boldsymbol{k}=\boldsymbol{j}$$

where the cyclic order is important; see also **Scalar Product of Vectors**.

Vector Quantity A *Physics* term; a physical quantity which has both magnitude and direction; a few examples include *displacement*, *velocity*, *acceleration*, *force*, and *moment*; see also **Scalar Quantity**.

VED Analysis An *Operations Management* term related to *inventory*; a *selective inventory control technique* which classifies the inventories based on the criticality of components into three categories: *V* (*vital*) items, *E* (*essential*) items and *D* (*desirable*) items; *V items* are absolutely required items whose absence will make the system inoperative (e.g. *boilers* for power plants and *heart* for a human body); *E items* are required for the efficient running of the system, whose presence will increase efficiency and its absence will decrease efficiency (e.g. *heaters* and *warm clothes* in winter season); *D items* are desirable which neither stop nor reduce the efficiency of a system but their availability increases efficiency and reduces failure (e.g. *boiler accessories* for power plants).

Velocity Head A *Fluid Mechanics* term; also called **Kinetic Head**; an energy parameter which appears in the *Bernoulli equation*; represents the elevation needed for a fluid to reach the velocity v from the rest during a frictionless free fall; expressed as $v^2/2\,g$; (v = velocity of the fluid, and g = acceleration due to gravity = 9.81 m/s^2); see also **Pressure Head** and **Elevation Head**.

Velocity of Light A *Physics* term; the distance travelled by the light in vacuum in unit time; equals to 3×10^8 m/s.

Velocity Polygon A *Fluid Mechanics* term; see **Velocity Triangle**.

Velocity Potential A *Fluid Mechanics* term; also called **Potential Function**; a *scalar function* of *space* and *time*; the negative partial derivative of velocity potential with respect to any direction gives the velocity component of the fluid flow in that direction; denoted by φ; given as;

$$V_x=-\frac{\partial\varphi}{\partial x}$$
$$V_y=-\frac{\partial\varphi}{\partial y}$$

$$V_z = -\frac{\partial \varphi}{\partial z}$$

Negative sign associated with φ indicates that φ decreases in the direction of fluid flow; existence of velocity potential for a flow implies that it is an *irrotational flow*; see also **Stream Function.**

Velocity Ratio A *Theory of Machines* term related to a *gear*; abbreviated as **VR**; in a weight lifting machine (*screw jack*), a ratio of the distance moved by *effort* to the distance moved by *load*; in a gearset, a ratio of the output speed (*angular velocity of gear*) to the input speed (*angular velocity of pinion*); also in a gearset, a ratio of the number of teeth on the pinion to the number of teeth on the gear; velocity ratio is the reciprocal of gear ratio; see also **Gear Ratio.**

Velocity Triangle A *Fluid Mechanics* term; also called **Velocity Polygon**; a triangle whose sides are used to represent the three velocities used in the analysis of turbo-machines (*turbines, pumps,* etc.); the three velocities include the *blade speed*, the *absolute fluid velocity*, and the *relative velocity*, where the *fluid velocity* is the vector sum of the *relative velocity* and the *blade velocity.*

Velocity A *Physics* term; a ratio of the displacement of an object to the corresponding time taken by the object; a *vector quantity*; the SI unit is m/s; see also **Speed.**

Vena Contracta A *Fluid Mechanics* term; the minimum area of a fluid flow in a sudden contraction which results when the converging streamlines begin to expand to fill up the downstream area, where pressure is minimum.

Vendor Managed Inventory An *Operations Management* term, abbreviated as **VMI**; a system of inventory management in which the supplier takes the responsibility of managing the inventories (raw materials, parts, components, etc.) for the manufacturer (buyer) at its location/premises; see also **Co-Managed Inventory.**

Vent Rod A *Manufacturing* term; also called **Vent Wire**; a sharp-pointed *foundry tool* used to produce small holes in the rammed sand of the mould to allow escape of gases generated during pouring of the molten metal in the mould so as to prevent the casting defects.

Vent Wire A *Manufacturing* term; see **Vent Rod.**

Vent A *Manufacturing* term related to *casting*; a large number of *small openings* made in the *sand mould* to carry off *air* and *gases* generated when the molten metal comes into contact with the sand in the mould cavity and core.

Venturimeter A *Fluid Mechanics* term; a *metering device* named after the *Italian* physicist *Giovanni B. Venturi* (1746–1822) who investigated its principle in 1791; used to *measure the flow rate* of both compressible and incompressible fluids, using Bernoulli's equation; resembles to a converging–diverging nozzle; the coefficient of discharge of a venturimeter is close to unity (≈ 0.99).

Vernier Calliper A *Metrology* term; also called **Calliper Gauge**; an *end measuring instrument* used to measure internal or external dimensions of an object.

Vertical Centroidal Axis A *Strength of Materials* term; the axis which is vertical to the cross section of a body; see also **Horizontal Centroidal Axis**.

Vibration A *Physics* term; the oscillation of high frequency of a body; for example, *the vibration of a string of a musical instrument.*

Vibratory Motion A *Physics* term; see **Oscillatory Motion**.

Vibrograph A *Physics* term; a device used to measure *amplitude* of vibration in a given direction.

Vickers Test A *Strength of Materials* term; also called **Diamond Pyramid Hardness Test**; a *destructive method* of hardness testing of a material which uses a 136° pyramid-shaped diamond indenter to be pressed against the test specimen by applying a load varying between 1 and 120 kg; uses a number, called Vickers pyramid number (VPN) to define the hardness, given by

$$\text{VPN} = \frac{\text{Load applied in } Kgf}{\text{Surface area of the pyramidal indentation in mm}^2} = \frac{1.854P}{D^2}$$

where

P = load applied on the test specimen
D = diagonal of indentation produced.

Victor H. Vroom (1932-) A well-known *motivation personality* who proposed '*expectancy theory of motivation*'; a business school professor at the Yale School of Management who is an authority on the psychological analysis of behaviour in organizations, particularly on leadership and decision-making. His most well-known books include '*Work and Motivation*', '*Leadership and Decision Making*', and '*The New Leadership*'.

Virtual Organization A *Management* term; see **Boundaryless Organization**.

Viscoelastic Fluid A *Fluid Mechanics* term; a fluid which recovers its deformed original shape partially on the removal of the applied shear stress.

Viscoelasticity A *Fluid Mechanics* term; a time-dependent deformation process which results from both *viscosity* and *elasticity*.

Viscometer A *Fluid Mechanics* term; a device used to measure the *viscosity* of a fluid.

Viscosity A *Fluid Mechanics* term; also called **Fluid Friction** or **Perfect Friction** or **Internal Friction** or **Viscous Drag**; the internal fluid friction which occurs on account of relative motion between every two layers of a fluid; an important property of a fluid which causes resistance to the flow of the fluid; a measure of fluid's resistance to shear or angular deformation; the viscosities of all liquids decrease, while viscosities of all gases increase with rise in temperature.

Viscous Drag A *Fluid Mechanics* term; see **Viscosity**.

Visible Light A *Physics* term; the portion of the *electromagnetic radiation* which can be perceived by the human eye; the wavelength of visible light varies between 400 and 700 nm.

Vitreosil A *Chemistry* term; a *translucent variety of glass* obtained by fritting sand with an electrically heated carbon rod or plate.

Vitreous Silica A *Chemistry* term; a commercial glass which is nearly pure SiO_2.

VMI An *Operations Management* term; see **Vendor Managed Inventory**.

V-Notch Weir A *Fluid Mechanics* term; see **Triangular Weir**.

VOC A *Quality Management* term; see **Voice of the Customer**.

Voice of the Customer A *Quality Management* term, abbreviated as **VOC**; a process of capturing customer needs and preferences; *quality function deployment* (QFD) is used as a tool for translating voice of the customer into specific technical requirements.

Volatility A *Physics* term; a *physical property* of a substance which enables it to get easily vaporized; a low melting point metal is more volatile than a higher melting point metal.

Volt An *Electrical* term; the force which causes the electrical current to flow; the SI unit of potential difference, named after an *Italian* physicist *Alessandro Volta* (1745–1827).

Voltage An *Electrical* term; a difference in electric potential.

Volume Flexibility An *Operations Management* term; the ability of a production system to quickly increase or reduce the volume of products or services produced; see also **Product Flexibility**.

Volume Flow Rate A *Fluid Mechanics* and a *Thermodynamics* term; also called **Discharge** or simply **Flow Rate**; the volume of a fluid that flows through a cross section per unit time; denoted by Q; the SI unit is m^3/s.

Volume Modulus of Elasticity A *Strength of Materials* term; see **Bulk Modulus of Elasticity**.

Volume The amount of space occupied by a three-dimensional object or enclosed within a container.

Volumetric Analysis A *Chemistry* term; the analysis of a mixture of gases based on the moles (or volumes) of each component; see also **Gravimetric Analysis**.

Volumetric Efficiency A *Fluid Mechanics* term usually used in case of centrifugal pumps and especially fans; a ratio of the *actual discharge* to the *ideal discharge*, which is the sum of the actual discharge and the loss of fluid due to leakage in the impeller casing clearances.

Volumetric Strain A *Physics* and a *Strength of Materials* term; also called **Dilatation**; the strain produced on account of change in volume of a body by a deforming force; calculated as a ratio of the change in volume to the original volume of a body.

Volute A *Fluid Mechanics* term related to a *pump*; see **Shroud**.

von Mises Yield Criterion A *Strength of Materials* term related to *theory of material's failure*; see **Maximum Distortion Energy Theory**.

Vortex Flow A *Fluid Mechanics* term; also called **Vortex Motion**; a fluid flow in which the streamlines are concentric circles; classified into *free vortex* and *forced vortex*.

Vortex Motion A *Fluid Mechanics* term; see **Vortex Flow**.

Vorticity A *Fluid Mechanics* term; a *vector quantity* defined as a *curl of the velocity vector* (V) of a fluid; used as a measure of rotation of a fluid particle as it moves in the flow field; equals to twice the angular velocity of a fluid particle; zero for an irrotational flow; denoted by ζ (zeta); expressed as:

$$\zeta = \nabla \times V = 2\omega$$

where

ω = angular velocity of the fluid particle
V = linear velocity of the fluid particle (shown in bold to indicate its vector nature).

Vroom's Expectancy Theory A *Management* term related to *motivation*; also called **Expectancy Theory of Motivation**; a motivation theory proposed by *Victor H. Vroom* (1932–) in 1964; states that a person's motivation to achieve a goal is dependent on the value which the person places on achieving the goal and on the expectancy of attaining it; the theory is more near to the realistic situations.

Vulcanization A *Chemistry* term; the transformation of a polymer with a linear structure into one with a network structure by means of *crosslinking*; a process in which crude rubber (called *latex*, obtainable from a rubber tree) is mixed with small amounts of *sulphur* and then heated; *imparts good mechanical properties to rubber*; addition of *filler materials* such as *carbon black* and *zinc oxide* to crude rubber before vulcanization improves its *wear resistance*.

W

W. Edwards Deming (1900–1993) An *American* quality expert who is widely known as '*father of quality control*'; introduced the concepts of '*total quality control*'; widely credited for placing *Japan* in its world leadership position in the quality of its manufactured products; also expanded the use of *plan-do-check-act* (PDCA) cycle, originally developed by *Walter A. Shewhart* (1891–1967); the *Deming award* for quality is named after him.

Wafer An *Electronics* term; a thin slice made from a cylindrical single crystal of high-purity material, usually *silicon.*

Wake A *Fluid Mechanics* term; a region of fluid flow which trails a body where the effects of the body on the fluid velocity can be felt.

Walter Andrew Shewhart (1891–1967) An *American* quality expert who is widely known as '*father of statistical quality control*'; one of the quality gurus and pioneers of *quality management* who developed '*control charts*' which formed the basis of statistical quality control; also introduced the term '*quality assurance*' to improve the quality through the use of statistical control methods; originally developed the *PDCA* (Plan-do-check-act) cycle; wrote a book '*Economic Control of Quality of Manufactured Product* (1931)', which contained most of his thinking on quality.

Warding File A *Manufacturing* term; one of the types of a file used in the *fitting shop* which resembles a *flat file*, but has a sharply tapered width towards point and reduced thickness; used in the filing of *narrow slots.*

Warm Forging A *Manufacturing* term; an intermediate stage of forging lying between cold forging and hot forging in which a workpiece is heated at a moderate temperature before hammering; see also **Cold Forging** and **Hot Forging**.

Warren Truss A *Strength of Materials* term related to a *truss*; a commonly used standard truss usually consisting of *equilateral triangles.*

Wash A *Casting* term; a type of *casting defect* which occurs on a *casting's drag surface* near the gate of the mould *due to high velocity of the molten metal.*

D. K. Singh, *Dictionary of Mechanical Engineering*,
https://doi.org/10.1007/978-981-99-1722-8_23

Water Gas A *Chemistry* term; an important *industrial fuel* which is a mixture of *carbon monoxide* (CO) and *hydrogen* (H_2), that is, ($CO + H_2$); produced when steam is passed over red hot coke; see also **Producer Gas**.

Water Hammer A *Fluid Mechanics* term; an undesirable phenomenon caused by the acoustic waves which propagate and reflect in a confined liquid, and produces noise similar to hammering on a pipe; a phenomenon of excess rise in pressure of a liquid (water), when its flow is suddenly decreased by manipulating a valve; for example, occurs when a valve is closed abruptly; causes knocking of a pipe system and results in even failure, if the pipe system is not properly designed; the hydroelectric plants incorporate *safety devices* to reduce the adverse effect of water hammer which may occur due to rapid change in water flow with load change on the turbine.

Water Jet Machining A *Nonconventional Machining* term, abbreviated as **WJM**; also called **Hydrodynamic Machining**; a nonconventional machining process which uses an extremely high pressure water jet for cutting soft materials like *wood, plastic*, and *aluminium*.

Water Level Indicator A *Power Plant* term related to a *boiler*; a *boiler mounting* used to indicate the safe level of water to be always maintained inside the boiler to prevent its damage due to overheating; a boiler usually consists of two water level indicators.

Water Table A *Hydrology* term; the top of the saturated groundwater zone.

Water Tube Boiler A *Power Plant* term related to a *boiler*; a type of boiler in which water flows through a number of inclined straight tubes surrounded by hot flue gases; important features of a water tube boiler include fast steam generation, high evaporative capacity, and high steam pressure range; suitable for large power plants and ships; examples include *Babcock and Wilcox boiler*, *Stirling boiler*, *La-Mont boiler*, and *Bension boiler*; Babcock and Wilcox boiler is the most commonly used water tube boiler; see also **Fire Tube Boiler**.

Water Vapour A *Thermodynamics* term; also called **Moisture**; the vapour of water.

Water A *Chemistry* term; a chemical compound of *hydrogen* and *oxygen* having the molecular formula H_2O and molecular weight 18; an incompressible substance which exists in three phases, namely solid (ice), liquid (water) and gas (vapour or steam); pure water has the pH value of 7 (neither acidic nor alkaline); density of water at standard atmospheric pressure and at 15.5 °C is assumed to be 1000 kg/m^3.

Water-Cooled Condenser A *Power Plant* term; a condenser which uses water as the condensing medium; see also **Air-Cooled Condenser** and **Evaporative Condenser**.

watt A *Physics* term; the SI unit of power named after *James Watt* (1736–1819), who invented the *steam engine*; abbreviated as *W*; equivalent to joule per second (J/s).

WBD A *Thermodynamics* term related to *psychrometry*; see **Wet Bulb Depression.**

WBT A *Thermodynamics* term related to *psychrometry*; see **Wet Bulb Temperature.**

Wealth of Nations A book authored by the famous economist *Adam Smith* in 1776 which deals with the economic advantages which can be realized due to the *division of labour* introduced for the workers.

Wear A process of removal of surface materials from a workpiece as a result of a mechanical action; classified into *abrasive wear*, *adhesive wear*, and *corrosive wear*.

Weber Number A *Fluid Mechanics* term; a *dimensionless* (*nondimensional*) group, named after *Moritz Weber* (1871–1951), who worked on problems involving *capillary* effects and developed the modern *laws of similitude*; defined differently for a moving liquid and a liquid at rest; in case of a *moving liquid*, it represents a ratio of *inertia* force to *surface tension* force; in case of a liquid at rest, it represents a ratio of *gravitational* force to *surface tension* force; given as:

$$\mathrm{We} = \frac{\rho V^2 L}{\sigma} \text{ (for a moving liquid)}$$

$$\mathrm{We} = \frac{\rho g L^2}{\sigma} \text{ (for a liquid at rest)}$$

where

ρ = density of the fluid
V = velocity of fluid flow
L = characteristic length
σ = surface tension
g = acceleration due to gravity.

Surface tension is important for flows having We << 1, but is usually neglected; for two systems to be *dynamically* similar, their *Weber number* should be same; useful in *model analysis* involving *capillary* waves in channels, very low weir head, air entrainment, and motion of liquid in the form of *jets* and *sprays*.

Wedge A *Physics* term; a means of producing small movements in the position of a body or a means of applying large forces; a *metallic* or *wooden* piece of *triangular* or *trapezoidal* cross section which can be used for lifting loads or for slight adjustments in the position of an object, for example, used for tightening *fits* or *keys* for shafts.

Weight A *Physics* term; the gravitational force exerted on a body by the earth; a concentrated force acting through the centre of gravity of the body; a *vector quantity*, which is directed towards the centre of the earth; denoted by W; quantified as the product of mass (m) of the body and the acceleration due to gravity (g), that is, *mg*; the weight of a body at the centre of the earth is zero, but its mass is not zero and changes with the change in the value of *g*; a body at moon has weight equal to *one-sixth* of its weight at earth; the SI unit is *newton* (N); has dimension [MLT^{-2}].

Weighted Moving Average Method An *Operations Management* term related to *forecasting*; a *quantitative* method of demand forecasting; a *modified* and *improved* method over moving average method; based on assigning *differential weighting* to data where *more weights* are assigned to *recent data* such as the weight W_1 is assigned to the most recent data, W_2 to the second most recent, and so on, where the sum of all the weights is equal to 1; the forecast value $F(t)$ in time t is given as:

$$F(t) = W_1 V_1 + W_2 V_2 + \cdots + W_n V_n$$

where $V_1, V_2 \ldots V_n$ represent past demand data; see also **Moving Average Method.**

Weightlessness A *Physics* term; a condition in which the observed weight of a body becomes zero; occurs when the body is under *free fall*, that is, the acceleration due to gravity (*g*) is zero; every object inside a satellite orbiting around the earth is in a state of weightlessness, as its acceleration is equal to the acceleration due to gravity directed towards the centre of the earth.

Wein's Displacement Law A *Heat Transfer* term; see **Wien's Law**.

Weir A *Fluid Mechanics* term; a *flow measuring* as well as a *controlling device*; usually a *simple dam* (narrow or broad) constructed across an *open channel*, where water to be metered is allowed to flow over its top or crest; the rate of flow is determined by measuring the depth of upstream flow over the crest of the weir; classified into *broad-crested weir* and *sharp-crested weir*; see also **Underflow Gate.**

Weld Bead A *Welding* term; a weld deposit which results from a pass during a welding operation.

Weld Crack A *Welding* term; a *welding defect*; occurs in the form of a crack in the weld metal.

Weld Joint A *Welding* term; a union of two or more members produced by using a welding operation.

Weld Metal A *Welding* term; the portion of the base metal which has been melted during welding.

Welder A *Welding* term; a person who is capable of performing a welding operation.

Welding Current A *Welding* term; the current flowing through a welding circuit during a welding operation.

Welding Procedure A *Welding* term; the detailed *methods* and *practices* including joint welding procedures involved in the production of a weldment.

Welding Shop A *Manufacturing* term; a part of the central workshop where workpieces are joined together using various welding processes.

Welding Technique A *Welding* term; the details of a manual, machine, or semi-automatic welding which a welder or welding operator follows and uses during a welding operation.

Welding Torch A *Welding* term; a device used to mix *oxygen* and *acetylene* in the required proportion to carry out the welding operation in *gas welding* or *torch brazing*; see also **Cutting Torch**.

Welding A *Manufacturing* term; a *manufacturing* process used to join two parts (mostly *similar metals*) permanently by heating them to suitable temperature with or without the application of *pressure* and with or without using *filler metal* or *flux*; produces the strongest possible joint which equals to the strength of parent metal; classified into *fusion* and *plastic* welding processes.

Weldment A *Welding* term; an assembly whose component parts are joined by welding.

Wet Bulb Depression A *Thermodynamics* term related to *psychrometry*, abbreviated as **WBD**; the difference between the *dry bulb temperature* (DBT) and the *wet bulb temperature* (WBT); used as an indirect measure of the *dryness of the air*; for saturated air, WBD is zero, as both DBT and WBT are equal; see also **Dry Bulb Temperature** and **Wet Bulb Temperature**.

Wet Bulb Temperature A *Thermodynamics* term related to *psychrometry*, abbreviated as **WBT**; the temperature of the *liquid water* in the *moist air*; see also **Dry Bulb Temperature**.

Wetness Fraction A *Thermodynamics* term related to *steam*; a second measure of quality of the wet steam; calculated as a ratio of the mass of water vapour (m_f) to the mass of total wet steam ($m_g + m_f$) where m_g is the mass of dry steam; expressed as $(1 - x)$, where x is the dryness fraction of the wet steam; see also **Dryness Fraction**.

Wetted Area A *Fluid Mechanics* term; the total surface area of a body in contact with a fluid; finds use in *drag analysis*.

Wetted Perimeter A *Fluid Mechanics* term; the length of a solid channel cross-section surface in contact with a liquid; used in a fluid flow through *non-circular conduits*.

Wetting A *Welding* term; the *bonding* or *spreading* of a liquid filler metal or flux on a solid base metal.

Whip A *Theory of Machines* term; an *undesirable phenomenon* which occurs in an *open belt drive* in which the belt vibrates in a direction normal to the direction of the belt motion; occurs when the centre-to-centre distance between the driving and driven pulleys exceeds 15 m; see also **Slip**.

Whipping A *Welding* term; a *manual welding technique* in which an arc or a gas flame is moved back and forth in the direction of the weld path (with or without lengthening the arc or gas flame).

Whistle-Blower A *Management* term; an employee who reports *illegal* or *wrongful activities* of his employer or fellow employees.

Whistle-Blowing A *Management* term; a process of making known to outside agencies the unethical company practices; see also **Whistle-Blower**.

White Cast Iron A *Materials Engineering* term; also called **Chilled Cast Iron** because of *fast cooling rate* during its formation; a *hard, very wear resistant, brittle* form of cast iron which has a characteristic white crystalline fracture surface; contains carbon in the combined state as *cementite* (Fe_3C); finds use where high abrasion resistance is required such as in *hammer mills, crusher jaws, crushing rollers,* and *wear plates.*

White Metal A *Materials Engineering* term; an alloy of *tin (Sn), antimony (Sb),* and *copper (Cu)* consisting of 82% *Sn,* 12% *Sb* and 6% *Cu*; used in making *table wares.*

White Zinc A *Chemistry* term; also known as **Zinc Oxide (ZnO)**; finds use in making *paints,* and is superior to white lead as it does not blacken.

Whiteware A *Materials Engineering* term; a commercial wired *ceramic* which has a typical white and fine-grained microstructure; a few examples include *tile* and *pottery.*

Wien's Law A *Heat Transfer* term; also called the **Wein's Displacement Law**; a radiation heat transfer law developed by a Nobel laureate *German* physicist *Wilhelm Wien* (1864–1928) in 1891; states that the product of the maximum wavelength of radiation and its temperature remains constant, that is, $\lambda_{max}T$ = constant = 2898 μm K; the locus of the maxima of the radiation emission curves.

William John MacQuorn Rankine (1820–1872) A *Scottish* engineer and a professor of civil engineering at the University of Glasgow, who made several outstanding contributions in the field of *thermodynamics* along with contributing in *mechanics, canals, steam engines,* and *water supply systems.*

Wilson Line A *Thermodynamics* term related to *steam*; the locus of points on the *h–s diagram* of steam where condensation of steam takes place regardless of the initial temperature and pressure at the entrance of a high-velocity nozzle; positioned between the 4 and 5% moisture curves (often approximated by the 4% moisture line) in the saturation region on the *h–s diagram.*

Wind Turbine A *Fluid Mechanics* term; also called **Windmill**; a *turbomachine* which uses air as a working fluid which extracts energy from the wind; used to pump water or generate electricity.

Windmill A *Fluid Mechanics* term; see **Wind Turbine**.

Window Glass A *Chemistry* term; see **Ordinary Glass**.

Winglet A *Fluid Mechanics* term; a short aerodynamically contoured wing which is set perpendicular to the wing of an aircraft at the tip; used to reduce the strength of the trailing vortex system and the induced drag; also produces a small component of force in the flight direction, which has the effect of further reducing the overall drag of the aircraft.

Wink Counter A *Work Study* term; see **Microchronometer**.

Wink An *Operations Management* term related to *work study*; a unit of *measuring time* which equals to 1/2000 of a minute (called *one wink*); finds use in the *SIMO chart* to record the time of elements of motion, called *therbligs*.

Wire Gauge A *Measurement* term; an instrument used to measure the *wire diameter*; consists of a thin circular disc having many slots on its circumference, which are numbered according to the size of the wire; high gauge number means thin wire and lower gauge number indicates thick wire.

Wire Rope A *Design* term; see **Cable**.

Wire-EDM A *Nonconventional Machining* term; a *variation of electrical discharge machining* (EDM) in which a continuously moving conductive wire is used as a tool electrode in place of a contour-shaped tool electrode used in conventional EDM; used to produce very *fine* and *intricate die openings in hard materials*; used for both *machining* and *cutting*.

Withdrawal Kanban An *Operations Management* term; also called **Conveyance Kanban** or **C-Kanban**; a card physically attached to a container or cart, which authorizes the *movement of goods* in a production system; a distinctive feature of JIT production system; see also **Production Kanban**.

WJM A *Nonconventional Machining* term; see **Water Jet Machining**.

Wood Charcoal A *Chemistry* term; an *amorphous* allotropic form of carbon obtained by strong heating of wood in a limited supply of air.

Wood A *Materials Engineering* term; a *natural, fibre-reinforced composite* which is soft enough to be given any shape; a *non-metallic material* which is lightweight; can be subjected to moderate shock loading; can absorb moisture and get distorted; a widely used *pattern material* for *sand casting*; finds extensive use in *furniture industry* in making various furniture items such as *chair*, *table*, *bed*, etc.

Woodruff Key A *Design* term related to a *key*; *a type of sunk key* in the form of a segment of a circular disc of uniform thickness; mainly used on the tapered shafts of machine tools and automobiles.

Work Centre An *Operations Management* term; see **Workstation.**

Work Group A *Management* term; a group which interacts primarily to share information and to make decisions which will help each other; see also **Work Team**.

Work Measurement An *Operations Management* term related to *work study*; a process of estimating the amount of worker time required to produce one unit of output (say product); useful in the development of labour standards (standard time involved in a task) required for planning and controlling operations.

Work Ratio An *IC Engine* term; a ratio of the *net work output* to the *positive work output* of a heat engine (the net work output is the difference between positive work output and input work); used as an *index of performance* for a heat engine; see also **Efficiency Ratio**.

Work Sampling A *Work Study* term; a work measurement method of taking random qualitative observations of a work to determine the percentage of its productive and unproductive time; commonly used in computing *allowance time* for a worker, in estimating costs of certain activities and in investigating work methods.

Work Study An *Operations Management* term; the systematic investigation of all the factors of human work which affect the efficiency and economy of a production system in order to improve them; a tool used to analyze and measure operations and to devise ways and means of improving productivity; classified into *motion study* and *time study*.

Work Team A *Management* term; a group which engages in *collective work* which requires joint effort, and generates a positive energy; see also **Work Group**.

Work A *Physics* term; a *form of energy*; the work is said to be done, if a force acting on a body moves it through a certain distance in the direction of force; a *scalar quantity* denoted by W; a scalar product of the *force* (F) and *displacement* (d), given as: $W = F.d \cos\theta$, where θ is the angle between the direction of force and the displacement; if the force is at right angles to the displacement, the work is zero; can be negative also; the CGS unit of work is *erg* which is equivalent to *dyne-cm*; the SI unit is *joule* (J) which is equivalent to N-m; $1 \text{ J} = 10^7$ erg.

Work-Energy Principle A *Physics* term; see **Work-Energy Theorem**.

Work-Energy Theorem A *Physics* term; also called **Work-Energy Principle**; states that the work done in displacing a body is equal to the change in kinetic energy of the body.

Worker-Machine Chart A *Motion Study* term; see **Activity Chart**.

Workforce A *Management* term; all non-managerial employees of an organization.

Working Capital A *Financial Management* term; see **Current Assets**.

Working Drawing An *Engineering Drawing*; see **Production Drawing.**

Working Fluid A *Thermodynamics* term; a fluid used in heat engines and other cyclic devices to transfer heat through it while undergoing a cycle; air is often used as an important working fluid in thermodynamic cycles.

Working Stress A *Strength of Materials* term; also called **Design Stress** or **Permissible Stress**; a safe limit stress that is obtained by dividing the ultimate (maximum) stress of a material by the factor of safety; finds use in the design of engineering components.

Working Stroke A *Thermodynamics* term related to an *IC engine*; see **Power Stroke**.

Working Substance A *Fluid Mechanics* and a *Thermodynamics* term; a substance usually a fluid, which is capable of undergoing deformation during its expansion or compression in a fluid system or in a thermodynamic cycle and takes part in energy transfer; can receive or reject heat energy or can be the means by which the work is done; *air* and *steam* are the common working substances used in a thermodynamic system.

Work-in-Process (WIP) An *Operations Management* term; also called **Work-in-Progress (WIP)**; components or assemblies needed for a final product in manufacturing; semi-finished components or parts over which some operations have already been carried out, but some more operations are still left to be performed.

Work-in-Progress (WIP) An *Operations Management* term; see **Work-in-Process (WIP)**.

Workpiece A *Manufacturing* term; the part or component on which a manufacturing operation is carried out.

Workstation An *Operations Management* term; also called **Work Centre**; the place where production activities may be carried out to transform raw materials into finished products using *processing machines* and involving *labours*.

World-Class Company See **World-Class Organization**.

World-Class Organization Also called **World-Class Company**; an organization or a company which produces each product and service of such a high standard that would be considered best-in-class by its customers.

Worm and Worm Gear A *Theory of Machines* term related to a *gear*; also called **Worm and Worm Wheel**; a special case of a pair of spiral gears in mesh in which the axes of the driving and the driven shafts are at *right angles* to each other; used to transmit rotary motion between *non-parallel* and *intersecting* shafts; the *worm is of smaller diameter* and acts as a *pinion* (*driving gear*) and is made of a *case hardened steel*, whereas *the worm gear is of bigger diameter* which is attached to a load wheel and is made of *phosphor bronze* centrifugally cast rim.

Worm and Worm Wheel A *Theory of Machines* term related to a *gear*; see **Worm and Worm Gear**.

Worm Gearing A *Theory of Machines* term related to a *gear*; a combination of *worm* and *worm gear*; a form of spiral gearing in which the axes of the driving and the driven shafts are usually at right angles; a power or motion transmitting device which has high velocity ratio; suitable for higher load carrying capacity because of line contact of the gear teeth.

Wright's Law A *Management* term; see **Learning Curve**.

Wrought Iron A *Materials Engineering* term; the *purest form of iron* which contains not more than 0.5% of carbon and other impurities; a mechanical mixture of *very pure iron* and a *silicate slag* distributed uniformly throughout the metal; its *high ductility* is used in hot working operations like *forging* to give various shapes to many products; its ultimate strength can be increased considerably by cold working (process is called *strain hardening*); has *good weldability* and is also *corrosion resistant*; *can never be cast as it becomes pasty on heating rather than melting*, hence all the shapes are given by *hammering, pressing* or *forging*; can resist *sudden* and excessive *shock loads* without undergoing permanent deformation, and this property finds use in making *chains, crane, hooks,* and *railway couplings*; principally used in the production of pipes and other products to be protected from rusting as it resists rusting effectively.

X

X-Bar Chart A *Quality Control* term; a *control chart for variables* which is used to measure the mean of the process distribution; see also **R-Chart**.

X-Radiography A *Materials Engineering* term; also called **Radiographic Testing** or **Radiographic Inspection**; a *nondestructive testing* in which defects are inspected by using the *attenuation* (reducing intensity) of X-rays.

XYZ Analysis An *Operations Management* term related to *inventory*; a *selective inventory control technique* which classifies the inventories based on its closing value in storage into three categories: *X items* having largest closing value (e.g. *gold*), *Y items* having moderate closing value (e.g. *renovated houses*), and *Z items* having lowest closing value (e.g. *old electronic gadgets*).

D. K. Singh, *Dictionary of Mechanical Engineering*,
https://doi.org/10.1007/978-981-99-1722-8_24

Y

Yield Point A *Strength of Materials* term; a stage in a *tension test*, when a material loses its resistance to counter the external applied force; that is, it yields to external force; a material loses elasticity and becomes a plastic body after the yield point; not well defined in case of *brittle materials*; the stress corresponding to this point is called *yield stress*, which is a *design stress* in engineering applications.

Yield Strength A *Strength of Materials* term; see **Yield Stress**.

Yield Stress A *Strength of Materials* term; also called **Yield Strength**; a stress which corresponds to the yield point; a material loses its elastic nature when it attains yield stress; an important design parameter which defines the maximum stress for ductile materials.

Young's Modulus of Elasticity A *Strength of Materials* term; also called simply **Modulus of Elasticity** or **Young's Modulus** or **Elastic Modulus**; a solid material's property and an *elastic constant*; a ratio of the *longitudinal stress* to the *longitudinal strain* within elastic limit; denoted by E (mostly) or Y; has higher values for most of the engineering materials; the same value in tension or compression; a measure of the stiffness of a material; the greater the value of E, the larger is the elasticity of the material; for a perfectly rigid body, the value of E is infinity; the SI unit is pascal (Pa) which is equivalent to N/m^2; the values of E for some materials are tabulated below:

Materials	**E(GPa)**
Steel	200
Wrought iron	190
Copper	120
Aluminium	70; see also **Modulus of Rigidity** and **Bulk Modulus of Elasticity**

Young's Modulus A *Strength of Materials* term; see **Young's Modulus of Elasticity**.

D. K. Singh, *Dictionary of Mechanical Engineering*,
https://doi.org/10.1007/978-981-99-1722-8_25

Z

Zero Defects A *Quality Management* term proposed by the quality guru *Philip B. Crosby* (1926–2001); a concept which ensures that defects are eliminated to a maximum extent, preferably completely; a programme which tends to change workers' attitudes about quality by stressing error-free performance; signifies that a production system and quality control measures be so designed and coordinated that defects are totally avoided as far as possible, with no scope to produce any scrap.

Zero-Emission Vehicle A vehicle which produces no pollutants; an *electric vehicle* is a zero-emission vehicle.

Zero Inventory System An *Operations Management* term; see **JIT Manufacturing**.

Zero Vector A *Physics* term; see **Null Vector**.

Zero-Force Member A *Strength of Materials* term related to a *truss*; a truss member which does not carry any force; simply used to give stability to a truss; a truss may contain one or more *zero-force members*.

Zeroth Law of Thermodynamics A *Thermodynamics* term, formulated by *Ralph H. Fowler* (1889–1944) in 1931; recognized more than half a century after the formulation of the first and second laws of thermodynamics; named zeroth law as it should have preceded the first and second laws of thermodynamics; states that if a body is in thermal equilibrium with other two bodies separately, then all the three bodies are in thermal equilibrium with each other; forms the basis of all the temperature measurements.

Zinc Chromate A *Chemistry* term; a *zinc compound* used as a yellow pigment in making paints.

Zinc Oxide A *Chemistry* term having the chemical formula ZnO; see **White Zinc**.

D. K. Singh, *Dictionary of Mechanical Engineering*,
https://doi.org/10.1007/978-981-99-1722-8_26

Zinc A *Chemistry* term; a *metallic element* having symbol *Zn* with atomic number 30 and atomic weight 65.37; an important *nonferrous metal*; its important ores include *zinc blende* (ZnS), *calamine* ($ZnCO_3$), and *zincite* (ZnO), but zinc is mainly extracted from *zinc blende*; a *low melting point metal* which melts at 421 °C; its *mechanical properties change with temperature* due to *formation of different allotropic forms*; shows brittle nature at room temperature, becomes malleable and ductile when heated to about 155 °C, and is again brittle at 220 °C; finds extensive use in the preparation of alloys, such as *Brass, Delta metal*, and *German silver*; used to coat iron because of its resistance to rusting and corrosion (process is called *Galvanizing*); also used in *Parke's process* for the extraction of silver from lead; zinc oxide (white zinc) finds use in making paints, and zinc chromate is used as a yellow pigment.